Progress in Mathematics
Volume 168

European Congress of Mathematics

Budapest, July 22–26, 1996
Volume I

A. Balog
G.O.H. Katona
A. Recski
D. Sza'sz
Editors

Birkhäuser Verlag
Basel · Boston · Berlin

Editors:

A. Balog
Mathematical Institute
Hungarian Academy of Sciences
Reáltanoda str. 13–15
H-1053 Budapest Hungary

G.O.H. Katona
Mathematical Institute
Hungarian Academy of Sciences
Reáltanoda str. 13–15
H-1053 Budapest Hungary

A. Recski
Mathematical Institute
Technical University of Budapest
H-1521 Budapest Hungary

D. Sza'sz
Mathematical Institute
Hungarian Academy of Sciences
Reáltanoda str. 13–15
H-1053 Budapest Hungary

1991 Mathematics Subject Classification 00B25

A CIP catalogue record for this book is available from the Library of Congress, Washington D.C., USA

Deutsche Bibliothek Cataloging-in-Publication Data

European Congress of Mathematics <2, 1996, Budapest>:
European Congress of Mathematics: Budapest, July 22 - 26, 1996 / A. Balog... ed. - Basel ; Boston ; Berlin : Birkhäuser.
ISBN 3-7643-5496-8 (Basel...)
ISBN 0-8176-5496-8 (Boston)

Vol. 1. - (1998
(Progress in mathematics ; Vol. 168)
ISBN 3-7643-5497-6 (Basel...)
ISBN 0-8176-5497-6 (Boston)

Vol. 2. - (1998
(Progress in mathematics ; Vol.169)
ISBN 3-7643-5498-4 (Basel...)
ISBN 0-8176-5498-4 (Boston)

Printed on acid-free paper produced of chlorine-free pulp. TCF ∞
Printed in Germany
ISBN 3-7643-5497-6 (Vol. I/PM 168)
ISBN 3-7643-5498-4 (Vol. II/PM 169)
ISBN 3-7643-5496-8 (Set)
ISBN 0-8176-5497-6 (Vol. I/PM 168)
ISBN 0-8176-5498-4 (Vol. II/PM 169)
ISBN 0-8176-5496-8 (Set)

9 8 7 6 5 4 3 2 1

Table of Contents of Volume I

Speeches

Address by J.-P. Bourgignon ix

Address by G. Demszky xi

Address by A. Goncz xii

Address by G. Katona xiii

List of talks xv

Prizes of the European Mathematical Society xvii

Contributions

N. Alon
Randomness and pseudo-randomness in discrete mathematics 1

L. Ambrosio
Free discontinuity problems and special functions with bounded variation 15

K. Astala
Recent connections and applications of planar quasiconformal mappings 36

R. Benedetti
A combinatorial approach to combings and framings of 3-manifolds 52

Ch. Bessenrodt
Algebra and combinatorics 64

F. Bethuel
Some recent results for the Ginzburg-Landau equation 92

P. Bjørstad
Mathematics, parallel computing and reservoir simulation 100

E. Bolthausen
Large deviations and perturbations of random walks and random surfaces 108

J. Bricmont, A. Kupiainen
Renormalization group for fronts and patterns 121

D. Burago
Geometry of tori: Riemannian versus Finsler? 131

L. Caporaso
Counting curves on surfaces: A guide to new techniques and results 136

U. Dierkes
Minimal surfaces in singular spaces 150

I. Dynnikov
Surfaces in 3-torus: Geometry of plane sections 162

L.H. Eliasson
One-dimensional quasi-periodic Schrödinger operators – dynamical systems and spectral theory 178

W.T. Gowers
Banach spaces with few operators 191

H. Hedenmalm
Recent developments in the function of the Bergman space 202

A. Huber
Extensions of motives 218

J. Kaczorowski
Boundary values of Dirichlet series and the distribution of primes 237

J. Kollár
Low degree polynomial equations: Arithmetic, geometry and topology 255

D.O. Kramkov, A.N. Shiryaev
Sufficient conditions of the uniform integrability of exponential martingales 289

C. Lescop
On the Casson invariant 296

R. März
EXTRA-ordinary differential equations: Attempts to an analysis of differential-algebraic systems 313

Table of Contents of Volume II

Contributions (continued)

J. Matousek
Geometric set systems 1

D. McDuff
Recent developments in symplectic topology 28

A.S. Merkurjev
K-theory and algebraic groups 43

V. Milman
Surprising geometric phenomena in high-dimensional convexity theory 73

St. Müller
Microstructures, phase transition and geometry 92

T. Nowicki
Different types of non-uniform hyperbolicity for interval maps are equivalent 116

E. Olivieri, E. Scoppola
Metastability and typical exit paths in stochastic dynamics 124

V.P. Platonov
Rationality problems for group varieties 151

L. Polterovich
Precise measurements in symplectic topology 159

J. Pöschel
Nonlinear partial differential equations, Birkhoff normal forms, and KAM theory 167

L. Pyber
Group enumeration and where it leads us 187

N. Simányi
Studying dynamical systems with algebraic tools 200

J.P. Solovej
Mathematical results on the structure of large atoms 211

A. Stipsicz
Geography of irreducible 4-manifolds 221

G. Tardos
Transversals of d-intervals – comparing three approaches 234

J.-P. Tignol
Algebras with involution and classical groups 244

A.P. Veselov
Huygens' principle and integrability 259

E. Zuazua
Some problems and results on the controllability of partial differential equations 276

Round Tables

(A) Electronic literature in mathematics
B. Wegner (chair); A. DeKemp, A. Bardelloni, J.-P. Allouche 315

(B) Mathematical Games
D. Singmaster (chair); A. Fraenkel, M.E. Larsen, T. Szentiványi 338

(D) Women and mathematics
K. Hag (chair); S. Paycha, R. Piene, D. McDuff, R. März 347

(E) Public image of mathematics
R. Bulirsch (chair); M. Chaleyat-Maurel, Gy. Staar, St. Deligeorges 376

(G) Education
V.L. Hansen (chair); Ch. Mauduit, J.-P. Boudine, M. Laczkovich, L. Pósa 380

Speeches

Address by Jean-Pierre Bourguignon, President of the European Mathematical Society

Monsieur le Maire de Budapest,
Monsieur le Secrétaire d'Etat l'Education,
Monsieur le Représentant de la Commission Européenne en Hongrie,
Messieurs les Présidents,
Mesdames et Messieurs,
Chères et chers collègues,

Au nom de la Société Mathématique Européenne, je viens vous remercier d'honorer de votre présence la cérémonie d'ouverture du Deuxième Congrès Européen de Mathématiques ici à Budapest, dans un pays de grande tradition mathématique.

Notre hôte est la Société Mathématique János Bolyai. Nous remercions ses membres pour leur confiance dans la S.M.E. en mettant sur pied ce congrès et pour tout le travail qu'ils y ont investi.

The Bolyai Society is named after one of the founders, some 175 years ago, of a completely new geometry. In his own words (which remarkably encapsulate the power of Mathematics), János Bolyai "created a universe out of nothing". Besides the new mathematical fields that were now revealed, a possible model was created which has now become a standard tool for statistical physicists in their study of disordered media. This model is also relevant for the architecture of computers of the latest generation. This illustrates the versatility of mathematical concepts, and the long-range value of investment in this field.

Mathematics is diverse as Europe itself. It struggles and sometimes achieves a unity as I would wish for Europe. This quest and this achievement, you will have plenty of opportunities of seeing at work during the week because the international Scientific Committee of this congress, chaired by Professor Jürgen Moser, which selected the speakers, chose the "Unity of Mathematics" as theme of the Congress.

Mathematics has become a key for the harmonious development of modern societies for at least three independent reasons:

1) Mathematical concepts lie at the heart of many different techniques, those on which high-technology is based, such as numerical windtunnels (whose use is now widespread in the aeronautics industry), scanners for medical imaging, telecommunications codes (on which the quality and security of data transmission rely), more generally models, either deterministic or stochastic, used in economy, in banks and insurance companies, but also in meteorology, epidemiology and in the environmental sciences, to name but a few.

2) Mathematics is now present in our daily life, not only in our regular use of many technical devices, but also in the constant reference made to statistics and polls. Democracy requires that citizens be properly trained to detect inadequate uses of such data. We, if we believe in democracy, must make sure that our citizens are indeed comfortable with mathematics; this is a necessity in our fast evolving world where decentralized centers of decision-making are a must.
3) Finally, and this third reason is particularly dear to my heart, mathematics remains, as it has been throughout history, a privileged path to critical reasoning, in school and more generally in cultural circles. To learn how to think independently is both a shield against authoritarianism and a passport to innovation.

You will therefore understand that I feel it my responsibility

- to call upon the attention of people in charge of running society to the conditions necessary to meet these challenges: positions for mathematicians (be they teachers, engineers, technicians or researchers) belong to strategic resources; means to develop research and innovation, and to make contacts between industry and the academic world fruitful belong to investments with the highest returns. Failing to recognize this may prove an expensive mistake;
- to appeal to my fellow mathematicians to be receptive to calls from society for training (at all levels, keeping in mind that students have to become ambassadors of our discipline no matter what their future professions are going to be), for innovation (this requires curiosity for applications), and for what may prove the decisive point, the cultural dimension of our discipline.

Mathematicians are here in one of the cities lying at the heart of historical Europe, in a century fast approaching its conclusion which has witnessed so many dramatic changes. Mathematics is indeed one of the major construction sites for tomorrow's world. All of us have to be convinced of this. We all have our share of work in this endeavour.

I thank you for your attention.

Jean-Pierre Bourguignon
President of the
European Mathematical Society

* * *

Address by Gábor Demszky, Mayor of Budapest

Ladies and Gentlemen:

Welcome to Budapest! As the mayor of Budapest, I am most happy that the Second European Mathematical Congress is organized in our city.

Mathematics has played a decisive role in the development of the European spirit, and the unity of Europe is based on this European spirit. All mathematicians understand each other in the language of mathematics, and so mathematics is perhaps the most universal, most European mediator between the various European nations.

The other basis of modern European unity is the unified information network which is likewise the product of the application of mathematics. Hence you, participants of the European Mathematical Congress, are in fact engineers of European unity.

Thus the goals of your previous and present congresses – to discuss scientific achievements, their applications and widespread propagation, to study the problems of mathematics education, to improve the popularization of mathematics — these are extremely important for the whole of European society.

I think it is no accident that Budapest received the right to organize this congress. The choice reflects the appreciation of the entire scientific world towards Budapest as a metropolis of dynamic development, as a city of gifted people and as one of the capitals of international mathematics.

Hungarian mathematicians and physicists have played an outstanding role since János Bolyai in the universal development of science. Several of them studied in the schools and universities of Budapest. János Neumann was a student at Fasor High School in Budapest.

A few days ago the Hungarian team achieved third place at the International Mathematical Olympiad in Bombay in the competition of young mathematicians of 75 nations. The Hungarian students won three gold, two silver and one bronze medals. Four out of the six Hungarian participants are studying in Budapest, three of them learn in a special high school supervised directly by the capital.

The Mathematical and Physical Society founded in 1891 and its successor today, the János Bolyai Mathematical Society, have always represented and helped the talented. The work of the Society is indispensable today, as well. The city of Budapest will make all efforts in order to support the goals of the Mathematical Society.

The Municipality of Budapest is aware that the quality and continuity of a city's life are determined to a great extent by the training of future scientists. In the Mihály Fazekas Elementary and High School, which is supervised directly by the capital, there have been special mathematical classes for the most gifted students for the last three and half decades. One of the initiators of this program was Professor János Surányi, Honorary President of the Mathematical Society. Many former students of this school have already achieved international fame in the mathematical scientific community.

Ladies and Gentlemen!

This is already the second occasion that the European Mathematical Society awards prizes to the 10 best young mathematicians. It is a great privilege for me that I can hand over these prizes after the introductory addresses. Hereby we would like to follow the example of the previous congress and the then mayor of Paris. I think that the city of Budapest should definitely support forming such an important tradition. Therefore we were happy to contribute to the financial and moral backing of the prizes.

I do hope that these prizes are not just gestures for the recognition of the talented, but will also inspire young mathematicians to take their increasing share of scientific research and development.

I am confident that scientific conferences promote the rapprochement of cities and nations in the long run. With this belief I wish fruitful work to the Congress and further successes in their professional careers to the prizewinners!

* * *

Address by Arpad Goncz, President of Hungary

Dear Guests, Dear Friends from Abroad and from Hungary!

It is my real pleasure that, on behalf of the European Mathematical Society and with the support of the European Community, our country could organize the Second European Congress of Mathematicians.

This honour is certainly due to the international prestige of Hungarian mathematics. The Hungarian team has always finished among the top ten of the International Mathematical Olympiads – our youngsters contribute to the great reputation of the Hungarian mathematics, but their widely acknowledged results and their success are due – to a considerable extent – to the János Bolyai Mathematical Society. Under this name the Society is "only" 50 years old, but its predecessor, the Mathematical and Physical Society, was founded in 1891, and its goals and course were set by eminent physicists and mathematicians. These goals, to integrate the scientific achievements and education, and to find and help the gifted students by competitions, periodicals, conferences have not changed since. We are happy to see that these Hungarian ideas and methods flourish in harmony with the European efforts. Our achievements show that.

I have learned from the program of the congress, that mathematical meetings fill the period of time from mid-July to the end of August. Beside the Congress in Budapest, the centerpoint of the series of these scientific events, conferences on various topics are organized in various places: Austria, Romania, the Czech Republic, Slovakia, Italy, and in Szeged, Miskolc, Eger, just to mention a few of them. Not knowing if the representatives of these events are present here in this session, I ask you to forward my greeting words to them as well.

I extend my heartfelt greetings to all the young mathematicians participating in this Congress and I greet with all respect the eminent scientists, who will be giving lectures and popularizing mathematics at this Congress. I express my gratitude to those who, with incessant enthusiasm, teach our students and to those who help and organize the competitions and the scientific meetings. Research, education, caring for the gifted are all united in an exemplary manner in these meetings.

As a private person and a one-time instructor I look with a genuine envy at the masters, who in the course of their lectures find followers from the above-average, gifted and motivated pupils, since I have come to realize that this is the first step toward immortality.

With the warmest regards

Arpad Goncz

* * *

Address of Gyula O.H. Katona, Chairman of the Organizing Committee to the Congress

Ladies and Gentlemen, may I welcome you to the Second European Congress of Mathematics.

As the Chairman of the Organizing Committee, it is my privilege to introduce the members of the platform party: Gábor Demszky, Mayor of Budapest, Zoltán Szabó, Vice-Minister of Culture and Education, Dr. Hans Beck, Head of Delegation, European Commission in Hungary, Professor Jean-Pierre Bourguignon, President of the European Mathematical Society, Professor László Márki, Vice-President of the European Mathematical Society, Professor Ákos Császár, President of the Division of Mathematics of the Hungarian Academy of Sciences, Professor Pál Révész, President of the János Bolyai Mathematical Society, László Lovász, Chairman of the Prize Committee and Tamás Székely, representing the Motorola Company.

Let me say a few words on behalf of the Organizers.

Four years ago a very important initiative was started in France. The first European Congress of Mathematics was organized in Paris in 1992. It is a great privilege for Hungarian Mathematics that we are the organizers of the second item of the sequence of these.

It was a long and difficult work. We could not have succeeded without the extensive help of the European Mathematical Society, the European Union, the City of Budapest and the different Hungarian government organizations, not to mention the very many Hungarian enthusiasts. Let me express our profound gratitude to all the above mentioned people and organizations.

But we are also extremely grateful to you, who have come to this celebration of European Mathematics. Yes, this is indeed a celebration, but also a demonstration; a demonstration of the vital unity of mathematical life in Europe.

When preparing this short speech, it occurred to me for a second that I could say a welcoming sentence in each European language. Unfortunately, as time is pressing, this would take too long. So, perhaps I could do it in the major languages, only. But I was not able to decide what the major languages were. Finally I came to the conclusion that there is no need for that, we have a common language, which may be obscure to many outsiders, but for us this is the language of mathematics. This is sufficient for us ... and necessary, too.

So I would like to wish you a happy mathematical experience. I hope that you will learn many deep theories, that you will prove exciting theorems and pose intriguing conjectures.

Now may I introduce another language which is, in some way, related to our own, the language of music. The Sonatore Pannoniae brass ensemble will perform both classical and preclassical compositions.

* * *

List of Talks

N. Alon: Randomness and pseudo-randomness in discrete mathematics

G. Ben Arous: Large deviations as a common probabilistic tool for some problems of analysis, geometry and physics

D. McDuff: Recent progress in symplectic topology

B. Dubrovin: Reflection groups, quantum cohomologies, and Painlevé's equation

J. Kollár: Low degree polynomial equations: arithmetic, geometry and topology

J. Laskar: The stability of the solar system

A.S. Merkurjev: K-theory and algebraic groups

V. Milman: Surprising geometric phenomena of high-dimensional convexity theory

St. Müller: Microstructures, geometry and the calculus of variations

J.-P. Serre: Correspondences and dictionaries in geometry and number theory

Parallel lectures

L. Ambrosio: Special functions with bounded variation and free discontinuity problems

K. Astala: Planar quasiconformal mappings – recent connections and applications

R. Benedetti: A combinatorial calculus for combed and framed 3-manifolds

Ch. Bessenrodt: Algebra and combinatorics

F. Bethuel: Some results on the Ginzburg-Landau equation

P. Bjørstad: Mathematics, parallel computations and oil reservoirs

E. Bolthausen: Large deviations and perturbations of random walks and random surfaces

J. Bricmont: The renormalization group: from statistical mechanics to partial differential equations

L. Caporaso: Enumerating curves on rational surfaces

J. de Jong: Families of curves and alterations

U. Dierkes: Minimal surfaces in singular spaces

I. Dynnikov: Surfaces in 3-torus: geometry of plane sections

H. Eliasson: One dimensional quasi-periodic Schrödinger operators

H. Hedenmalm: Function theory in Bergman spaces: a playground for complex analysis, operator theory and elliptic PDE

E. Hrushovski: Highly symmetric finite structures

J. Kaczorowski: Boundary values of Dirichlet series and the distribution of primes

C. Lescop: On the Casson invariant

R. März: EXTRA-ordinary differential equations – attempts to an analysis of differential-algebraic systems

J. Matousek: Geometric set systems

L. Merel: Arithmetic of elliptic curves and diophantine equations

T. Nowicki: Different types of non-uniform hyperbolicity

A. Pastur: Random matrices and related problem of analysis, probability and mathematical physics

R. Pérez-Marco: Beyond linearization

V.P. Platonov: Rationality problems for group varieties

J. Pöschel: On quasi- and almost-periodic solutions of nonlinear partial differential equations

L. Pyber; Group enumeration and where it leads us

H.P. Schlickewei: Exponential diophantine equations

E. Scopolla: Metastability for stochastic dynamics of interacting particle systems

A.N. Shiryaev: Towards stochastic calculus and stochastics in finance

N. Simányi: Studying dynamical systems with algebraic tools

J.Ph. Solovej: Mathematical results on the structure of large atoms

A. Stipsicz: Geography of irreducible 4-manifolds

G. Tardos: Homotopical methods in combinatorial optimization

J.-P. Tignol: Algebras with involution and classical groups

A. Veselov: Huygens' principle and integrability

E. Zuazua: Recent results on the controllability of partial differential equations

Round Tables

(A) Electronic literature in mathematics;
B. Wegner (chair); A. DeKemp, A. Bardelloni, J.-P. Allouche

(B) Mathematical Games;
D. Singmaster (chair); A. Fraenkel, M.E. Larsen, T. Szentiványi

(C) Demography of mathematicians;
J.-P. Bourguignon (chair); D. Wallace, L. Lemaire, Ch. Berg

(D) Women and mathematics;
K. Hag (chair); S. Paycha, R. Piene, D. McDuff, R. März

(E) Public image of mathematics;
R. Bulirsch (chair); M. Chaleyat-Maurel, Gy. Staar, St. Deligeorges

(F) Mathematics and Eastern Europe;
D. Cioranescu (chair); Vl. A. Molchanov, W. Jäger, J.F. Rodriguez, M. Niezgodka, Ch. Duhamel, Fl. Topsoe, D. Miklós

(G) Education;
V.L. Hansen (chair); Ch. Mauduit, J.-P. Boudine, M. Laczkovich, L. Pósa

Prize Winners
2nd European Congress of Mathematics

Alexis Bonnet works on a broad spectrum of problems in applied analysis. His results on the Mumford-Sha conjecture in the theory of computer vision meant a breakthrough. This problem deals with a variational problem with a singular boundary set, and proposes a finite representation of the optimum solution. Bonnet obtained the first finiteness result under additional assumptions, which is a major step in understanding this difficult free boundary value problem. In a different direction, his results on partial differential equations, in particular on flame propagation and combustion, are very significant.

Willaim Timothy Gowers' work has made the geometry of Banach spaces look completely different. To mention some of his spectacular results: he solved the notorious Banach hyperplane problem, to find a Banach space which is not isomorphic to any of its hyperplanes. He gave a counterexample to the Schroeder-Bernstein theorem for Banach spaces. He proved a deep dichotomy principle for Banach spaces which if combined with a result of Komorowski and Tomczak-Jaegermann shows that if all closed infinite-dimensional subspaces of a Banach space are isomorphic to the space, then it is a Hilbert space. He gave (jointly with Maurey) an example of a Banach space such that every bounded operator from the space to itself is a Fredholm operator. His mathematics is both very original and technically very strong. The techniques he uses are highly individual; in particular, he makes very clever use of infinite Ramsey theory.

Annette Huber developed a difficult and important theory, the theory of the derived category of mixed motivic realisations. The theory of motives was discovered by Alexander Grothendieck in the 60's. This important topic is still largely conjectural. The definition of mixed motives is one of the central problems of this theory. Annette Huber defines a derived category of the category of mixed realisations defined by Jannsen. She constructs a functor from the category of simplicial varieties to this derived category, whose cohomology objects are precisely the mixed realisations of the variety. She then defines an absolute cohomology theory, over which the usual absolute theories – absolute Hodge-Deligne and continuous étale cohomology – naturally factorise.

Aise Johan de Jong has produced in a large variety of deep results on various aspects of arithmetic algebraic geometry. His personal influence on the work in the field is impressive. His work is characterized by a truly geometric approach and a abundance of new ideas. Among others, his results include the resolution of a conjecture of Vyes and the answer to a long-standing question of Mumford on moduli spaces. Resolution of singularities by modification is difficult and unknown in most cases; in a recent outstanding work, de Jong found an elegant method for the resolution of singularities by alterations, which is a slightly weaker question but sufficient for most applications. This basic method combines geometric insight and technical knowledge.

Dmitri Kramkov has important results in statistics and the mathematics of finance. He did fundamental work in filtered experiments. In particular, he obtained a deep result on the structure of Le Cam's distance between two filtered statistical experiments, and proved very general theorems about the structure of the limit experiments which cover many results in the asymptotic mathematical statistics of stochastic processes. Recently he proved a remarkable "Optional decomposition of supermartingales" which is an extension of the fundamental Doob-Meyer decomposition for the case of many probability measures. This unexpected result is rather difficult and refined technically, and, from the conceptual point of view, very important. In the direction of mathematical finance, Kramkov obtained impressive results on pricing formulas for certain classes of "exotic" options based on geometric Brownian motion. He succeeded in computing explicit solutions for "Asian options" where the pay-off is given by a time-average of geometric Brownian motion.

Jiri Matousek's achievements have combinatorial and geometric flavor; his research is characterized by its breadth, by its algorithmic motivation, as well as the difficulty of the problems he attacks. He gave constructions of epsilon-nets in computational geometry, which provides tools for derandomization of geometric algorithms. He obtained the best result on several key problems in computational and combinatorial geometry and optimization, such as linear programming algorithms and range searching. He solved several long-standing problems (going back to the work of K.F. Roth) in geometric discrepancy theory, in particular on the discrepancy of halfplanes and of arithmetical progressions. He solved a problem by Johnson and Lindenstrass on embeddings of finite metric spaces into Banach spaces. He also obtained sharp results on almost isometric embeddings of finite dimensional Banach spaces using uniform distributions of points on spheres. In mathematical logic, he found a striking example of a combinatorial unprovable statement.

Loic Merel proved an absolute bound for the torsion of elliptic curves. Thereby he gave a solution to a long-standing problem, open for more than 30 years, that has resisted the efforts of the greatest specialists of elliptic curves. The group of torsion points of an elliptic curve over a number field is finite. Merel found a bound of the order of this group in terms of the degree of the number field; such a bound was known in a very few cases only (the case of the rational numbers (Mazur 1976), number fields of degree less than 8 (Kamieny-Mazur 1992), and number fields of degree less than 14 (Abramovitch 1993).)

Grigory Perelman's work played a major role in the development of the theory of Alexandrov spaces of curvature bounded from below, giving new insight into to what extent results of Riemannian geometry rely on the smoothness of the structure. Now, mainly due to Perelman, the theory is rather complete. His results include a structure theory of these spaces, a stability theorem (new even for Riemannian manifolds), and a synthetic geometry à la Aleksandrov. He proved a conjecture of Gromov concerning an estimation of the product of weights, and the Cheeger-Gromov conjecture. This last problem attracted the attention and efforts of many geometers for more than 20 years, and the method developed by Perelman yielded an astonishingly short solution.

Ricardo Perez-Marco solved several outstanding problems, and obtained basic results, in the theory of dynamics of non-linearizable germs and non-linearizable analytic diffeomorphisms of the circle, and in the theory of centralizers, a natural complement of non-linearizability. He discovered a new arithmetic condition under which a germ without periodic orbits is linearizable. He gave a negative answer to a question of Arnold on the linearizability of analytic diffeomorphisms of the circle without accumulating periodic orbits. Perez-Marco developed a theory of analytic non-linearizable germs based on an important and useful compact invariant.

Leonid Polterovich contributed in a most important way to several domains of geometry and dynamical systems, in particular to symplectic geometry. Polterovich ties together complex analytic and dynamical ideas in a unique way, leading to significant progress in both directions. In particular, he brings complex analysis into the realm of Hamiltonian mechanics, which marks a principally new step in this classical field. Among others, he established (with Bialy) an anti-KAM estimate in terms of the Hofer displacement of a Hamiltonian flow. Polterovich found the first non-trivial restriction on the Maslov class of an embedded Lagrangian torus, and (with Eliashberg) completely solved the Lagrangian knot problem in the real 4-space.

Progress in Mathematics, Vol. 168, © 1998 Birkhäuser Verlag Basel/Switzerland

Randomness and Pseudo-Randomness in Discrete Mathematics

Noga Alon*

Department of Mathematics
Raymond and Beverly Sackler Faculty of Exact Sciences
Tel Aviv University, Tel Aviv, Israel
Email: noga@math.tau.ac.il.

The discovery, demonstrated in the early work of Paley, Zygmund, Erdős, Kac, Turán, Shannon, Szele and others, that *deterministic* statements can be proved by *probabilistic* reasoning, led in the first half of the century to several striking results in Analysis, Number Theory, Combinatorics and Information Theory. It soon became clear that the method, which is now called *the probabilistic method*, is a very powerful tool for proving results in Discrete Mathematics. The early results combined combinatorial arguments with fairly elementary probabilistic techniques, whereas the development of the method in recent years required the application of more sophisticated tools from probability. The books [10], [54] are two recent texts dealing with the subject.

Most probabilistic proofs are existence, non-constructive arguments. The rapid development of theoretical Computer Science, and its tight connection to Combinatorics, stimulated the study of the algorithmic aspects of these proofs. In a typical probabilistic proof, one establishes the existence of a combinatorial structure satisfying certain properties by considering an appropriate probability space of structures, and by showing that a randomly chosen point of this space is, with positive probability, a structure satisfying the required properties. Can we find such a structure *efficiently*, that is, by a (deterministic or randomized) polynomial time algorithm? In several cases the probabilistic proof provides such a randomized efficient algorithm, and in other cases the task of finding such an algorithm requires additional ideas. Once an efficient randomized algorithm is found, it is sometimes possible to *derandomize* it and convert it into an efficient deterministic one. To this end, certain explicit *pseudo-random* structures are needed, and their construction often requires tools from a wide variety of mathematical areas including Group Theory, Number Theory and Algebraic Geometry.

The application of probabilistic techniques for proving deterministic theorems, and the application of deterministic theorems for derandomizing probabilistic existence proofs, form an interesting combination of mathematical ideas from various areas, whose intensive study in recent years has led to the development

*Research supported in part by a USA Israeli BSF grant and by the Fund for Basic Research administered by the Israel Academy of Sciences.

of fascinating techniques. In this paper I survey some of these developments and mention several related open problems.

1. Probabilistic methods

The applications of probabilistic techniques in Discrete Mathematics, initiated by Paul Erdős who contributed to the development of the method more than anyone else, can be classified into three groups. The first one deals with the study of certain classes of random combinatorial objects, like random graphs or random matrices. The results here are essentially results in Probability Theory, although most of them are motivated by problems in Combinatorics. The second group consists of applications of probabilistic arguments in order to prove the existence of combinatorial structures which satisfy a list of prescribed properties. Existence proofs of this type often supply extremal examples to various questions in Discrete Mathematics. The third group, which contains some of the most striking examples, focuses on the application of probabilistic reasoning in the proofs of deterministic statements whose formulation does not give any indication that randomness may be helpful in their study.

The above classification is, of course, somewhat arbitrary, and there are results that can fit more than one of the above groups. Most of the combinatorial results obtained by applying probabilistic arguments belong, however, naturally to one of these groups.

There has been recent interesting progress in all three groups. This chapter contains a brief description of several typical results in each of them.

1.1. Random structures Although there have been several papers by various researchers in the late 50's that dealt with the statistical aspects of graphs, the systematic study of Random Graphs was initiated by Erdős and Rényi whose first two papers on the subject are [21], [22]. Formally, $G(n,p)$ denotes the probability space whose points are graphs on a fixed set of n labelled vertices, where each pair of vertices forms an edge, randomly and independently, with probability p. The term "the random graph $G(n,p)$" means, in this context, a random point chosen in this probability space. Each graph property A (that is, a family of graphs closed under graph isomorphism) is an event in this probability space, and one may study its probability $Pr[A]$, that is, the probability that the random graph $G(n,p)$ lies in this family. In particular, we say that A holds *almost surely* if the probability that $G(n,p)$ satisfies A tends to 1 as n tends to infinity. There are numerous papers dealing with random graphs, and the book of Bollobás [13] is an excellent extensive account of the known results in the subject proved before its publication in 1985.

One of the most important discoveries of Erdős and Rényi was the discovery of *threshold functions*. A function $r(n)$ is called a threshold function for a graph property A, if when $p(n)/r(n)$ tends to 0, then $G(n,p(n))$ does not satisfy A almost surely, whereas when $p(n)/r(n)$ tends to infinity, then $G(n,p(n))$ satisfies A almost surely. Thus, for example, it is shown in [21] that the function $r(n) = \ln n/n$ is a threshold function for the property "G is connected." (In fact, a much more precise estimate follows from the results in [21]: if $p(n) = \frac{\ln n}{n} + \frac{c}{n}$, then, as n tends to infinity, the probability that $G(n, p(n))$ is connected tends to $e^{-e^{-c}}$.)

A graph property is *monotone* if it is closed under the addition of edges. Note that many interesting graph properties, like hamiltonicity, non-planarity, connectivity or containing at least 10 vertex disjoint triangles are monotone.

Bollobás and Thomason [15] proved that *any* monotone graph property has a threshold function. Their proof applies to any monotone family of subsets of a finite set, and relies on the Kruskal-Katona Theorem that describes the possible number of subsets of each cardinality in a monotone family. By viewing a monotone graph property as a family of subsets of the set of all potential edges, this yields the result for random graphs. Their theorem shows that for any monotone property A, if the probability that a random graph $G(n,p)$ satisfies A exceeds ϵ, then for $q \geq C(\epsilon)p$, the probability that $G(n,q)$ satisfies A is at least $1-\epsilon$. This result applies even without the assumption that the property A is closed under graph isomorphism. In fact, if one is not interested in the precise behaviour of $C(\epsilon)$ this can be deduced simply by observing that if $(1-\epsilon)^k < \epsilon$ then the probability is more than $1-\epsilon$ that at least one of k graphs G_i chosen independently according to the distribution $G(n,p)$ satisfies A. The probability that their union satisfies A is therefore also at least $1-\epsilon$.

Friedgut and Kalai showed that the symmetry of graph properties can be applied to obtain a sharper result, as follows.

THEOREM 1.1 ([24]) *For any monotone graph property A, if $G(n,p)$ satisfies A with probability at least ϵ, then $G(n,q)$ satisfies A with probability at least $1-\epsilon$, for $q = p + O(\log(1/2\epsilon)/\log n)$.*

The proof follows by combining two results. The first is a simple but fundamental lemma of Margulis [41] and Russo [52], which is useful in Percolation Theory. This lemma can be used to express the derivative with respect to p of the probability that $G(n,p)$ satisfies A as a sum of contributions associated with the single potential edges. The second result is a theorem of [17] that asserts that at least one such contribution is always large. The symmetry implies that all contributions are the same and the result follows.

Another interesting early discovery in the study of Random Graphs was that many interesting graph invariants are highly concentrated. A striking result of this type was first proved by Matula [40] and strengthened by various researchers; for fixed values of p almost all graphs $G(n,p)$ have the same *clique number*. The clique number of a graph is the maximum number of vertices in a clique, that is, in a subgraph in which any two vertices are adjacent. It turns out that for every fixed positive value of $p < 1$ and every n, there is a real number $r_0 = r_0(n,p)$ which is roughly $2\log n/\log(1/p)$, such that the clique number of $G(n,p)$ is either $\lfloor r_0 \rfloor$ or $\lceil r_0 \rceil$ almost surely. Moreover, $r_0(n,p)$ can be chosen to be an integer for most values of n and p. The proof of this result is not difficult, and is based on the second moment method. One estimates the expectation and the variance of the number of cliques of a given size contained in $G(n, p)$ and applies the inequalities of Markov and Chebyshev.

An *independent set* of vertices in a graph G is a set of vertices no two of which are adjacent. The *chromatic number* $\chi(G)$ of G is the minimum number of independent sets needed to cover all its vertices. This is a more complicated quantity than the clique number, and its behaviour for the random graph $G(n, p)$ is much less understood than the corresponding behaviour of the clique number.

Answering a problem suggested by Erdős and Rényi, Bollobás [14] showed that the chromatic number of $G(n, 0.5)$ is almost surely $(1 + o(1))n/2\log_2 n$. His proof applies a Martingale Inequality to show that almost surely, every set of at least, say, $n/\log^2 n$ vertices of $G(n, 0.5)$ contains an independent subset of size nearly as large as the maximum independent set in the whole graph, implying that a greedy approach of omitting maximum independent sets from the graph one by one yields a nearly optimal coloring.

How concentrated is the chromatic number of $G(n, p)$? Shamir and Spencer [53] proved that there is always a choice of an interval $I = I(n, p)$ of length roughly $\sqrt{n}$, such that the chromatic number of $G(n, p)$ lies, almost surely, in I. More surprisingly, if $p(n) < n^{-5/6-\epsilon}$, then there is always such an interval containing only four distinct values. This was improved by Łuczak [38], who showed that for such values of $p(n)$ the chromatic number is actually, almost surely, one of two consecutive values. In a very recent joint work of the author and Krivelevich it is shown that this is the case whenever $p(n) \leq n^{-1/2-\epsilon}$. This implies the following.

PROPOSITION 1.2 *For every $\alpha < 1/2$ and every integer valued function $r(n) < n^{\alpha}$, there exists a function $p(n)$ such that the chromatic number of $G(n, p(n))$ is precisely $r(n)$ almost surely.*

Therefore, for such values of $p(n)$, almost all graphs $G(n, p(n))$ have the same chromatic number ! The proofs of all these results start by applying a Martingale Inequality to show that if $\delta > 0$ is an arbitrarily small real, and t is the smallest integer for which the chromatic number of $G(n, p(n))$ is at least t with probability that exceeds δ, then one can omit, with probability at least $1 - \delta$, a set of at most $C(\delta)\sqrt{n}$ vertices from $G(n, p(n))$ to get a t-colorable subgraph. This can be combined with several additional combinatorial and probabilistic tools to deduce the above results.

1.2. Probabilistic constructions The *Ramsey number* $R(k, t)$ is the minimum number n such that every graph on n vertices contains either a clique of size k or an independent set of size t. By a special case of the celebrated theorem of Ramsey (cf., e.g., [28]), $R(k, t)$ is finite for every positive integers k and t, and satisfies $R(k, t) \leq \binom{k+t-2}{k-1}$. In particular, $R(k, k) < 4^k$. The problem of determining or estimating the numbers $R(k, t)$ received a considerable amount of attention, and seems to be very difficult in general.

In one of the first applications of the probabilistic method in Combinatorics, Erdős [18] proved that if $\binom{n}{k}2^{1-\binom{k}{2}} < 1$ then $R(k, k) > n$, that is, there exists a graph on n vertices containing neither a clique of size k nor an independent set of size k. The proof is extremely simple; every fixed set of k vertices in the random graph $G(n, 0.5)$ is a clique or an independent set with probability $2^{1-\binom{k}{2}}$. Thus $\binom{n}{k}2^{1-\binom{k}{2}}$ (< 1) is an upper bound for the probability that the random graph $G(n, 0.5)$ contains a clique or an independent set of size k. Despite the simplicity of this proof, there is no constructive version of it, in the sense that there is no known deterministic algorithm that constructs a graph on $n > (1+\epsilon)^k$ vertices with neither a clique nor an independent set of size k, in time which is polynomial in n, where $\epsilon > 0$ is any positive absolute constant.

Ajtai, Komlós and Szemerédi [1] showed, using a probabilistic proof, that $R(3, t) \leq O(t^2/\log t)$. In a recent paper, Kim [33] proves that this is tight, up to a constant factor. This provides the correct asymptotic behaviour of $R(3, t)$:

THEOREM 1.3 ([1], [33]) *There are two positive constants c_1, c_2 such that*

$$c_1 \frac{t^2}{\log t} \leq R(3, t) \leq c_2 \frac{t^2}{\log t},$$

for every t.

Kim's proof is based on a clever "semi-random" construction and proceeds in stages. Starting from the empty graph on n vertices, in each stage choose every potential edge which does not form a triangle with two of the edges picked so far, randomly and independently, with probability $1/(\sqrt{n}\log^2 n)$. If triangles are formed, omit a maximal collection of pairwise edge disjoint triangles, thus completing the stage. The process, which clearly generates a triangle-free graph, terminates after some n^δ stages. It is shown in [33], by combining subtle combinatorial and probabilistic arguments, that with positive probability this process produces a graph whose independence number does not exceed $t = c\sqrt{n \log n}$ for an appropriate choice of an absolute positive constant c. Therefore, $R(3, t) > n = \Omega(t^2/\log t)$, as needed. As is the case with the Ramsey numbers $R(k, k)$, there is no known deterministic efficient algorithm that constructs a triangle-free graph on n vertices which contains no independent sets of size $n^{1/2+o(1)}$.

The above mentioned semi-random approach for constructing the required combinatorial structure in stages, where in each stage some correction may be applied, is influenced by a method developed by Rödl in [51], following similar ideas that appeared in [1]. This technique, which is sometimes called the "Rödl Nibble", was initiated by Rödl in order to solve a packing and covering problem of Erdős and Hanani [19]. His result forms another interesting example of a probabilistic construction. It asserts that for every fixed $k \geq l \geq 2$, there is a collection of at most $\binom{n}{l}/\binom{k}{l} + o(n^l)$ subsets of cardinality k of an n-element set, so that each l-element subset is contained in at least one k-tuple. Note that this means that most l-subsets are covered precisely once, that is, are contained in exactly one of the k-tuples in the collection. The proof is obtained by repeatedly picking a small random subset of the k-tuples that does not intersect any of the ones picked already by more than $l-1$ points. By a careful analysis it can be shown that this produces, with positive probability, a collection of at most $\binom{n}{l}/\binom{k}{l} + o(n^l)$ k-

tuples that cover all but at most $o(n^l)$ l-subsets. Covering the uncovered l-subsets by additional k-sets, one obtains a collection with the desired properties.

The main part of the proof here, as well as in [33], is to maintain certain regularity properties of the combinatorial structure which is being constructed in stages, during the whole process.

This technique has been developed by several researchers, who applied it to prove various interesting results about packing, covering and coloring problems for hypergraphs. Some of these results are mentioned in the next subsection.

Probabilistic constructions have been used extensively in Combinatorial Geometry and Combinatorial Number Theory. A recent geometric example, answering a question of Füredi and Stanley [27], appears in [11], where it is shown that for every k and d there are collections of at least $d^{\Omega(\log(k+2)/\log\log(k+2))}$ nonzero vectors in R^d, in which any $k+1$ members contain an orthogonal pair.

1.3. Proving deterministic theorems A *hypergraph* H is a pair (V, E), where V is a finite set whose members are called *vertices* and E is a finite collection of subsets of V, called *edges*. If every edge contains precisely k vertices, the hypergraph is *k-uniform*. Thus, 2-uniform hypergraphs are graphs. A *matching* in H is a subset of its edges no two of which share a vertex. A *proper coloring* of the edges of H is an assignment of colors to the edges of H so that each color class forms a matching. The *chromatic index* of H is the smallest number of colors used in a proper edge coloring of it.

Several researchers noticed that the Nibble technique developed in [51] can be applied for tackling various packing, covering and coloring problems for hypergraphs. See [25], [47] and [32] for some interesting examples. The results in all these papers are deterministic theorems about hypergraphs, and therefore belong to this subsection. The strongest result of this type, due to Kahn, deals with proper edge colorings of hypergraphs.

THEOREM 1.4 ([32]) *For every $\epsilon > 0$ and every k there is a finite $D_0 = D_0(k, \epsilon)$ with the following property. Let H be a k-uniform hypergraph with maximum degree D, where $D > D_0$. If no two vertices of H share more than ϵD common edges, then for any assignment of a list of at least $D(1+\epsilon)$ colors for each edge of H, there is a proper edge coloring of H assigning to each edge a color from its list.*

In particular, this implies that the chromatic index of H does not exceed $(1+\epsilon)D$, as proved already in [47].

The proofs in the above mentioned papers and in several related ones are based on the Nibble technique, and usually combine it with several Martingale Inequalities or other large deviation inequalities like the one of Talagrand in [58].

A *proper k-coloring* of a graph is an assignment of a color from a set of k colors to each of its vertices so that adjacent vertices get distinct colors. Such a coloring is *acyclic* if there is no two-colored cycle. The *acyclic chromatic number* of a graph is the minimum number of colors in an acyclic coloring of it. The Four Color Theorem, which is the best known result in Graph Theory, asserts that the chromatic number of every planar graph is at most 4. Answering a problem of Grünbaum and improving results of various authors, Borodin [16] showed that every planar graph has an acyclic 5-coloring. He conjectured that for any surface but the plane, the maximum possible chromatic number of a graph embeddable on the surface, is equal to the maximum possible acyclic chromatic number of a graph embeddable on it. The Map Color Theorem (see [50]) determines precisely the maximum possible chromatic number of any graph embeddable on a surface of genus g and shows this maximum is

$$\lfloor \frac{7+\sqrt{1+48g}}{2} \rfloor = \Theta(g^{1/2}).$$

The following result shows that the maximum possible acyclic chromatic number of a graph on such a surface is asymptotically different.

THEOREM 1.5 ([9]) *The acyclic chromatic number of any graph embeddable on a surface of genus g is at most $O(g^{4/7})$. This is nearly tight in the sense that for every $g > 0$ there is a graph embeddable on a surface of genus g whose acyclic chromatic number is at least $\Omega(g^{4/7}/(\log g)^{1/7})$.*

Therefore, the above mentioned conjecture of Borodin is false for all surfaces with large genus.

The proof of the $O(g^{4/7})$ upper bound is probabilistic, and combines some combinatorial arguments with the Lovász Local Lemma. This Lemma, first proved in [20], is a tool for proving that under suitable conditions, with positive probability, none of a large finite collection of nearly independent, low probability events in a probability space holds. This positive probability is often exponentially small, and yet the Local Lemma can be used to show it is positive. The proof of the $\Omega(g^{4/7}/(\log g)^{1/7})$ lower bound is also probabilistic, and is based on an appropriate random construction.

Among the deterministic theorems proved by probabilistic arguments, there are examples of probability theorems. An interesting example of this type is a derivation of a large deviation inequality of Janson ([30], [29], see also [10]). Another example is the 123-theorem proved in [12]; for every two independent identically distributed real random variables X and Y

$$Pr[|X-Y| \leq 2] < 3Pr[|X-Y| \leq 1].$$

2. Pseudo-randomness

The rapid development of theoretical Computer Science and its tight connection to Discrete Mathematics motivated the study of the algorithmic aspects of probabilistic techniques. Can a combinatorial structure whose existence is proved by probabilistic means be constructed *explicitly* (that is, by an efficient deterministic algorithm)? Can the algorithmic problems corresponding to existence probabilistic proofs be solved by efficient procedures? The area of *randomized algorithms* has been developed tremendously during the last decade, when it has been realized that for numerous computational problems, the simplest and fastest algorithms are often randomized ones. Can such algorithms be derandomized, that is, can they be converted into efficient deterministic ones? The investigation of these questions in recent years led to fascinating techniques which are often related to other branches of Mathematics. In this section I briefly describe some of the highlights.

2.1. Expanders An (n, d, c)-*expander* is a d-regular graph on n vertices, such that every set X of at most $n/2$ of its vertices has at least $c|X|$ neighbors outside the set. Infinite families of such graphs with fixed positive values of d and c and growing number of vertices have numerous applications in Combinatorics and Theoretical Computer Science. The simplest way of proving the existence of such families is by a probabilistic construction first described by Pinsker [46]; for every $d \geq 3$ there is some $c = c(d) > 0$ such that a random bipartite graph obtained by choosing d random permutations between the two parts is a $(2n, d, c)$-expander almost surely.

The problem of constructing such families of graphs explicitly is more complicated. Most known constructions rely on the tight relationship between the expansion properties of a graph and the ratio between its largest and second largest eigenvalues. The *adjacency matrix* of a graph $G = (V, E)$ is the matrix $A = (a_{u,v} : u, v \in V)$ in which $a_{u,v}$ is the number of edges between u and v. This is a symmetric matrix, and thus it has real eigenvalues and an orthonormal basis of eigenvectors. If the graph is d-regular, then the largest eigenvalue is d, and the second largest eigenvalue, which is denoted by $\lambda(G)$, is strictly smaller than d iff the graph is connected. It is not too difficult to see that any d regular graph with n vertices and second eigenvalue λ is an (n, d, c)-expander for $c = (d - \lambda)/(2d)$. This (in a slightly stronger form) has been proved, independently, by Tanner in [57] and by the author and Milman in [8]. The proof is simple and applies the variational definition of the second eigenvalue to an appropriate test function.

The converse is more complicated, but is also true, and has been proved in [6].

THEOREM 2.1 *For any (n, d, c)-expander G, $\lambda(G) \leq d - \frac{c^2}{4+2c^2}$.*

Therefore, a d regular graph is highly expanding iff its second eigenvalue is far from the first. Combining this fact with some known results about Kazhdan's Property T of group representations, it is possible to give some explicit families of expanders. These are not, however, the best known constructions.

It is known (see [6]) that for any infinite family of d-regular graphs, the limsup of the second largest eigenvalue is at least $2\sqrt{d-1}$. Lubotzky, Phillips and Sarnak [37], and independently, Margulis [42], constructed, for every $d = p + 1$ where p is a prime congruent to 1 modulo 4, explicit infinite families of d-regular graphs in which the second largest eigenvalue is at most $2\sqrt{d-1}$. Thus, at least in terms of the second eigenvalue, these expanders are best possible. Moreover, in these graphs all the eigenvalues, besides the first, are bounded in absolute value by $2\sqrt{d-1}$. This fact implies certain strong pseudo-random properties, which are useful for some of the applications.

The graphs of [37] and [42] are Cayley graphs of factor groups of the group of all 2 by 2 matrices over a finite field. Their spectral properties are proved by applying results of Eichler and Igusa on the Ramanujan conjectures concerning the number of ways an integer can be represented as a sum of four squares of some special form. Eichler's proof is based on Weil's famous theorem known as the Riemann Hypothesis for curves. More details can be found in [36].

Expanders have numerous applications. They form the basic building blocks of various interconnection and sorting networks, including the sorting network of Ajtai, Komlós and Szemerédi [2] that sorts n elements in $O(\log n)$ parallel steps. They are useful for parallel sorting, merging and selection, and for various variants of the sorting problem, like the "nuts and bolts sorting problem" considered in several papers including [4], [34]. Expanders have recently been used in the construction of Spielman [55] of linear time encodable and decodable error-correcting codes which correct a linear number of errors. Considered as (finite) metric spaces, such graphs cannot be embedded in the Banach spaces ℓ_p with low distortion, as shown by Matoušek [39]. They are also useful in amplification of probabilities, as the random walks on them converge quickly to a uniform distribution. The connection between the expansion properties of graphs and the rate of convergence of random walks on them forms the basis for several algorithms for approximating difficult combinatorial quantities using rapidly mixing Markov chains, developed by Jerrum and Sinclair, see, e.g., [56].

2.2. Derandomization The tremendous recent development of randomized algorithms, described, among other places, in the comprehensive recent book of Motwani and Raghavan [43], motivated the study of the possibility to convert such algorithms into deterministic ones. Although this is not known in many cases, there are several general techniques that often supply the desired derandomization.

One of the general techniques is the *method of conditional probabilities*. An early instance of this method is implicit in a paper of Erdős and Selfridge [23], but the explicit description of the method is due to Spencer (see, e.g., [54] or [10]), and further developments are due to Raghavan [49]. The basic approach is the following; given a random variable X defined on a finite probability space, the objective is to find deterministically and efficiently a point s of the sample space in which the value of X does not exceed its expectation $E(X)$. To do so, assume the points of the sample space are represented by binary vectors, and try to determine the bits of an appropriate point s one by one, where each bit is chosen in a way that ensures that the conditional expectation of X given the bits chosen so far does not exceed $E(X)$. This process, which can be viewed as a variant of binary search, is possible only when the required conditional expectations can be computed efficiently. In some cases precise computation is difficult, and one may use estimates that satisfy certain requirements. These estimates, introduced in [49] and called *pessimistic estimators*, are often useful in applications of this method. Several illustrations of the method appear, among other places, in [49], [10], [54].

Another general technique relies on the fact that many randomized algorithms run successfully even when the random choices they utilize are not fully independent. For the analysis some limited amount of independence, like *k-wise independence* for some fixed k, often suffices. In these cases, it is possible to replace the appropriate exponentially large sample spaces required to simulate all random choices of the algorithms by ones of polynomial size. The algorithms can then be converted into deterministic ones by searching the relatively small sample spaces deterministically.

A simple construction of small sample spaces supporting k-wise independent random variables, appears in [31]. For the case of binary, uniform random variables this is treated under the name *orthogonal arrays* in the Coding Theory literature, see, e.g., [44]. These constructions, as well as some others, are based on some simple properties of polynomials over a finite field or on certain explicit error-correcting codes.

Several researchers realized that constructions of this type are useful for derandomizing *parallel* algorithms, since one may simply check all points of the sample space in parallel. The following simple result supplies a lower bound for the size of any sample space supporting n k-wise independent nonconstant random variables.

PROPOSITION 2.2 *Let $\mathcal{S}$ be a sample space supporting n nontrivial k-wise independent random variables. Then, if k is even, $\mathcal{S}$ has at least $\sum_{i=0}^{k/2}\binom{n}{i}$ points, and if k is odd $\mathcal{S}$ has at least $\sum_{i=0}^{(k-1)/2}\binom{n}{i}+\binom{n-1}{(k-1)/2}$ points.*

Note that this implies that for fixed k and large n, the size of $\mathcal{S}$ is $\Omega(n^{\lfloor k/2 \rfloor})$. For the binary uniform case this proposition is essentially the Rao bound [48], whereas for the general case it is shown in [3], where it is also observed that this is nearly tight in several cases including the binary uniform one. It follows that polynomial size sample spaces suffice only for handling k-wise independence for fixed k. There are, however, several ways to achieve a higher amount of independence. The most promising way, initiated by Naor and Naor in [45] and improved in [5], constructs sample spaces that support random variables any k of which are *nearly* independent. The constructions here are based on certain error-correcting codes together with some simple properties of the Fourier transform of a distribution on an Abelian group.

The above techniques have been applied in numerous papers dealing with derandomization of parallel as well as sequential algorithms and I make no attempt to include a comprehensive list of references here.

There are several additional derandomization techniques, including ones that rely on cryptographic assumptions to generate pseudo-random sequences and including more specific methods, that are not described here.

2.3. Explicit constructions There have been many attempts to convert some known probabilistic proofs of existence of combinatorial structures into explicit constructions. To consider these problems systematically, the notion of an *explicit construction* should first be defined precisely. There are several definitions of this notion and the most natural one is probably the existence of an algorithm for constructing the desired structure in time which is polynomial in its size.

Since the early work of Shannon it has been known that randomly chosen codes have powerful error-correcting properties. A major part of the work in the theory of error-correcting codes is focused on attempts to try and construct explicit codes that are (nearly) as good as random ones. It is still an open problem to determine or estimate the maximum number of vectors of length n over an alphabet of size q so that the Hamming distance between any two vectors is at least d. Let $A_q(n, d)$ denote this maximum. There is, of course, a large number of known upper and lower bounds for $A_q(n, d)$ (cf., e.g., [44]), but even the correct asymptotic behaviour of its logarithm in the binary case is not known. The problem of finding explicit large collections of vectors providing lower bounds for $A_q(n, d)$ is also very difficult, and there are several explicit constructions that rely on some simple properties of polynomials over finite fields as well as on certain deep estimates of character sums. The most exciting explicit constructions are the Algebraic-Geometric codes introduced by Goppa in 1981. Tsfasman, Vladut and Zink proved in [59] that for alphabets that are even powers of primes and exceed 49, these codes yield explicit collections of vectors providing lower bounds for $A_q(n, d)$ which are exponentially better than the best bounds obtained by a random construction (or, equivalently, by the Gilbert-Varshamov bound). Therefore, in coding theory there are interesting cases where explicit constructions beat the best known random ones.

Another example of explicit constructions which are better than the best known random ones is the construction of dense graphs without short cycles see, e.g., [37]. A more recent example, due to Kollár, Rónyai and Szabó [35], is a construction of dense bipartite graphs that do not contain some fixed complete bipartite subgraph. The properties of these graphs are proved by applying some basic tools from Algebraic Geometry.

The best known problem of finding an explicit construction of a combinatorial structure is probably that of constructing explicit Ramsey graphs. As described in subsection 1.2, it is very simple to prove, by a probabilistic argument, the existence of graphs with at least $2^{k/2}$ vertices which contain neither a clique nor an independent set of size k. Yet, the largest known explicit graphs with this property contain only $2^{\Omega(\log^2 k/ \log\log k)}$ vertices. These graphs have been constructed by Frankl and Wilson [26], using certain results on intersections of finite sets, which are proved by applying some linear algebra techniques.

Another Ramsey-type question mentioned in subsection 1.2 deals with the existence of large triangle-free graphs with no large independent sets. Kim [33] proved by an appropriate random construction that there are triangle-free graphs on n vertices whose largest independent sets are of size $O(\sqrt{n}\sqrt{\log n})$. There is no known explicit construction of such a graph. The best known explicit construction, described in [7], gives explicit triangle-free graphs on n vertices whose largest independent set is of size $O(n^{2/3})$. The properties of these graphs, which are Cayley graphs of Abelian groups, are deduced from their spectral properties. These in turn are proved by applying some estimates on character sums.

Combinatorial examples like the last two, in which random constructions give much better results than explicit ones, seem to be much more frequent than examples in which the constructive approach wins. This could be viewed as a victory of the probabilistic method and a sign of its power in the study of problems in Discrete Mathematics, or as a sign of our lack of imagination and ability to find more constructive solutions. In any case, I am convinced that the study and application of probabilistic arguments, and the related study of pseudo-random structures, will keep playing a crucial role in the development of Combinatorics and Theoretical Computer Science in the future.

References

[1] M. Ajtai, J. Komlós and E. Szemerédi, A note on Ramsey numbers, *J. Combinatorial Theory Ser. A* 29 (1980), 354–360.
[2] M. Ajtai, J. Komlós and E. Szemerédi, Sorting in $c \log n$ parallel steps, *Combinatorica* 3 (1983), 1–19.
[3] N. Alon, L. Babai and A. Itai, A fast and simple randomized parallel algorithm for the maximal independent set problem, *J. Alg.* 7 (1986), 567–583.
[4] N. Alon, M. Blum, A. Fiat, S. K. Kannan, M. Naor and R. Ostrovsky, Matching nuts and bolts, *Proc. of the Fifth Annual ACM-SIAM SODA* (1994), ACM Press, 690–696.
[5] N. Alon, O. Goldreich, J. Håstad and R. Peralta, Simple constructions of almost k–wise independent random variables, *Random Structures and Algorithms* 3 (1992), 289–303.
[6] N. Alon, Eigenvalues and expanders, *Combinatorica* 6 (1986), 83–96.
[7] N. Alon, Explicit Ramsey graphs and orthonormal labelings, *The Electronic J. Combinatorics* 1 (1994), R12, 8pp.
[8] N. Alon and V. D. Milman, Eigenvalues, expanders and superconcentrators, *Proc. 25th Annual Symp. on Foundations of Computer Science*, Singer Island, Florida, IEEE (1984), 320–322. (Also: λ_1, isoperimetric inequalities for graphs and superconcentrators, *J. Combinatorial Theory, Ser. B* 38 (1985), 73–88.)
[9] N. Alon, B. Mohar and D. P. Sanders, On acyclic colorings of graphs on surfaces, *Israel J. Math.* 94 (1996), 273–283.
[10] N. Alon and J. H. Spencer, *The Probabilistic Method*, Wiley, New York, 1992.
[11] N. Alon and M. Szegedy, Large sets of nearly orthogonal vectors, to appear.
[12] N. Alon and R. Yuster, The 123 Theorem and its extensions, *J. Combinatorial Theory Ser. A* 72 (1995), 322–331.
[13] B. Bollobás, *Random Graphs*, Academic Press, London, 1985.
[14] B. Bollobás, The chromatic number of random graphs, *Combinatorica* 8 (1988), 49–55.
[15] B. Bollobás and A. Thomason, Threshold functions, *Combinatorica* 7 (1987), 35–38.
[16] O.V. Borodin, On acyclic colorings of planar graphs, *Discrete Math.* 25 (1979), 211–236.
[17] J. Bourgain, J. Kahn, G. Kalai, Y. Katznelson and N. Linial, The influence of variables in product spaces, *Israel J. Math.* 77 (1992), 55–64.
[18] P. Erdős, Some remarks on the theory of graphs, *Bulletin of the Amer. Math. Soc.* 53 (1947), 292–294.
[19] P. Erdős and H. Hanani, On a limit theorem in combinatorial analysis, *Publ. Math. Debrecen*, 10 (1963), 10–13.
[20] P. Erdős and L. Lovász, Problems and results on 3-chromatic hypergraphs and some related questions, in *Infinite and Finite Sets*, A. Hajnal et. al. eds, North Holland (1975), 609–628.

[21] P. Erdős and A. Rényi, On random graphs I, *Publ. Math. Debrecen* 6 (1959), 290–297.
[22] P. Erdős and A. Rényi, On the evolution of random graphs, *Publ. Math. Inst. Hungar. Acad. Sci.* 5 (1960), 17–61.
[23] P. Erdős and J. L. Selfridge, On a combinatorial game, *J. Combinatorial Theory, Ser. A* 14 (1973), 298–301.
[24] E. Friedgut and G. Kalai, Every monotone graph property has a sharp threshold, Proc. AMS, to appear.
[25] P. Frankl and V. Rödl, Near perfect coverings in graphs and hypergraphs, *Europ. J. Combinatorics* 6 (1985), 317–326.
[26] P. Frankl and R. M. Wilson, Intersection theorems with geometric consequences, *Combinatorica* 1 (1981), 357–368.
[27] Z. Füredi and R. Stanley, Sets of vectors with many nearly orthogonal pairs (Research Problem), *Graphs and Combinatorics* 8 (1992), 391–394.
[28] R. L. Graham, B. L. Rothschild and J. H. Spencer, *Ramsey Theory*, Second Edition, Wiley, New York, 1990.
[29] S. Janson, Poisson approximation for large deviations, *Random Structures and Algorithms* 1 (1990), 221–230.
[30] S. Janson, T. Łuczak and A. Ruciński, An exponential bound for the probability of nonexistence of a specified subgraph in a random graph, in *Random Graphs 87* (M. Karonski et. al. eds.), Wiley (1990), 73–87.
[31] A. Joffe, On a set of almost deterministic k-independent random variables, *Annals of Probability* 2 (1974), 161–162.
[32] J. Kahn, Asymptotically good list-colorings, *J. Combinatorial Theory, Ser. A* 73 (1996), 1–59.
[33] J. H. Kim, The Ramsey number $R(3,t)$ has order of magnitude $t^2/\log t$, *Random Structures and Algorithms* 7 (1995), 173–207.
[34] J. Komlós, Y. Ma and E. Szemerédi, Matching nuts and bolts in $O(n \log n)$ time, *Proc. of the 7th Annual ACM-SIAM SODA* (1996), ACM Press, 232–241.
[35] J. Kollár, L. Rónyai and T. Szabó, Norm-graphs and bipartite Turán numbers, *Combinatorica*, to appear.
[36] A. Lubotzky, *Discrete Groups, Expanding Graphs and Invariant Measures,* Birkhäuser Verlag, 1994.
[37] A. Lubotzky, R. Phillips and P. Sarnak, Explicit expanders and the Ramanujan conjectures, *Proc. of the 18th ACM Symp. on the Theory of Computing*, (1986), 240–246; (Also: Ramanujan graphs, *Combinatorica* 8 (1988), 261–277).
[38] T. Łuczak, A note on the sharp concentration of the chromatic number of random graphs, *Combinatorica* 11 (1991), 295–297.
[39] J. Matoušek, On embedding expanders into ℓ_p spaces, to appear.
[40] D. W. Matula, On the complete subgraph of a random graph, *Combinatory Mathematics and its Applications*, Chapel Hill, North Carolina (1970), 356–369.
[41] G. A. Margulis, Probabilistic characteristics of graphs with large connectivity, *Prob. Peredachi Inform.* 10 (1974), 101–108.
[42] G. A. Margulis, Explicit group-theoretical constructions of combinatorial schemes and their application to the design of expanders and superconcentrators, *Prob. Peredachi Inform.*, 24 (1988), 51–60 (in Russian). (English translation in *Problems of Information Transmission*, 24 (1988), 39–46).
[43] R. Motwani and P. Raghavan, *Randomized Algorithms*, Cambridge University Press, New York, 1995.
[44] F. J. MacWilliams and N. J. A. Sloane, *The Theory of Error-Correcting Codes*, North Holland, Amsterdam, 1977.

[45] J. Naor and M. Naor, Small–bias probability spaces: efficient constructions and applications, *SIAM J. Comput.* 22 (1993), 838–856.

[46] M. Pinsker, On the complexity of a concentrator, 7th *Internat. Teletraffic Conf.*, (1973), Stockholm, 318/1–318/4.

[47] N. Pippenger and J. H. Spencer, Asymptotic behaviour of the chromatic index for hypergraphs, *J. Combinatorial Theory, Ser. A* 51 (1989), 24–42.

[48] C. R. Rao, Factorial experiments derivable from combinatorial arrangements of arrays, *J. Royal Stat. Soc.* 9 (1947), 128–139.

[49] P. Raghavan, Probabilistic construction of deterministic algorithms: approximating packing integer programs, *J. Comput. Syst. Sci.*, 37 (1988), 130–143.

[50] G. Ringel and J. W. T. Youngs, Solution of the Heawood map coloring problem, *Proc. Nat. Acad. Sci. U.S.A.* 60 (1968), 438–445.

[51] V. Rödl, On a packing and covering problem, *European Journal of Combinatorics* 5 (1985), 69–78.

[52] L. Russo, On the critical percolation probabilities, *Z. Wahrsch. verw. Gebiete* 43 (1978), 39–48.

[53] E. Shamir and J. H. Spencer, Sharp concentration of the chromatic number on random graphs $G_{n,p}$, *Combinatorica* 7 (1987), 124–129.

[54] J. H. Spencer, *Ten lectures on the Probabilistic Method*, Second Edition, SIAM, Philadelphia, 1994.

[55] D. Spielman, Linear-Time Encodable and Decodable Error-Correcting Codes, *Proc. of the* 27th *ACM Symp. on the Theory of Computing* 1995, ACM Press, 388–397.

[56] A. Sinclair and M. R. Jerrum, Approximate counting, uniform generation and rapidly mixing Markov chains, *Information and Computation* 82 (1989), 93–133.

[57] R. M. Tanner, Explicit construction of concentrators from generalized N-gons, *SIAM J. Alg. Disc. Meth.* 5 (1984), 287–293.

[58] M. Talagrand, A new isoperimetric inequality for product measure and the tails of sums of independent random variables, *Geometric and Functional Analysis* 1 (1991), 211–223.

[59] M. A. Tsfasman, S. G. Vladut and T. Zink, Modular curves, Shimura curves and Goppa codes, better than the Varshamov-Gilbert bound, *Math. Nachr.* 104 (1982), 13–28.

Progress in Mathematics, Vol. 168, © 1998 Birkhäuser Verlag Basel/Switzerland

Free Discontinuity Problems and Special Functions with Bounded Variation

LUIGI AMBROSIO

Dipartimento di Matematica, Università di Pavia
Via Abbiategrasso 215, 27100 Pavia, Italy

Introduction

Many problems in the Calculus of Variations are characterized by a competition between volume and surface energies. A simple example is the so-called *prescribed curvature problem* (see for instance [58])

$$\min\left\{\int_E g(x)\,dx + \mathcal{H}^{N-1}(\partial E) : E \subseteq \mathbf{R}^N\right\} \tag{1}$$

where $g \in L^1(\mathbf{R}^N)$ is a given function and $\mathcal{H}^{N-1}$ is the Hausdorff $(N-1)$-dimensional measure in $\mathbf{R}^N$. In this problem, if $g << 0$ in some (not too irregular) region, it is convenient to include it in E, paying the perimeter of the region. The terminology for problem (1) is justified by the first variation, performed by changes in the independent variable: if g is continuous at a regular point x of ∂E, the following holds

$$\mathbf{H}(x) = g(x)\nu_E(x) \tag{2}$$

where $\mathbf{H}$ is the mean curvature vector of ∂E and ν_E is the outer normal to E.

According to the terminology introduced in [44], by a *free discontinuity problem* we mean a variational problem with a competition between volume and surface energies, with the second ones supported on sets which are not fixed a priori, as in (1). However, unlike (1), these sets are not necessarily boundaries.

The most popular, and in some sense canonical, example of a free discontinuity problem is the Mumford–Shah minimization problem

$$\inf\left\{F(u,K) : K \subseteq \overline{\Omega} \text{ compact}, \;\; u \in C^1(\Omega \setminus K)\right\} \tag{3}$$

where F is given by

$$F(u,K) := \int_{\Omega\setminus K} \left[|\nabla u|^2 + \alpha(u-g)^2\right] dx + \beta\mathcal{H}^{N-1}(K \cap \Omega). \tag{4}$$

Here Ω is a bounded open set in $\mathbf{R}^N$, α, $\beta > 0$ are fixed parameters and $g \in L^\infty(\Omega)$. By minimizing F one looks for a "piecewise smooth" approximation of g. Even though F depends on two variables, the most important unknown is K; indeed, given K, u is the unique solution of the Neumann boundary value problem

$$\begin{cases} \Delta u = \alpha(u-g) & \text{in } \Omega \setminus K; \\ \dfrac{\partial u}{\partial n} = 0 & \text{in } \partial\Omega \cup K. \end{cases} \tag{5}$$

If $K = \emptyset$, problem (5) corresponds to the classical globally smooth approximation of g, as α goes to ∞.

In the two-dimensional case ($N = 2$), problem (3) has been suggested by Mumford and Shah in [63] in connection with a variational approach to the edge detection problem. In this case (see Figure 1), typically Ω is a rectangle in the plane and the function g represents the grey level of an image seen by a camera. By minimizing F one tries to eliminate noise effects, smoothing g out of a discontinuity set K representing the edges of the objects seen by the camera. The advantage of the piecewise smooth approximation, as opposed to a globally smooth one, is that, since u may be discontinuous across K, smoothing is prevented near K and this allows a more precise definition of the edge set in the applications.

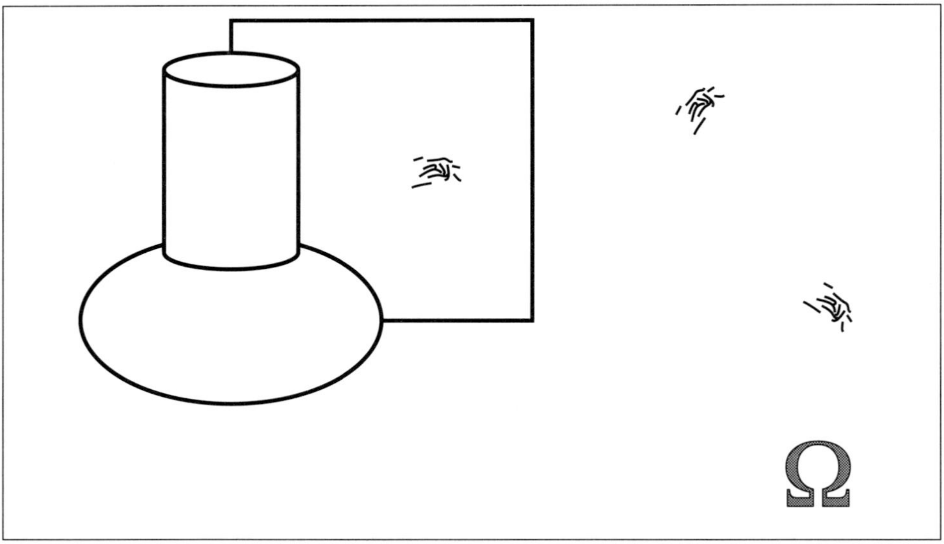

Fig. 1

Besides the Mumford–Shah problem, it is worth mentioning a different example of a free discontinuity problem, related to the mathematical theory of nematic liquid crystals. For any bounded domain $D \subseteq \mathbf{R}^3$ with Lipschitz boundary and any function $\mathbf{n} \in H^1(D, \mathbf{S}^2)$ we define

$$E(\mathbf{n}, D) := \int_D W(\mathbf{n}, \nabla\mathbf{n})\, dx + \int_{\partial D} f(\mathbf{n}, \nu_D)\, d\mathcal{H}^2 \tag{6}$$

where ν_D is the outer normal to D. Interpreting D as a drop of liquid crystal and $\mathbf{n}$ as the optic axis of the crystal, the functional above represents the sum of the volume energy and the surface contact energy associated with the crystal; typical choices of W and f are

$$W_F(\mathbf{n},\nabla\mathbf{n}) := \kappa_1\big(\operatorname{div}\mathbf{n}\big)^2 + \kappa_2\big(\langle\mathbf{n},\operatorname{curl}\mathbf{n}\rangle\big)^2 + \kappa_3\big|\mathbf{n}\wedge\operatorname{curl}\mathbf{n}\big|^2$$
$$+ (\kappa_2+\kappa_4)\big(\operatorname{tr}(\nabla\mathbf{n})^2 - (\operatorname{div}\mathbf{n})^2\big) \quad \text{and} \quad f(\mathbf{n},\nu) := g(\langle\mathbf{n},\nu\rangle).$$

W_F, the so-called Oseen-Frank energy, reduces to $\kappa|\nabla\mathbf{n}|^2$ for a particular choice of κ_i. Even though initially the surface energy is supported on a boundary, the minimization of $E(\mathbf{n},D)$ with suitable boundary conditions and constraints can be included in the category of free discontinuity problems, because D might develop an interior boundary with surface energy supported on it. Usually, the minimization of (6) is achieved either by fixing D a priori, or by putting constraints on D (e.g., convexity) which prevent the formation of interior boundaries (see for instance [23], [56]).

The aim of this lecture is to give a brief survey of the current research in the field of free discontinuity problems, focusing on the following issues:

- Existence of weak solutions;
- Regularity of solutions;
- Computation of (approximate) solutions.

1. The space SBV and existence of weak solutions

The main difficulty in proving the existence of solutions of problem (3), even in the two-dimensional case, is due to the fact that no topology on compact sets ensures at the same time compactness of minimizing sequences and lower semicontinuity of

$$K \mapsto \mathcal{H}^{N-1}(K\cap\Omega).$$

Hence, the direct method cannot be applied to this formulation of the problem and a weak formulation seems to be necessary. Since the problem involves functions which are discontinuous across K, it is natural to look for weak solutions in the space $BV(\Omega)$ of functions with *bounded variation* in Ω.

Definition 1.1. We will denote by $Du := (D_1u,\dots,D_Nu)$ the vector measure representing the distributional derivative of $u\in BV(\Omega)$, that is

$$\int_\Omega u\frac{\partial\phi}{\partial x_i}\,dx = -\int_\Omega \phi\,dD_iu \qquad \forall\phi\in C_0^\infty(\Omega). \tag{1.1}$$

Moreover, we write Du as D^au+D^su with D^au absolutely continuous with respect to Lebesgue measure $\mathcal{L}^N$, D^su singular with respect to $\mathcal{L}^N$, and we denote by ∇u the density of D^au with respect to $\mathcal{L}^N$. Finally, Su will denote the *discontinuity set* of u, i.e., the set of all points $x\in\Omega$ which are not Lebesgue points of u.

With these definitions at hand, a naive interpretation of (4) would be

$$\inf_{u\in BV(\Omega)}\int_\Omega\big[|\nabla u|^2+\alpha(u-g)^2\big]\,dx+\beta\mathcal{H}^{N-1}(Su), \tag{1.2}$$

noticing that the integral can be extended to the whole of Ω because Su is Lebesgue negligible. However, the space $BV(\Omega)$ is too large for the minimization of the Mumford–Shah functional. To show this fact, let us consider the class $\mathcal{C}$ of continuous functions $u \in BV(\Omega)$ such that $\nabla u = 0$. The class $\mathcal{C}$ contains nonconstant functions (the Cantor-Vitali function, for instance) and it is not hard to see that $\mathcal{C}$ is dense in $L^2(\Omega)$. Since on $\mathcal{C}$ the functional in (1.2) reduces to

$$\alpha \int_\Omega |u-g|^2\,dx$$

we obtain the surprising result that the infimum in (1.2) is zero, attained if and only if u is (equivalent to) a function in $\mathcal{C}$. This pathological behaviour can be overcome by a more detailed analysis of the distributional derivative of a BV function:

Definition 1.2. Let $u \in BV(\Omega)$. The restriction $D^j u$ of $D^s u$ to Su is called the *jump* part of the derivative, i.e.,

$$D^j u(B) := (D^s u \llcorner Su)(B) = D^s u(B \cap Su) \qquad \forall B \in \mathcal{B}(\Omega).$$

The restriction $D^c u$ of $D^s u$ to $\Omega \setminus Su$ is called the *Cantor* part of the derivative, i.e.,

$$D^c u(B) := D^s u \llcorner (\Omega \setminus Su)(B) = D^s u(B \setminus Su) \qquad \forall B \in \mathcal{B}(\Omega).$$

In particular

$$Du = D^a u + D^j u + D^c u. \tag{1.3}$$

Now we recall the main properties of the components $D^a u$, $D^j u$, $D^c u$ of Du, noticing that the first two, unlike the third, can be both identified by blow-up arguments.

Theorem 1.3. ([47], [31], [72], [73]) *Let $u \in BV(\Omega)$. Then*
(i) for $\mathcal{L}^N$-almost every $x \in \Omega$ the functions

$$u_{x,\varrho}(y) := \frac{u(x+\varrho y) - u(x)}{\varrho}$$

converge in $L^1_{\rm loc}(\mathbf{R}^N)$ as $\varrho \to 0$ to the linear function $\langle \nabla u(x), y\rangle$, where ∇u is the density of $D^a u$;
(ii) for $\mathcal{H}^{N-1}$-almost every $x \in Su$ the functions

$$v_{x,\varrho}(y) := u(x+\varrho y) \qquad y \in \frac{\Omega - x}{\varrho}$$

converge in $L^1_{\rm loc}(\mathbf{R}^N)$ as $\varrho \to 0^+$ to the function v_x defined by

$$v_x(y) := \begin{cases} u^+(x) & \text{if } \langle y, \nu_u(x)\rangle > 0; \\ u^-(x) & \text{if } \langle y, \nu_u(x)\rangle < 0 \end{cases}$$

for suitable $u^+(x)$, $u^-(x) \in \mathbf{R}$, $\nu_u(x) \in \mathbf{S}^{N-1}$. *In addition,*

$$D^j u(B) = \int_{B\cap Su} (u^+ - u^-)\nu_u \, d\mathcal{H}^{N-1} \qquad \forall B \in \mathcal{B}(\Omega);$$

(iii) $|D^c u|(B) = 0$ *for any Borel set* $B \subseteq \Omega$ *such that* $\mathcal{H}^{N-1}(B) < +\infty$.

The functions ∇u, (u^+, u^-), ν_u in Theorem 1.3 are called respectively the *approximate differential* of u, the *approximate one-sided limits* of u, the *approximate normal* to Su. The word "approximate" means that the limits are understood only in a measure theoretic sense.

Since free discontinuity problems deal with volume and surface energies, it is natural to allow in these problems only functions $u \in BV(\Omega)$ whose distributional derivative has the same structure. Since the Cantor part of the derivative is singular with respect to $\mathcal{L}^N$ and is not supported in $(N-1)$-dimensional sets, this leads to the following definition:

Definition 1.4. ([45], [6]) Let $u \in BV(\Omega)$; we say that u is a *special function with bounded variation*, and we write $u \in SBV(\Omega)$, if $D^c u = 0$. By Theorem 1.3 it follows that $u \in SBV(\Omega)$ if and only if

$$Du(B) = \int_B \nabla u \, dx + \int_{B\cap Su} (u^+ - u^-)\nu_u \, d\mathcal{H}^{N-1} \qquad \forall B \in \mathcal{B}(\Omega).$$

The fundamental result concerning the space $SBV(\Omega)$ is the compactness–lower semicontinuity theorem stated below. We denote by $\phi : [0,\infty) \to [0,\infty)$, $\theta : (0,\infty) \to (0,\infty)$ a convex and a concave function respectively such that

$$\lim_{t\to\infty} \frac{\phi(t)}{t} = \infty, \qquad \lim_{t\to 0} \frac{\theta(t)}{t} = \infty. \tag{1.4}$$

Theorem 1.5. *Let* $\Omega \subseteq \mathbf{R}^N$ *be a bounded open set and let* $(u_n) \subseteq SBV(\Omega)$ *be a sequence satisfying*

$$\int_\Omega \phi(|\nabla u_n|) \, dx + \int_{Su_n} \theta(|u_n^+ - u_n^-|) \, d\mathcal{H}^{N-1} + \|u_n\|_\infty \le C \tag{1.5}$$

for some real constant C. *Then,* (u_n) *has a subsequence* $(u_{n(k)})$ *converging in* $L^1(\Omega)$ *to* $u \in SBV(\Omega)$. *Moreover,* $\nabla u_{n(k)}$ *weakly converges to* ∇u *in* $L^1(\Omega; \mathbf{R}^N)$ *and*

$$\int_\Omega \phi(|\nabla u|) \, dx \le \liminf_{k\to\infty} \int_\Omega \phi(|\nabla u_{n(k)}|) \, dx \tag{1.6}$$

$$\int_{Su} \theta(|u^+ - u^-|) \, d\mathcal{H}^{N-1} \le \liminf_{k\to\infty} \int_{Su_{n(k)}} \theta(|u^+_{n(k)} - u^-_{n(k)}|) \, d\mathcal{H}^{N-1}. \tag{1.7}$$

The proof of Theorem 1.5 was achieved in [6] using a reduction, via an integral-geometric approach, to the one-dimensional case. This approach was based on a useful *structure* theorem for BV functions, showing that the decomposition (1.3) is preserved by the one-dimensional sections of u.

We notice that the growth assumptions on ϕ and θ cannot be weakened; indeed, *any* BV function can be approximated by smooth functions (v_n) satisfying

$$\sup_n \int_\Omega |\nabla v_n|\,dx < +\infty$$

and by piecewise constant functions (w_n) satisfying

$$\sup_n \int_{Jw_n} |w_n^+ - w_n^-|\,d\mathcal{H}^{N-1} < +\infty.$$

Roughly speaking, the growth conditions on ϕ and θ together with (1.5) prevent the possibility of creating a Cantor part of the derivative either with $D^a u_n$ or with $D^j u_n$, and force a separate convergence of $D^a u_n$ to $D^a u$ and of $D^j u_n$ to $D^j u$.

We also notice that the L^∞ bound on u_n can be removed working in the generalized space (see [6], [8], [64], [65])

$$GSBV(\Omega) := \big\{u : \Omega \to \mathbf{R} : K \wedge u \vee -K \in SBV(\Omega)\ \forall K > 0\big\}.$$

The proof of the structure theorem is rather technical; recent and simpler proofs of the SBV compactness theorem have been given in [10] for the case $\theta \equiv 1$ and in [4] for the general case. These proofs are based on the following criterion for membership to SBV:

Proposition 1.6. *Let $u \in BV(\Omega)$ and for any function $\psi \in C_0^1(\mathbf{R})$ define*

$$\|\psi\|_\theta = \sup\left\{\frac{|\psi(t) - \psi(s)|}{\theta(t-s)} : s,\, t \in \mathbf{R},\ \ s < t\right\}.$$

Then, $u \in SBV(\Omega)$ and

$$I_\theta(u) = \int_{Su} \theta(|u^+ - u^-|)\,d\mathcal{H}^{N-1} < +\infty$$

if and only if there exist a constant M and a function $w \in L^1(\Omega; \mathbf{R}^N)$ such that

$$|D\psi(u) - \psi'(u) w \mathcal{L}^N|(\Omega) \le M\|\psi\|_\theta \qquad \forall \psi \in C_0^1(\mathbf{R}). \tag{1.8}$$

Moreover, if (1.8) holds, then $w = \nabla u$ and the least constant M is $I_\theta(u)$.

The idea behind Proposition 1.6 is that the chain rule enables the jump part $D^j u$ to be discriminated from the diffuse part $D^a u + D^c u$ of the derivative of u; indeed, for any $u \in BV(\Omega)$ and for any function $\psi \in C_0^1(\mathbf{R})$ the following holds (see for instance [73], [10])

$$D\psi(u) = \psi'(u) D^a u + \big(\psi(u^+) - \psi(u^-)\big)\nu_u \mathcal{H}^{N-1} \llcorner Su + \psi'(\tilde{u}) D^c u \tag{1.9}$$

where $\tilde{u}$ is the Lebesgue representative of u, defined at any point in the complement of Su. In particular, the measure valued map

$$\psi \mapsto D\psi(u) - \psi'(u)D^a u$$

discriminates between the jump part and the Cantor part of the derivative. Clearly, (1.9) shows that any SBV function fulfils (1.8) with $w = \nabla u$ and $M \geq I_\theta(u)$. The opposite implication, proved in [4] (in [10] for $\theta \equiv 1$), is based on (1.4) and on the possibility of choosing functions ψ with very large derivatives and such that $\|\psi\|_\theta \leq 1$.

If $\theta \equiv 1$, (1.9) becomes

$$\left| \int_\Omega \psi(u)\nabla\phi - \phi\psi'(u)w\,dx \right| \leq M \operatorname{osc}(\psi) \qquad \forall\phi \in C_0^1(\Omega),\ \psi \in C_0^1(\mathbf{R}). \tag{1.10}$$

As explained in [10], a geometric interpretation of (1.10) in terms of the representability by integration of the boundary (in the sense of currents) of the vertical part of the graph of u can be given.

Proof of Theorem 1.5. From (1.4) and (1.5) it follows that

$$\int_\Omega |\nabla u_n|\,dx + \int_{Su_n} |u_n^+ - u_n^-|\,d\mathcal{H}^{N-1} \leq C' < +\infty.$$

Hence, $|Du_n|(\Omega) \leq C'$ and the BV compactness theorem yields a subsequence (which we still denote by (u_n)) converging in $L^1(\Omega)$ to $u \in BV(\Omega)$.

To prove that $u \in SBV(\Omega)$, we notice that the Dunford–Pettis theorem implies that the sequence $w_n = \nabla u_n$ is weakly relatively compact in $L^1(\Omega; \mathbf{R}^N)$, so that it is not restrictive to assume that w_n weakly converges as $n \to \infty$ to some w. Applying (1.8) to the functions u_n, and using the convergence of the measures $D\psi(u_n) - \psi'(u_n)w_n\mathcal{L}^N$ to $D\psi(u) - \psi'(u)w\mathcal{L}^N$, we infer that u satisfies (1.8) with

$$M = \liminf_{n\to\infty} I_\theta(u_n).$$

Then, $u \in SBV(\Omega)$ and $w = \nabla u$, that is, ∇u_n weakly converges to ∇u. Since ϕ is convex, (1.6) follows. Finally, (1.7) follows by the inequality $I_\theta(u) \leq M$.

Coming back to the Mumford–Shah problem (3) and to its weak formulation (1.2), we can restrict the domain of the functional to $SBV(\Omega)$; moreover, by a truncation argument, we can look for minimizers in the class

$$\{u \in SBV(\Omega) \cap L^\infty(\Omega) : \|u\|_\infty \leq \|g\|_\infty\}.$$

Applying the compactness-lower semicontinuity Theorem 1.5 to a minimizing sequence (with $\phi(t) := t^2$, $\theta \equiv 1$), the direct method yields the following result:

Theorem 1.7. *The functional*

$$u \in SBV(\Omega) \mapsto \int_\Omega \left[|\nabla u|^2 + \alpha(u-g)^2\right] dx + \beta\mathcal{H}^{N-1}(Su) \tag{1.11}$$

has minimizers in $SBV(\Omega)$. Moreover,

$$\min_{SBV(\Omega)} E(u) \leq \inf_{u,K} F(u,K). \tag{1.12}$$

The inequality (1.12) follows from the fact that $u \in SBV(\Omega)$ and $Su \subseteq K$ whenever (u,K) satisfies the properties in (3) and $F(u,K) < +\infty$. We will see in the next section that actually equality holds in (1.12).

Now, let us show how a similar existence result can be stated for a weak formulation of problem (6). In order to exclude trival solutions, we will minimize the functional $E(\mathbf{n},D)$ in (6) in the class of sets D satisfying $D \subseteq K$, $\mathcal{L}^3(D) = \gamma$, with $K \subseteq \mathbf{R}^N$ compact and $\gamma > 0$ given. In this case, we will look for solutions in a class of vector valued SBV functions, noticing that the definitions of Su, Du, $D^a u$, $D^j u$, $D^c u$ as well as Theorem 1.3 can be extended to this case (see [72], [73], [8]).

More precisely, we set

$$\mathcal{A} := \left\{\mathbf{n} \in [SBV(\mathbf{R}^3)]^3 : \mathbf{n}(x) \in \mathbf{S}^2 \cup \{0\} \quad \text{for } \mathcal{L}^3\text{-a.e. } x \in \mathbf{R}^3\right\}.$$

In addition, for any function $\mathbf{n} \in \mathcal{A}$ we define

$$E^*(\mathbf{n}) := \int_{\mathbf{R}^3} W^*(\mathbf{n}, \nabla\mathbf{n})\,dx + \int_{S\mathbf{n}} [f^*(\mathbf{n}^+, \nu_{\mathbf{n}}) + f^*(\mathbf{n}^-, -\nu_{\mathbf{n}})]\,d\mathcal{H}^2$$

with

$$f^*(\mathbf{n},\nu) := \begin{cases} f(\mathbf{n},\nu) & \text{if } |\mathbf{n}| = 1; \\ 0 & \text{if } \mathbf{n} = 0, \end{cases} \qquad W^*(\mathbf{n},z) := \begin{cases} W(\mathbf{n},z) & \text{if } |\mathbf{n}| = 1; \\ |z|^2 & \text{if } \mathbf{n} = 0. \end{cases} \tag{1.13}$$

Then, denoting by $D_{\mathbf{n}}$ the set $\{x \in \mathbf{R}^3 : \mathbf{n}(x) \in S^2\}$, we can consider the minimum problem

$$\min\left\{E^*(\mathbf{n}) : \mathbf{n} \in \mathcal{A},\ D_{\mathbf{n}} \subseteq K,\ \mathcal{L}^3(D_{\mathbf{n}}) = \gamma\right\}. \tag{1.14}$$

In this setting, $D_{\mathbf{n}}$ represents the drop of liquid crystal and the surface energy term contains an additional energy contribution due to "interior boundary points", i.e., points $x \in S\mathbf{n}$ such that $\mathbf{n}^+(x)$, $\mathbf{n}^-(x)$ both belong to $\mathbf{S}^2$.

Theorem 1.8. *Assume that $W(\mathbf{n},z)$ is continuous, convex in z and that*

$$W(\mathbf{n},z) \geq c|z|^2 \qquad \forall \mathbf{n} \in \mathbf{S}^2,\ z \in \mathbf{R}^{3\times 3}$$

for a suitable constant $c > 0$. Assume in addition that $f(\mathbf{n},\nu)$ is continuous, strictly positive and that

$$f_{\mathbf{n}}(p) := \begin{cases} |p| f\left(\mathbf{n}, \dfrac{p}{|p|}\right) & \text{if } p \neq 0; \\ 0 & \text{if } p = 0 \end{cases} \tag{1.15}$$

is convex in $\mathbf{R}^3$ for any $\mathbf{n} \in \mathbf{S}^2$. Then, the minimum problem (1.14) has solution.

The proof of Theorem 1.8 (see [8]) is mainly based on Theorem 1.5, which can be applied to each component of a minimizing sequence. However, we notice that the surface energy term in (1.14) is not isotropic in general, so we cannot apply (1.7) to obtain the lower semicontinuity of surface energy.

The lower semicontinuity of the volume and surface energies follow by the two theorems stated below:

Theorem 1.9. *Let $f(x, s, z) : \Omega \times \mathbf{R}^p \times \mathbf{R}^q \to [0, \infty]$ be a Borel function satisfying:*
(i) $(s, z) \mapsto f(x, s, z)$ is lower semicontinuous in $\mathbf{R}^p \times \mathbf{R}^q$ for any $x \in \Omega$;
(ii) $z \mapsto f(x, s, z)$ is convex in $\mathbf{R}^q$ for any $x \in \Omega$, $s \in \mathbf{R}^p$.
Then, the following holds

$$\int_\Omega f\big(x, u(x), z(x)\big)\, dx \leq \liminf_{n\to\infty} \int_\Omega f\big(x, u_n(x), z_n(x)\big)\, dx$$

for any sequence $(u_n) \subseteq L^1(\Omega; \mathbf{R}^p)$ converging to u in the strong L^1 topology and any sequence $(z_n) \subseteq L^1(\Omega; \mathbf{R}^q)$ converging to z in the weak L^1 topology.

The above result was proved by Ioffe (see [55]) in the framework of optimal control theory and applies, of course, to the function $W^*(\mathbf{n}, z)$ in (1.13), suitably extended to $\mathbf{R}^3 \times \mathbf{R}^{3\times 3}$.

We notice that, under the convexity assumption (ii) on f, the equality $z = \nabla u$ (and $q = Np$) is not relevant for lower semicontinuity. As we will see, this information is crucial in estabilishing lower semicontinuity results for the volume energy under weaker assumptions on f, such as polyconvexity or quasi-convexity (see [19], [1], [38], [9]).

Theorem 1.10. *Let $K \subseteq \mathbf{R}^p$ be a compact set and let $\varphi(a, b, \nu) : K \times K \times \mathbf{S}^{N-1} \to [0, \infty]$ be a function such that*

$$\varphi(a, b, \nu) = \varphi(b, a, -\nu) \qquad \forall a, b \in K,\ \nu \in \mathbf{S}^{N-1}. \tag{1.16}$$

Denoting by $\mathcal{M}$ the space of $\mathbf{R}^N$-valued Radon measures in K, let us assume the existence of a positively 1-homogeneous, convex and lower semicontinuous map $\Phi : \mathcal{M} \to [0, \infty]$ such that

$$\varphi(a, b, \nu) = \Phi\big((\delta_a - \delta_b)\nu\big) \qquad \forall a, b \in K,\ \nu \in \mathbf{S}^{N-1}. \tag{1.17}$$

Then,

$$\int_{Su} \varphi(u^+, u^-, \nu_u)\, d\mathcal{H}^{N-1} \leq \liminf_{n\to\infty} \int_{Su_n} \varphi(u_n^+, u_n^-, \nu_{u_n})\, d\mathcal{H}^{N-1}$$

for any sequence $(u_n) \subseteq SBV(\Omega; K)$ satisfying, component by component, the assumptions of Theorem 1.5.

The proof of Theorem 1.10 is based on the Hahn–Banach theorem, which enables φ to be approximated from below with functions φ_V of the form

$$\varphi_V(a, b, \nu) = \langle V(a) - V(b), \nu\rangle \qquad V : K \to \mathbf{R}^N \text{ continuous.}$$

The chain rule (1.9) shows that the surface energies induced by φ_V are lower semicontinuous and, by approximation, the same property is valid for φ. Setting $K = \mathbf{S}^2 \cup \{0\} \subseteq \mathbf{R}^3$, the function

$$\varphi(a,b,\nu) = f^*(a,\nu) + f^*(b,-\nu)$$

appearing in the surface energy term of E^* fulfils (1.16); in addition, the functional

$$\Phi(\mu) := \int_K f^*\left(s, \frac{d\mu}{d|\mu|}(s)\right) d|\mu|(s)$$

is convex, positively 1-homogeneous, weakly lower semicontinuous (see [30]) and satisfies (1.17), therefore Theorem 1.10 applies to φ.

We conclude this section with a short review of other free discontinuity problems for which the SBV theory provides existence of weak solutions:

- Problems involving vector valued functions and nonconvex bulk energies.

In the framework of SBV, the Acerbi–Fusco lower semicontinuity theorem can be reproduced (see [9]), proving lower semicontinuity of some functionals whose volume energy

$$\int_\Omega f(x,u,\nabla u)\, dx$$

is not convex in ∇u, the convex case being covered by Theorem 1.9. The main assumptions on the energy density f are the *quasi-convexity* and a standard p-growth condition for some $p > 1$. An example of a functional in this class is

$$\int_\Omega \left[|\nabla u|^2 + |\det \nabla u|\,\right] dx + \mathcal{H}^1(Su) + \text{ lower order terms} \tag{1.18}$$

with $\Omega \subseteq \mathbf{R}^2$, $u : \Omega \to \mathbf{R}^2$, $p = 2$. Regularity of minimizers of the above problem and the density lower bound for their jump sets have been studied in [2].

- Second order problems with discontinuities in the gradient.

In these problems, one looks for continuous solutions, allowing jumps of the gradient. The analytic formulation is based on the space

$$SBH(\Omega) := \left\{u \in W^{1,1}(\Omega) : \nabla u \in SBV(\Omega; \mathbf{R}^N)\right\}$$

(analogous to the $BH(\Omega)$ space introduced in [43]) and the model functional is

$$\int_\Omega |\nabla^2 u|^2\, dx + \mathcal{H}^{N-1}(S\nabla u) + \text{ lower order terms.} \tag{1.19}$$

Applications to the theory of elastic-plastic plates have been given in [33], [34]. Recently, problems with discontinuities both in u and in ∇u have also been considered in [35].

- Special functions with bounded deformation.

The space $BD(\Omega)$ of functions with *bounded deformation* was introduced in [59], [68], [69] to model plastic behaviour in linear elasticity. A function $u \in L^1(\Omega; \mathbf{R}^N)$ belongs to $BD(\Omega)$ if the symmetric part Eu of the distributional derivative Du is representable by a (symmetric matrix valued) measure. In other words, there exist measures $E_{ij}u$ such that

$$\frac{1}{2}\int_\Omega \Big[u^i \frac{\partial \phi}{\partial x_j} + u^j \frac{\partial \phi}{\partial x_i}\Big]\, dx = -\int_\Omega \phi\, dE_{ij}u \qquad \forall \phi \in C_0^\infty(\Omega),\ 1 \le i, j \le N.$$

The analogy with BV spaces is apparent, but $BD(\Omega)$ is stricly larger than $BV(\Omega; \mathbf{R}^N)$. In [21] and [12] the SBV theory has been extended to this context, decomposing Eu into an absolutely continuous part $E^a u$, a jump part $E^j u$ and a Cantor part $E^c u$. Accordingly, a space $SBD(\Omega)$ of *special* functions with bounded deformation can be defined, in which variational problems (analogous to (3))

$$\int_\Omega |\mathcal{E}u|^2\, dx + \mathcal{H}^{N-1}(Ju) + \text{ lower order terms} \tag{1.20}$$

can be formulated. Here the *approximate symmetric differential* $\mathcal{E}u$ is the density of $E^a u$ with respect to $\mathcal{L}^N$ and the *jump set* Ju of u is the set of points $x \in Su$ such that u has approximate one–sided limits at x, in the sense specified in Theorem 1.3(ii).

2. Regularity of solutions

We have seen in the preceding section that many free discontinuity problems have a weak formulation in SBV. The essential difference between the weak formulation and the original one is that the jump set Su of a SBV function u is not necessarily closed and, still being $(N-1)$-dimensional from the measure theoretic viewpoint, it may even be dense. Hence, the first step in the regularization process is to ascertain whether, at least for minimizers, a pair (u, K) with K closed and u continuously differentiable out of K can be recovered.

Let us confine our attention, to fix ideas, to a minimizer u of the Mumford–Shah functional in (1.11), and let $K = \overline{Su}$. We know that the measure Du is absolutely continuous outside K, hence $u \in W^{1,2}(\Omega \setminus K)$; moreover, the standard variation $u \to u + \varepsilon\phi$ with $\phi \in C_0^\infty(\Omega \setminus K)$ leads to the equation

$$\Delta u = \alpha(u - g) \in L^\infty \qquad \text{in } \Omega \setminus K,$$

(the equation must be understood in the sense of distributions) hence $u \in W^{2,p}_{\text{loc}}(\Omega \setminus K)$ for any $p < \infty$. In particular, $u \in C^1(\Omega \setminus K)$. In order to prove equality in (1.12) and that (u, K) is a strong solution, it is necessary to know that

$$\mathcal{H}^{N-1}(K \cap \Omega \setminus Su) = 0 \tag{2.1}$$

because, if this is true, we can replace $\mathcal{H}^{N-1}(Su)$ in (1.11) by $\mathcal{H}^{N-1}(K \cap \Omega)$ in (3). Equation (2.1) is a consequence, among other things, of the following *density lower bound*, proved in [46] for general N and in [40], [62] in the two-dimensional case.

Theorem 2.1. *There exists a constant $\theta = \theta(N) > 0$ such that*

$$\mathcal{H}^{N-1}\big(Su \cap B_\varrho(x)\big) \geq \theta \varrho^{N-1} \tag{2.2}$$

for any minimizer u of (1.11) and any ball $B_\varrho(x) \subseteq \Omega$ centered at $x \in \overline{Su}$ such that $\alpha \|g\|_\infty^2 \varrho \leq \beta$.

Equation (2.1) follows by (2.2) recalling that the $(N-1)$-dimensional density of Su is 0 for $\mathcal{H}^{N-1}$-almost every $x \notin Su$, while, by (2.2), the density is strictly positive for any $x \in K \cap \Omega$. Analogous density lower bounds for different free discontinuity problems have been obtained in [32], [49].

An equivalent but more appealing formulation of (2.2) is the *elimination property*:

$$\mathcal{H}^{N-1}\big(Su \cap B_\varrho(x)\big) < \frac{\theta}{2^{N-1}} \varrho^{N-1} \qquad \Longrightarrow \qquad Su \cap B_{\varrho/2}(x) = \emptyset$$

for balls $B_\varrho(x) \subseteq \Omega$ such that $\alpha \|g\|_\infty^2 \varrho \leq \beta$. In other words, if the Hausdorff measure of Su inside a ball $B_\varrho(x)$ is sufficiently small, then Su does not intersect the reduced ball $B_{\varrho/2}(x)$, i.e., u is regular in $B_{\varrho/2}(x)$. The implication is valid in sufficiently small balls, depending on the data α, β, g.

The density lower bound implies mild regularity properties of K; we mention, for instance, the agreement between the Hausdorff measure $\mathcal{H}^{N-1}$ and the Minkowski content $\mathcal{M}^{N-1}$, defined by

$$\mathcal{M}^{N-1}(K) := \lim_{\varrho \to 0+} \frac{\mathcal{L}^N\big(I_\varrho(K)\big)}{2\varrho}$$

where $I_\varrho(K)$ is the open ϱ neighbourhood of K (the above limit does not exist in general).

Proposition 2.2. [17] *Let u be a minimizer of (1.11). Then,*

$$\mathcal{H}^{N-1}(K) = \mathcal{M}^{N-1}(K) \tag{2.3}$$

for any compact set $K \subseteq \overline{Su} \cap \Omega$. If Ω is a rectangle, then (2.3) holds even with $K = \overline{Su}$.

Even though (2.3) is a very weak property (quite far from $C^{1,\alpha}$ regularity), it is very useful from the analytic viewpoint. For instance, (2.3) plays an important rôle in the approximation theorems of §3.

In the two-dimensional case the density lower bound and the elimination property can also be improved, getting concentration properties (see [40], [62]), uniform projection properties (see [48], [49]) and uniform rectifiability properties (see [41]):

Definition 2.3. We say that a closed set $K \subseteq \mathbf{R}^N$ satisfies the *concentration property* if for every $\varepsilon > 0$ there exists $\alpha_\varepsilon > 0$ such that, if B is any open ball centered at $x \in K$ with radius $\varrho \in (0,1)$, there exists a concentric open ball B' with radius at least $\alpha\varrho$ such that

$$\mathcal{H}^1\big(K \cap B'\big) \geq (1-\varepsilon)\mathrm{diam}\, B'.$$

We say that a closed set $K \subseteq \mathbf{R}^N$ is *uniformly rectifiable* if there exists an absolute constant C such that for any ball $B = B(x,r)$ there exists a curve Γ such that $K \cap B \subset \Gamma$ and

$$\mathcal{H}^1\big(\Gamma\big) \leq Cr.$$

The concentration property has been used in [48], [62] to get a new proof, independent of the SBV theory, of the existence of minimizing pairs (u, K) for the Mumford–Shah problem (3) in the two-dimensional case. The idea is to minimize the functional first with an upper bound on the number of connected components of K and then, using the fact that

$$K \mapsto \mathcal{H}^1(K \cap \Omega)$$

is lower semicontinuous along sequences converging in the Kuratowski topology and satisfying a uniform concentration property, a solution is found letting the bound go to ∞.

Actually, the *Mumford–Shah conjecture* states that any optimal segmentation $K = \overline{Su}$ is locally a finite union of $C^{1,1}$ arcs, so that the number of connected components should be locally finite. The Mumford–Shah conjecture is still open; however, in the last two years, considerable progress has been made towards its proof. Indeed, the author, Fusco and Pallara proved in [13] and [14] (see also [15]) a partial regularity theorem for K valid in any dimension, stated in Theorem 2.4 below. At the same time and independently, David obtained in [42] a similar result in dimension 2 and Bonnet proved in [26] (see also [24], [25]) the Mumford–Shah conjecture under the a priori assumption that K has finitely many connected components (see Theorem 2.5 below).

Theorem 2.4. *Let $u \in SBV(\Omega)$ be a minimizer of the Mumford–Shah functional and let $K = \overline{Su}$. Then, there exists a set $S \subseteq \Omega$ such that*

$$\overline{S} \cap \Omega = S \cap \Omega, \qquad \mathcal{H}^{N-1}\big(S\big) = 0 \tag{2.4}$$

and $K \cap \Omega \setminus S$ is locally a $C^{1,\alpha}$ hypersurface for any $\alpha < 1$. If $N = 2$, $K \cap \Omega \setminus S$ is locally a $C^{1,1}$ curve.

The regularity theory for free discontinuity problems is a challenging problem, because it mixes regularity theory for minimal surfaces with regularity theory for Neumann problems in nonsmooth domains. Indeed, confining our attention again to the Mumford–Shah problem, we have already seen that u solves a Neumann problem outside K; on the other hand, if K and g are sufficiently smooth in some region $A \subseteq \Omega$, making Hadamard's variations in the independent variable one finds the equation

$$\beta\mathbf{H} = \big[|\nabla u|^2 + \alpha(u-g)^2\big]_-^+ \nu_K \qquad \text{in } K \cap A \tag{2.5}$$

where $\mathbf{H}$ is the mean curvature vector of K, ν_K is the normal to K and $[\cdot]_-^+$ denotes the jump across K. Hence, if K is sufficiently smooth, higher regularity follows by the Neumann boundary value problem; on the other hand, if u is sufficiently regular (e.g., ∇u is bounded), then higher regularity of K comes from the prescribed mean curvature equation (2.5). The difficulty of the problem is due to the fact that in the beginning we have too little information to initiate the bootstrap argument.

The proof in [13], [14] (compare with [42], where the problem is studied with tools from harmonic analysis and potential theory) takes advantage, in particular, of the regularity theory for varifolds developed by Allard in [5] and by Brakke in [29], which is in turn reminiscent of the regularity techniques for minimal boundaries. In particular, a key quantity in the regularization process is the *tilt* of K, defined by

$$\mathbf{T}(x,\varrho) := \varrho^{1-N} \int_{B_\varrho(x)\cap K} |\nu_K(y) - \nu_K(x)|^2 \, d\mathcal{H}^{N-1}(y)$$

where ν_K is a unit normal to K, in a approximate sense. This quantity can be used in conjunction with the *scaled Dirichlet energy*, defined by

$$\mathbf{D}(x,\varrho) := \varrho^{1-N} \int_{B_\varrho(x)} |\nabla u|^2(y) \, dy.$$

In [14] it is proved that any point $x \in K$ where $\mathbf{T}(x,\varrho) + \mathbf{D}(x,\varrho)$ falls below a critical treshold $\varepsilon_0 = \varepsilon_0(N) > 0$ for a sufficiently small ϱ (depending on α, β, $\|g\|_\infty$) is a regular point of K. Hence, this leads to a constructive characterization of the singular set S

$$x \in S \qquad \Longleftrightarrow \qquad \limsup_{\varrho \to 0^+} \big[\mathbf{T}(x,\varrho) + \mathbf{D}(x,\varrho)\big] \geq \varepsilon_0. \tag{2.6}$$

The strategy of the proof is based on two decay properties: the first one, concerning $\mathbf{D}$, is true if $\mathbf{T} << \mathbf{D}$ and is related to classical elliptic regularity theory; the second one, concerning $\mathbf{T}$, is true if $\mathbf{D}$ is comparable with $\mathbf{T}$ and is related to the classical flatness improvement theorem of De Giorgi, Allard, Almgren. If $\mathbf{D}(x,\varrho) + \mathbf{T}(x,\varrho)$ is small enough a *joint* decay property can be proved, leading to regularity near x.

Theorem 2.5. *Let $u \in SBV(\Omega)$ be a minimizer of the two-dimensional Mumford–Shah functional, let $K = \overline{Su}$ and let $B \subset\subset \Omega$ be a ball such that $K \cap B$ is connected. Then, there exists a finite set $S \subseteq K \cap B$ such that $K \cap B \setminus S$ is locally a $C^{1,1}$ curve.*

The proof of Theorem 2.5 is based on a monotonicity formula (here the topological assumption on $K \cap B$ plays a fundamental rôle) and on blow-up arguments, which lead to the classification of singularities conjectured in [63], i.e., "triple junctions" where three curves meet with angles of $2\pi/3$ and "crack tips" where the curve ends (see Figure 2).
We notice that at triple junctions $\mathbf{D}(x,\varrho)$ is infinitesimal as ϱ tends to 0 while $\mathbf{T}(x,\varrho)$ is not; on the other hand, at crack tips the situation is reversed.

We remark that almost nothing is known for $N > 2$: even in dimension 3 there is no conjectured classification of singular points x in (2.6).

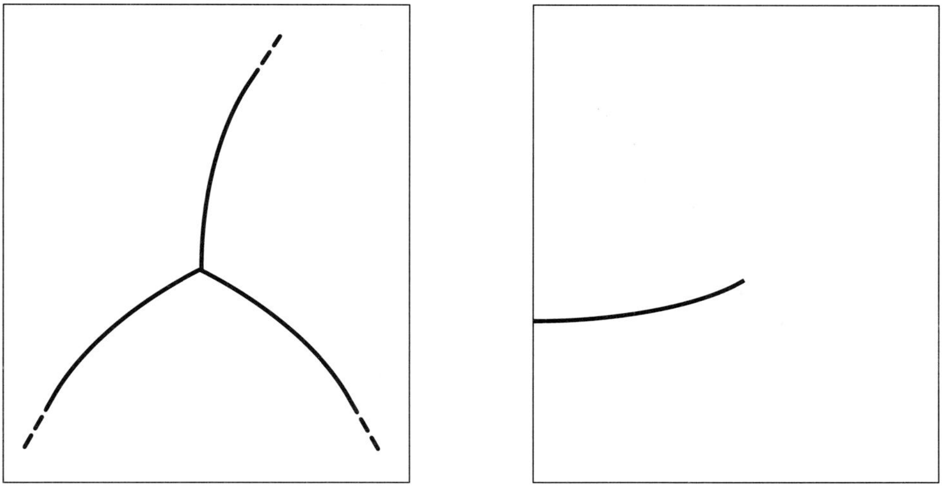

Fig. 2

3. Variational approximation of free discontinuity problems

Even though a general existence theory is now available, exact computations of solutions of free discontinuity problems can be very rarely performed, with the exception of situations in which symmetries allow a reduction to a one-dimensional problem. Hence, the computation of approximate solutions of free discontinuity problems is a crucial issue in applications; moreover, these computations can also be useful as a heuristic guide in the mathematical analysis of still open problems (e.g., the nature of singularities in dimensions higher than 2). Finally, we notice that Mumford and Shah have been inspired in the formulation of their continuous model by discrete models due to Blake and Zisserman and Geman and Geman, and the numeric solution of the Mumford–Shah problem depends, of course, on the connection between discrete and continuous models.

Assuming for the sake of simplicity $N = 1$ and $\Omega = (0, 1)$, the *weak membrane model* of Blake-Zisserman (see [22]) is based on the minimization of the functional

$$BZ_h(u) := h \sum_i W_h\Big(\frac{u_{i+1} - u_i}{h}\Big) + \alpha h \sum_i (u_i - g_i)^2 \tag{3.1}$$

where h is the mesh size and $W_h(t) := t^2 \wedge \beta/h$ is a truncated quadratic potential. Heuristically, near a discontinuity with jump J the cost is β for $h < J^2/\beta$ sufficiently small, while near regular points the cost is the square of the the difference quotient. Hence, the first sum in (3.1) is related, in the continuous model, to

$$\int_{\Omega \setminus K} |u'(t)|^2 \, dt + \beta \mathcal{H}^0(K)$$

and obviously the second one is an approximation of $\alpha \|u - g\|_2^2$.

A rigorous proof of the convergence of the Blake-Zisserman model to the Mumford–Shah model in the one-dimensional case was established in [36]. The convergence must be understood in the sense of Γ^-*–convergence*, which ensures convergence of extremal values to extremal values and of minimizers to minimizers. Let us recall its definition (see [39] for an exhaustive treatment of this subject):

Definition 3.1. Let X be a metric space and let $f,\ f_n : X \to [-\infty, +\infty]$ be functions. We say that (f_n) Γ^-–converges to f if the following two conditions hold:
(i) for any sequence $(x_n) \subseteq X$ converging to x, the following holds

$$\liminf_{n\to\infty} f_n(x_n) \geq f(x);$$

(ii) for any $x \in X$ there exists a sequence $(x_n) \subseteq X$ converging to x, such that

$$\limsup_{n\to\infty} f_n(x_n) \leq f(x).$$

The function f is uniquely determined by (i) and (ii) and it is denoted by $\Gamma^- - \lim\limits_{n\to\infty} f_n$.

We notice that Γ^-–convergence is not in general comparable with pointwise convergence, because we allow the argument x of f_n to depend on n. Therefore, this kind of convergence is well suited to the approximation of problems by other problems defined on completely different function spaces (as in Theorem 3.4 below).

The variational character of Γ^-–convergence is explained by the following proposition:

Proposition 3.2. *Assume that (f_n) Γ^-–converges to f and that there exists a compact set $K \subseteq X$ such that*

$$\inf_{x\in K} f_n(x) = \inf_{x\in X} f_n(x) \qquad \forall n \in \mathbf{N}. \tag{3.2}$$

Then, $\inf_X f_n$ converges as $n \to \infty$ to $\min_X f$ and any limit point of any sequence (x_n) such that

$$\lim_{n\to\infty}\left(f_n(x_n) - \inf_{x\in X} f_n(x)\right) = 0$$

is a minimizer of f.

Now, setting $X = L^2(\Omega)$, we can extend $BZ_h(u)$, initially defined on

$$\mathcal{P}_h := \{u \in C([0,1]) : u \text{ is linear on } [ih, (i+1)h] \text{ for any } i\}$$

to the whole of X, setting $BZ_h(u) = +\infty$ if $u \in X \setminus \mathcal{P}_h$. Analogously, we can define

$$F(u) := \int_\Omega \left[|u'|^2 + \alpha(u-g)^2\right] dt + \beta\mathcal{H}^0(Su)$$

if $u \in SBV(\Omega)$, $+\infty$ if $u \in X \setminus SBV(\Omega)$.

Theorem 3.3. [36] *The functionals BZ_h Γ^-–converge as $h \to 0$ to F. Moreover, assumption (3.2) is fulfilled for any infinitesimal sequence (h_n), hence*

$$\lim_{h\to 0}\min_{u\in X} BZ_h(u) = \min_{u\in X} F(u)$$

and any limit point of minimizers of BZ_h is a minimizer of F.

As explained in [36], the functionals $BZ_h(u)$ can be extended to the two-dimensional case by

$$BZ_h(u) := h \sum_{i,j} W_h\Big(\frac{u_{i+1,j} - u_{i,j}}{h}\Big) + W_h\Big(\frac{u_{i,j+1} - u_{i,j}}{h}\Big) + \alpha h \sum_{i,j} (u_{i,j} - g_{i,j})^2$$

but their Γ^-–limit, being necessarily anisotropic, does not coincide with the Mumford–Shah functional and is indeed an anisotropic version of it. To overcome this difficulty, a different approach, valid in any dimension, has been suggested in [28]. This approach is based on approximation by the *nonlocal* functionals

$$BD_\varepsilon(u) := \frac{1}{\varepsilon}\int_\Omega f\left(\varepsilon^{1-N}\int_{B_\varepsilon(x)\cap\Omega} |\nabla u|^2(y)\,dy\right)dx + \alpha\int_\Omega \big(u(x) - g(x)\big)^2\,dx$$

with $\varepsilon \to 0^+$. Here, f is a continuous increasing function satisfying

$$f(0) = 0, \qquad f'(0) = 1, \qquad f(+\infty) = \beta/2$$

(for instance $f(t) = t \wedge \beta/2$); this function plays the same rôle as the truncated quadratic potential in (3.1).

A different strategy, closer to [53], is based on the approximation of functionals depending on two variables (u, v), with the second one related to the set K. If K is the (measure theoretic) boundary $\partial^* E$ of a set E, the Modica-Mortola theorem allows variational approximation of $\mathcal{H}^{N-1}\big(\partial^* E \cap \Omega\big)$ by the quadratic elliptic functionals

$$MM_\varepsilon(v) := \int_\Omega \left(\varepsilon|\nabla v|^2 + \frac{W(v)}{\varepsilon}\right)dx \qquad\qquad v \in H^1(\Omega) \tag{3.3}$$

where $W(t)$ is a "double well" potential (see [60], [61]). For instance, choosing $W(t) = t^2(1-t)^2$, assuming that Ω has Lipschitz boundary and setting $MM_\varepsilon(v) = +\infty$ if $v \in L^2(\Omega) \setminus H^1(\Omega)$, the functionals $MM_\varepsilon(v)$ Γ^-–converge in $L^2(\Omega)$ to

$$F(v) := \begin{cases} \dfrac{1}{3}\mathcal{H}^{N-1}\big(\partial^* E \cap \Omega\big) & \text{if } v = \chi_E; \\[2ex] +\infty & \text{otherwise.} \end{cases}$$

Here $\partial^* E = S\chi_E$ is the *essential boundary* of E, i.e., the set of points where the density of E is neither 0 nor 1.

In the Mumford–Shah problem K need not be a boundary but we can still use a construction similar to (3.3), with the potential $W(t) = (1-t)^2$. Indeed, let $X = \big[L^2(\Omega)\big]^2$ and let us define

$$AT_\varepsilon(u, v) := \int_\Omega v^2\left[|\nabla u|^2 + \alpha(u-g)^2\right]dx + \frac{\beta}{2}\int_\Omega \left(\varepsilon|\nabla v|^2 + \frac{W(v)}{\varepsilon}\right)dx \tag{3.4}$$

if

$$v \in H^1(\Omega), \qquad uv \in H^1(\Omega), \qquad 0 \le v \le 1$$

$+\infty$ otherwise (here $v\nabla u$ is, by definition, $\nabla(uv) - u\nabla v$), and let

$$F(u,v) := \int_\Omega \left[|\nabla u|^2 + \alpha(u-g)^2\right] dx + \beta\mathcal{H}^{N-1}(Su)$$

if $u \in SBV(\Omega)$ and $v \equiv 1$, $+\infty$ otherwise. Then, the following result holds:

Theorem 3.4. ([17], [18]) *The functionals $AT_\varepsilon(u,v)$ Γ^-–converge in X to $F(u,v)$ and (3.2) is satisfied for any infinitesimal sequence (ε_n).*

The heuristic explanation of Γ^-–convergence is the following: on the one hand v_ε is forced to stay very close to 1 as $\varepsilon \to 0^+$, because the potential $W(t)$ is strictly positive and vanishes only for $t = 1$; on the other hand, since u_ε approximates discontinuous functions u, the factor v_ε^2 in front of $|\nabla u_\varepsilon|^2$ must go to zero near to discontinuities of u to keep the Dirichlet integral bounded. Hence, v_ε is forced to make transitions (which are sharper and sharper as $\varepsilon \to 0^+$) between 0 and 1 near to discontinuities of u. The balance between the W term and the Dirichlet integral in the second integral of (3.4) shows that the energy of cheapest transitions is proportional to $\mathcal{H}^{N-1}(Su)$. The typical behaviour of u_ε and v_ε near discontinuity points of u is shown in Figure 3.

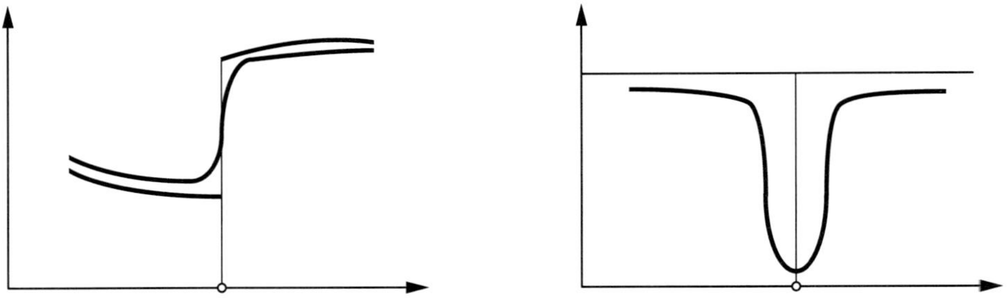

Fig. 3

Proposition 3.2 and Theorem 3.4 imply that any limit point of minimizers $(u_\varepsilon, v_\varepsilon)$ of AT_ε is a pair $(u,1)$, with $u \in SBV(\Omega)$ minimizer of the Mumford–Shah problem. Numerical experiments (see [57]) based on discretized versions $AT_{\varepsilon,h}$ of AT_ε show that, even for ε not too small, the functions $v_{\varepsilon,h}$ make sharp transitions which allow edge detection. In [20] it is proved that we still have Γ^-–convergence of the discretized functionals $AT_{\varepsilon,h(\varepsilon)}$ provided $h(\varepsilon) = o(\varepsilon)$.

We conclude this survey by observing that the approximation scheme developed in Theorem 3.4 might be extended to many other free discontinuity problems, at least those involving isotropic or anisotropic surface energies independent of the jump of u. For instance, this idea could be succesfully applied to variational problems involving the functionals in (1.18), (1.19), (1.20).

References

[1] E. Acerbi & N. Fusco, *Semicontinuity problems in the calculus of variations*, Arch. Rational Mech. & Anal., **86** (1984), 125–145.

[2] E. Acerbi, I. Fonseca & N. Fusco, *Regularity results for equilibria in a variational model for fracture*, Proc. Royal Society Edinburgh, to appear.

[3] G. Alberti, *Rank-one properties for derivatives of functions with bounded variation*, Proc. Royal Soc. Edinburgh, **123A** (1993), 239–274.

[4] G. Alberti & C. Mantegazza, *A note on the theory of SBV functions*, Boll. Un. Mat. Ital., to appear.

[5] W.K. Allard, *On the first variation of a varifold*, Ann. of Math., **95** (1972), 417–491.

[6] L. Ambrosio, *A compactness theorem for a new class of functions of bounded variation*, Boll. Un. Mat. Ital., **3-B** (1989), 857–881.

[7] L. Ambrosio, *Variational problems in SBV and image segmentation*, Acta Appl. Math., **17** (1989), 1–40.

[8] L. Ambrosio, *Existence theory for a new class of variational problems*, Arch. Rational Mech. & Anal., **111** (1990), 291–322.

[9] L. Ambrosio, *On the lower semicontinuity of quasi-convex integrals in SBV*, Nonlinear Anal., **23** (1994), 405–425.

[10] L. Ambrosio, *A new proof of the SBV compactness theorem*, Calc. Var., **3** (1995), 127–137.

[11] L. Ambrosio, A. Braides & A. Garroni, *Special functions with bounded variation and with weakly differentiable traces on the jump set*, submitted to Manuscripta Mathematica.

[12] L. Ambrosio, A. Coscia & G. Dal Maso, *Fine properties of functions in BD*, Arch. Rational Mech. & Anal., to appear.

[13] L. Ambrosio & D. Pallara, *Partial regularity of free discontinuity sets I*, submitted to Ann. Sc. Norm. Sup. Pisa.

[14] L. Ambrosio, N. Fusco & D. Pallara, *Partial regularity of free discontinuity sets II*, submitted to Ann. Sc. Norm. Sup. Pisa.

[15] L. Ambrosio, N. Fusco & D. Pallara, *Higher regularity of free discontinuity sets*, submitted to Diff. Int. Eq., (1996).

[16] L. Ambrosio & D. Pallara, *Partial regularity in free discontinuity problems*, Progress in partial differential equation: the Metz surveys 4, M. Chipot and I. Shafrir Eds., Pitman Research Notes in Mathematics, **345** (1996), Longman, Harlow, 3–17.

[17] L. Ambrosio & V. M. Tortorelli, *Approximation of functionals depending on jumps by elliptic functionals via* Γ*-convergence*, Comm. Pure Appl. Math., **43** (1990), 999–1036.

[18] L. Ambrosio & V. M. Tortorelli, *On the approximation of free discontinuity problems*, Boll. Un. Mat. Ital., **6**-B (1992), 105–123.

[19] J.M. Ball, *Convexity conditions and existence theorems in nonlinear elasticity*, Arch. Rational Mech. & Anal., **63** (1977), 337–403.

[20] G. Bellettini & A. Coscia, *Discrete approximation of a free discontinuity problem*, Numer. Funct. Anal. and Optimiz., **XV** (1994), 201–224.

[21] G. Bellettini, A. Coscia & G. Dal Maso, *Compactness and lower semicontinuity in SBD*, Preprint SISSA, Trieste (1996).

[22] A. Blake & A. Zisserman, *Visual Reconstruction*, MIT Press, 1987.

[23] H. Brezis, J.M. Coron & H. Lieb, *Harmonic maps with defects*, Comm. Math. Phys. **107**, 649–705.

[24] A. Bonnet, *Caractérisation des minima globaux de la fonctionelle de Mumford–Shah en segmentation d'images*, C.R. Acad. Sci. Paris, **321** (1995), 1121–1126.

[25] A. Bonnet, *Sur la régularité des bords des minima de la fonctionelle de Mumford–Shah*, C.R. Acad. Sci. Paris, **321** (1995), 1275–1279.

[26] A. Bonnet, *On the regularity of edges in image segmentation*, to appear in Ann. Inst. Poincaré.

[27] G. Bouchitte, A. Braides & G. Buttazzo. *Relaxation of Free Discontinuity Problems*, J. Reine Angew. Math, **458** (1995), 1–18.

[28] A. Braides & G. Dal Maso, *Nonlocal approximation of the Mumford Shah functional*, Preprint SISSA, (1996).

[29] K.A. Brakke, *The motion of a surface by its mean curvature*, Princeton U.P., 1978.

[30] G. Buttazzo, *Semicontinuity, Relaxation and Integral Representation in the Calculus of Variations*, Pitman Res. Notes Math. Ser., **207**, Longman, Harlow, 1989.

[31] A.P. Calderón & A. Zygmund, *On the differentiability of functions which are of bounded variation in Tonelli's sense*, Rev. Union Mat. Argentina, **20** (1960), 102–121.

[32] G. Carriero & A. Leaci, *S^k-valued maps minimizing the L^p-norm of the gradient with free discontinuities*, Ann. Scuola Norm. Sup. Pisa ser. IV, **18** (1991), 321–352.

[33] M. Carriero, A. Leaci & F. Tomarelli, *Strong solution for an elastic-plastic plate*, Calc. Var., **2** (1994), 219–240.

[34] M. Carriero, A. Leaci & F. Tomarelli, *Free gradient discontinuities*, Calculus of Variations, Homogenization and continuum mechanics, Ed. Buttazzo, Bouchitté, Suquet, World Scientific Singapore (1994), 131–147.

[35] M. Carriero, A. Leaci & F. Tomarelli, *A second order model in image segmentation: Blake & Zisserman functional*, Preprint Politecnico di Milano 177/P (1995).

[36] A. Chambolle, *Image segmentation by variational methods: Mumford and Shah functional and the discrete approximations*, SIAM J. Appl. Math., **55** (1995), 827–863.

[37] G. Congedo & I. Tamanini, *On the existence of solutions to a problem in multidimensional segmentation*, Ann. Inst. H. Poincaré Anal. Non Linéaire, **2** (1991), 175–195.

[38] B. Dacorogna, *Direct Methods in the Calculus of Variations*, Springer-Verlag, Berlin, 1989.

[39] G. Dal Maso, *An Introduction to Γ-convergence*, Birkhäuser, Boston, 1993.

[40] G. Dal Maso, J.M. Morel & S. Solimini, *A variational method in image segmentation: existence and approximation results*, Acta Math., **168** (1992), 89–151.

[41] G. David & S. Semmes, *On the singular set of minimizers of the Mumford–Shah functional*, J. Math. Pures et Appl., to appear.

[42] G. David, *C^1 arcs for the minimizers of the Mumford–Shah functional*, SIAM J. of Appl. Math. **56**, (1996).

[43] F. Demengel, *Fonctions a Hessian Borné*, Ann. Inst. Fourier, **34** (1985), 155–190.

[44] E. De Giorgi, *Free Discontinuity Problems in Calculus of Variations*, Frontiers in pure and applied Mathematics, a collection of papers dedicated to J.L.Lions on the occasion of his 60th birthday, R. Dautray ed., North Holland, 1991.

[45] E. De Giorgi & L. Ambrosio, *Un nuovo funzionale del calcolo delle variazioni*, Atti Accad. Naz. Lincei Rend. Cl. Sci. Fis. Mat. Natur., **82** (1988), 199–210.

[46] E. De Giorgi, G. Carriero & A. Leaci, *Existence theorem for a minimum problem with free discontinuity set*, Arch. Rational Mech. *&* Anal., **108** (1989), 195–218.

[47] H. Federer, *Geometric Measure Theory*. Springer-Verlag, New York, 1969.

[48] F. Dibos, *Uniform rectifiability of image segmentations obtained by a variational method*, J. Math. Pures Appliquées, to appear.

[49] F. Dibos & G. Koepfler, *Propriété de régularité des contours d'une image segmentée*, CRAS T.**313** (1991), 573–578.

[50] F. Dibos & E. Séré, *An approximation result for the minimizers of the Mumford-Shah functional*, Boll. Un. Mat. Ital., to appear.
[51] I. Fonseca & G. Francfort, *A model for the interaction between fracture and damage*, Calc. Var., **3** (1995), 407–446.
[52] I. Fonseca & N. Fusco, *Regularity results for anisotropic image segmentation models*, Ann. Sc. Norm. Sup. Pisa, to appear.
[53] S. Geman & D. Geman, *Stochastic relaxation, Gibbs distributions and the Bayesian restoration of images*, IEEE PAMI **6**, 1994.
[54] A.A. Griffith, *The phenomenon of rupture and flow in solids*, Phil. Trans. Royal Soc. London A, **221** (1920), 163–198.
[55] A.D. Ioffe, *On lower semicontinuity of integral functionals I, II*, Siam J. Cont. Optim., **15** (1977), 521–538 and 991–1000.
[56] F.H. Lin, *Liquid crystal droplets and free interfaces*, Diaz (ed.) et al., Free boundary problems: theory and applications. Proceedings of the international conference held in Toledo, Spain, June 21-26, 1993. Harlow: Longman Scientific & Technical. Pitman Res. Notes Math. Ser., **323** (1995), 156–161.
[57] R. March, *Visual reconstruction with discontinuities using variational methods*, Image and vision computing, **10** (1992), 30–38.
[58] U. Massari & M. Miranda: *Minimal surfaces of codimension one*, North Holland, Amsterdam, 1984.
[59] H. Matthies, G. Strang & E. Christiansen, *The saddle point of a differential program*, Energy methods in finite elements analysis, ed. by Glowinsky, Rodin and Zienkiewicz, John Wiley & Sons, 1979.
[60] L. Modica, *The gradient theory of phase transitions and the minimal interface criterion*, Arch. Rational Mech. *&* Anal., **98** (1987), 123–142.
[61] L. Modica & S. Mortola, *Un esempio di Γ-convergenza*, Boll. Un. Mat. Ital., **14-B** (1977), 285–299.
[62] J.M. Morel & S. Solimini, *Variational Models in Image Segmentation*, Birkhäuser, Boston, 1994.
[63] D. Mumford & J. Shah, *Optimal approximation by piecewise smooth functions and associated variational problems*, Comm. Pure Appl. Math., **17** (1989), 577–685.
[64] D. Pallara, *Nuovi teoremi sulle funzioni a variazione limitata*, Atti Accad. Naz. Lincei, Rend. Cl. Sci. Fis. Mat. Natur. Ser. 9, **1** (1990), 309–316.
[65] D. Pallara, *Some new results on functions of bounded variation*, Rend. Acc. Naz. delle Scienze dei XL, **14** (1990), 295–321.
[66] T.J. Richardson, *Limit theorems for a variational problem arising in computer vision*, Ann. Sc. Norm. Sup. Pisa, **19** (1992), 1–49.
[67] S. Solimini, *Simplified excision techniques for free discontinuity problems in several variables*, preprint Univ. Bari (1995).
[68] P. Suquet, *Existence et régularité des solutions des équations de la plasticité parfaite*, C. R. Acad. Sc. Paris, **286** (1978), 1201–1204.
[69] R. Temam, *Problèmes Mathématiques en Plasticité*, Gauthier-Villars, Paris, 1983.
[70] E.G. Virga, *Drops of nematic liquid crystals*, Arch. Rational Mech. *&* Anal., **107** (1989), 371–390.
[71] E.G. Virga & S. Faetti, *On a curvature surface energy for liquid crystals*, to appear in Arch. Rational Mech. *&* Anal., 1996.
[72] A.I. Vol'pert, *Spaces BV and quasi-linear equations*, Math. USSR Sb., **17** (1967), 225–267.
[73] A.I. Vol'pert & S.I. Hudjaev, *Analysis in Classes of Discontinuous Functions and Equations of Mathematical Physics*, Martinus Nijhoff Publisher, Dordrecht, 1985.

Progress in Mathematics, Vol. 168, © 1998 Birkhäuser Verlag Basel/Switzerland

Recent Connections and Applications of Planar Quasiconformal Mappings

Kari Astala

University of Jyväskylä, Department of Mathematics
P.O. Box 35, FIN-40351 Jyväskylä, Finland
Astala@math.jyu.fi

1. Introduction

Quasiconformal mappings are geometric deformations, homeomorphisms defined in subdomains of $\mathbf{R}^n$, $n \geq 2$, which in a quantitative manner behave like conformal mappings. The need for this notion and the methods of the field has arisen in a variety of different areas of analysis and geometry, recently even in group theory in connection with Gromov's theory of hyperbolic groups.

Describing first loosely the basic features of quasiconformality, these deformations preserve relative sizes of nearby objects with locally uniform bounds, they map positive angles to positive ones (with quantitative bounds), and similarly preserve the zero angles. There are two ways to define these properties precisely. A moment's thought shows that the former is conveniently described by the notion of quasisymmetry: a mapping $f : A \to B$, $A, B \subset \mathbf{R}^n$, is called *quasisymmetric* if

$$\frac{|f(x) - f(y)|}{|f(x) - f(z)|} \leq \eta\left(\frac{|x-y|}{|x-z|}\right) \tag{1}$$

for all points $x, y, z \in A$ and for some continuous strictly increasing function $\eta : \mathbf{R}_+ \to \mathbf{R}_+$ with $\eta(0) = 0$. On the other hand, the deformation of angles is best approached on the infinitesimal level. Let us therefore say that a homeomorphism $f : \Omega \to \Omega'$ between domains $\Omega, \Omega' \subset \mathbf{R}^n$ is called *K-quasiconformal* if it is contained in the Sobolev class $W^{1,n}_{loc}(\Omega)$ and its directional derivatives satisfy

$$\max_\alpha |\partial_\alpha f(x)| \leq K \min_\alpha |\partial_\alpha f(x)| \qquad a.e.\ \ x \in \Omega. \tag{2}$$

It then follows that a K-quasiconformal mapping satisfies (1) in a neighbourhood of every point $x \in \Omega$ with η depending only on K, n and $\operatorname{dist}(x, \partial\Omega)$. Furthermore, every quasiconformal mapping of the whole $\mathbf{R}^n$ is globally quasisymmetric with η depending only on K and n. Conversely, the requirement of quasisymmetry always implies (2) with $K = K(n, \eta)$.

In a sense these mappings, applicable also on manifolds with non-smooth structures or on suitable Lipschitz surfaces etc., are the most general class of homeomorphisms one can do analysis with.

In the special case of the plane, quasiconformality admits besides the geometric approach, in addition tools from other parts of analysis, in particular from differential equations, singular integrals and complex analysis. This adds new aspects to the theory on which we shall concentrate in the sequel. We shall mainly discuss some recent connections of planar quasiconformal mappings to topics such as complex dynamics, holomorphic deformations, solutions of elliptic PDE's, in particular their regularity and removability properties, singular integrals and the Beurling transform, nonlinear elasticity and homogenization.

2. Holomorphic deformations

The main reason for the existence of the various connections of quasiconformality to apparently distinct topics is that in the planar case (2) is equivalent to the well-known Beltrami differential equation

$$\overline{\partial} f(z) = \mu(z)\, \partial f(z), \tag{3}$$

where μ is the complex dilatation or the Beltrami coefficient of f with $\|\mu\|_\infty = \frac{K-1}{K+1} < 1$. This is a first order linear differential equation, solvable by singular integrals. In fact, by the *measurable Riemann mapping theorem* for any $\mu \in L^\infty(\mathbf{C})$ with $\|\mu\|_\infty < 1$ we have a homeomorphic $W^{1,2}_{loc}$-solution of (3), see [2] [8] [19], unique up to postcomposing with a Möbius transformation. Note also that if $\mu|_D \equiv 0$ in a subdomain D, then the solution $f = f^\mu$ is conformal there.

To show the existence of the solutions one must use the the classical Beurling operator, the two-dimensional counterpart of the Hilbert transform,

$$S\omega(z) = -\frac{1}{\pi} \int_{\mathbf{C}} \frac{\omega(\zeta)\, dm(\zeta)}{(\zeta - z)^2}.$$

It defines an isometry in $L^2(\mathbf{C})$, and what makes it useful in our setting is that S transforms $\overline{\partial}$-derivatives to ∂-derivatives,

$$S(\overline{\partial} u) = \partial u, \quad \forall u \in W^{1,2}(\mathbf{C}). \tag{4}$$

Assuming that we are first looking for a solution of (3) in the case where $\mu \in L^\infty(\mathbf{C})$ has a compact support and $f(z) = z + O(|z|^{-1})$, using the above identity for $u(z) = f(z) - z$ and combining with the equation $\overline{\partial} f(z) = \mu(z) \partial f(z)$ we have

$$\partial f(z) = 1 + (I - S\mu)^{-1} S(\mu). \tag{5}$$

Since as an operator on $L^2(\mathbf{C})$, the norm of $g \mapsto S(\mu g)$ is at most $\|\mu\|_\infty \|S\|_{L^2} < 1$, the expression is well-defined and shows that $\partial f(z)$ can be obtained as a Neumann series. The function itself has the representation

$$f(z) = z - \frac{1}{\pi} \int_{\mathbf{C}} \frac{\mu \partial f(\zeta)\, dm(\zeta)}{(\zeta - z)}.$$

The injectivity of f can be shown easily for smooth μ with the general case obtained by a limiting argument. A similar limiting argument also proves the existence for general μ's with non-compact support.

In any case the reasoning, originally due to Bojarski [8], shows that $f \equiv f^\mu$ depends *holomorphically* on μ. To study this dependence more systematically let us assume that the Beltrami coefficient is normalized so that μ with $\|\mu\|_\infty = 1$. Considering a subset $E \subset \mathbf{C}$ and the mapping

$$(\lambda, z) \mapsto \Phi(\lambda, z) = f^{\lambda\mu}(z), \quad \lambda \in \Delta \equiv \{\lambda \in \mathbf{C} : |\lambda| \leq 1\}, z \in E,$$

we see that f gives rise to what Mañé, Sad and Sullivan [22] call a holomorphic motion.

DEFINITION 2.1 *A function $\Phi : \Delta \times E \to \overline{\mathbf{C}}$ is called a holomorphic motion of a set $E \subset \overline{\mathbf{C}}$ if*

(i) *for any fixed $a \in E$, the map $\lambda \mapsto \Phi(\lambda, a)$ is holomorphic in Δ*
(ii) *for any fixed $\lambda \in \Delta$, the map $a \mapsto \Phi_\lambda(a) = \Phi(\lambda, a)$ is an injection, and*
(iii) *the mapping Φ_0 is the identity on E.*

Note in particular that this definition makes no assumptions on the set E or on the *a priori* continuity of Φ_λ. Therefore it is quite surprising that there is a converse to the above reasoning. This is a consequence of the remarkable generalized λ-lemma proved by Slodkowski.

THEOREM 2.2 (Slodkowski [28]) *Any holomorphic motion Φ of any set $E \subset \overline{\mathbf{C}}$ extends to a holomorphic motion of $\overline{\mathbf{C}}$.*

COROLLARY 2.3 (Mañé, Sad and Sullivan [22]) *For all holomorphic motions Φ the partial mappings Φ_λ extend quasiconformally to $\overline{\mathbf{C}}$ with*

$$K(\Phi_\lambda) \leq \frac{1 + |\lambda|}{1 - |\lambda|}.$$

Proof. According to Slodkowski's theorem we may assume that Φ is a holomorphic motion of the whole extended complex plane. Then by normalizing with a Möbius transformation we may further assume that for each λ, Φ_λ fixes the point ∞.

Considering now three different points $x, y, z \in \mathbf{C}$ the function

$$h_\lambda = \frac{\Phi_\lambda(x) - \Phi_\lambda(y)}{\Phi_\lambda(x) - \Phi_\lambda(z)}$$

is holomorphic in the open unit disk and omits the three values $0, 1$ and ∞. Since $\overline{\mathbf{C}} \setminus \{0, 1, \infty\}$ can be equipped with the hyperbolic metric ρ and since holomorphic mappings do not increase hyperbolic distances [20],

$$\rho(h(\lambda), h(0)) \leq \rho(\lambda, 0) = \log \frac{1 + |\lambda|}{1 - |\lambda|}. \tag{6}$$

On the other hand, in the hyperbolic metric $\overline{\mathbf{C}} \setminus \{0, 1, \infty\}$ is complete. This means that whenever z is fixed and $w \to 0$, necessarily $\rho(z, w) \to \infty$. Consequently we can interpret the inequality (6) in the form

$$|\frac{\Phi_\lambda(x) - \Phi_\lambda(y)}{\Phi_\lambda(x) - \Phi_\lambda(z)}| \leq \eta \left(|\frac{x-y}{x-z}| \right) \tag{7}$$

where $\eta = \eta_\lambda$ is a strictly increasing function, not depending on the choice of x, y or z, with $\eta_\lambda(0) = 0$. In other words, each Φ_λ is quasisymmetric. Finally, the bound on $K(\Phi_\lambda)$ is deduced from the Schwartz lemma and the holomorphic dependence of the Beltrami coefficient of Φ_λ on λ. □

Summing up the above reasoning, holomorphic motions and quasiconformal mappings are just different aspects of one and the same quantity!

Since typical holomorphic motions arise in perturbing a complex analytic dynamical systems, holomorphic motions explain in a very clear manner the appearence of quasiconformal phenomena in different complex dynamical systems. One of the best known examples is the quasiconformal stability of Julia sets.

EXAMPLE 2.4 *Let $P_c = z^2 + c$. Then, if c and γ are in the same component W of the interior of the Mandelbrot set, then their Julia sets are quasiconformally equivalent.*

Proof. We recall the argument only in the case where the component W is hyperbolic, i.e. in the case where for each $c \in W$ the iterated derivatives $|(P_c^n)'(z)|$ grow exponentially and uniformly on $z \in J(P_c)$; it is conjectured that all components of the Mandelbrot set are hyperbolic. Firstly recall that the Julia set of $P_c = z^2 + c$ is the closure of the repelling periodic points of P_c [12]. On the other hand for a fixed c the periodic points are precisely the zeroes of the functions $F(c, z) = P_c^n(z) - z$, $n \in \mathbf{N}$. Since by hyperbolicity if $P_c^n(z) = z$

$$\partial_z F(c, z) = (P_c^n)'(z) - 1 \neq 0, \quad z \in J(P_c) \tag{8}$$

and the implicit function theorem shows that each periodic point can be continued holomorphically and uniquely to neighbouring parameter values. Hence we obtain a holomorphic motion of the periodic points in the component W, i.e. as long as (7) holds for all n. By 2.2 and 2.3 the motion gives the quasiconformal equivalence of the corresponding Julia sets. □

Similar arguments work, for example, for Kleinian groups: Let Γ be a discrete group of Möbius transformations with limit set $L(\Gamma) \neq \overline{\mathbf{C}}$ and assume that Γ has no elliptic elements. Then if one changes the coefficients of the group elements holomorphically, this gives a holomorphic motion. Hence the corresponding limit sets are quasiconformally equivalent as long as the new groups Γ_λ obtained are discrete, the natural identification $(gh)_\lambda = g_\lambda h_\lambda$ remains an isomorphism and no accidental parabolics are formed.

Once the role of quasiconformality was realized in complex dynamics, the quasiconformal methods became basic tools in the area. The pioneering work was done by Sullivan and Douady and Hubbard; as a result Sullivan [29] obtained e.g. the non-existence of wandering domains of rational maps, a result basic to the classification of dynamics in the Fatou set. In their work [13] Douady and Hubbard established the theory of polynomial like mappings: If we are given two topological disks U and V with $\overline{U} \subset V$, then a holomorphic proper map $g : U \mapsto V$ is called *polynomial like.* A typical quasiconformal approach is given by their straightening theorem:

THEOREM 2.5 (Douady, Hubbard [13]) *If $g : U \mapsto V$ is a polynomial like mapping of degree d, then*

$$f \circ g = P \circ f$$

for some quasiconformal mapping f and some polynomial P with $\deg(P) = d$.

Moreover, the mapping f can chosen to be conformal in the interior of the filled-in Julia set $\bigcap_n g^{-n}V$.

Idea of proof. Without loss of generality we may assume that V is the disk $B(0, R)$ for some $R > 1$. The mapping g is now defined only in U but we may set $g(z) = z^d$ for $|z| \geq R$. We may then extend g to a quasiregular function on $V \setminus \overline{U}$. This means that $g \in W^{1,2}_{loc}$ and its derivatives satisfy the bounds (2), but g is not necessarily injective. It follows that g is quasiregular in the whole plane $\mathbf{C}$. In fact it is easily seen that here the dilatations $K(g^n) \leq K_{abs} < \infty$ are uniformly bounded in $n \in \mathbf{N}$.

According to the construction the values of the corresponding complex dilatations $\mu_{g^n}(z)$, as defined in (3), are uniformly bounded in the hyperbolic geometry of the unit disk Δ. We can define a new dilatation μ by choosing for each $z \in \Delta$ the value $\mu(z)$ to be the (hyperbolic) barycenter of the set

$$\{\mu_{g^n}(z) : n \in \mathbf{N}\}.$$

By the measurable Riemann mapping theorem the new dilatation μ admits a quasiconformal solution of (3), $f = f^\mu$. The transformation and uniqueness properties of the dilatation show that necessarily $f \circ g = P \circ f$ for some P holomorphic in $\mathbf{C}$.

Since the extended g is topologically a polynomial of degree d, this must be also true for P. □

The straightening theorem has become the basic tool in the quasiconformal surgery developed by Douady, Hubbard, Branner and others (see e.g. [10]), as well as in understanding, for example, the renormalization phenomena in complex dynamics or self-similarity properties of the Mandelbrot set.

One of the important open problems in the dynamics of rational functions is the question of whether the hyperbolic rational maps form a dense subset. For quadratic polynomials this can actually be phrased in terms of quasiconformal mappings; the density of hyperbolicity in the quadratic family is equivalent [22] to the following.

CONJECTURE 2.6 *The Julia set of a quadratic polynomial $P(z) = z^2 + c$ does not support an invariant linefield. That is, there does not exist a dilatation μ with*

$$0 \neq \mu(P(z)) = \mu(z)\frac{\overline{P'(z)}}{P'(z)} \quad \text{a.e. } z \in J(P).$$

3. Quasiconformal distortion

Let us study next how the ideas of holomorphic deformations can be useful in understanding quasiconformal mappings themselves, in particular in describing the manner these mappings distort the metric quantities like distance or area.

It was shown in the mid 50's, by the work of Ahlfors [1] and Mori [24] that K-quasiconformal mappings are Hölder continuous with exponent $1/K$. However, the following recent argument due to G.Martin [23] elegantly exposes the relations of this (sharp) distortion bound to holomorphicity. Let us denote

$$H_K(r) = \sup\{|f(z)| : |z| = r, f \text{ } K\text{-quasiconformal on } \mathbf{C} \text{ fixing } 0 \text{ and } 1\}\,. \qquad (9)$$

Let ρ be the density of the hyperbolic metric in $\mathbf{C} \setminus \{0, 1\}$. Then we have (as shown originally by Teichmüller) that

$$\int_r^{H_K(r)} \rho(-x)dx = \log(K).$$

Proof. (G. Martin) If we are given a Beltrami coefficient μ with $\|\mu\|_\infty = 1$ and a point $z \in \mathbf{C}$ let us write $\phi(\lambda) = f^{\lambda\mu}(z)$. We assume that the quasiconformal mapping $f^{\lambda\mu}$ with complex dilatation $\lambda\mu$ is normalized by Möbius transformations so that it fixes $0, 1$ and ∞. Then $\phi : \Delta \mapsto \mathbf{C} \setminus \{0, 1\}$ is holomorphic with $\phi(0) = z$. But by Theorems 2.2 and 2.3, a converse statement is also true. Hence

$$H_K(r) = \sup\{|\phi(\lambda)| : \phi : \Delta \mapsto \mathbf{C} \setminus \{0,1\} \text{ holomorphic, } |\phi(0)| = r \text{ and } K = \frac{1+|\lambda|}{1-|\lambda|}\}.$$

By a symmetrization argument [16], the value of $|\phi(\lambda)|$ is largest when $\phi(\lambda)$ and $\phi(0)$ both lie on the negative real axis. □

Finally, estimates for ρ [16] show that $H_K(r) \leq \frac{e^{\pi K}}{16} \max\{1, r\}^K$. Thus by composing with similarities we deduce for general quasiconformal mappings of $\mathbf{C}$ that they are uniformly $1/K$-Hölder continuous. The K-quasiconformal map

$$f_0(z) = z|z|^{\frac{1}{K}-1} \qquad (10)$$

shows that exponent $1/K$ cannot be improved.

In view of the above approach it is plausible that also many other quasiconformal properties are related to holomorphic deformations. A question basic to many quantitative bounds and applications is the optimal control of the distortion of area $|E|$ of subsets $E \subset \mathbf{C}$. The example (10) suggests that K-quasiconformal mappings also distort the area in a similar $1/K$-Hölder continuous manner. This, as conjectured by Gehring and Reich in mid 60's, is in fact the case.

THEOREM 3.1 (Astala, [3]) *For each K-quasiconformal mapping f of $\mathbf{C}$ fixing $0, 1$ and ∞ we have*

$$|f(E)| \le M_K |E|^{1/K}, \quad E \subset \mathbf{C}, \tag{11}$$

where M_K depends only on K.

Sketch of proof. The proof consists of two separate arguments, with the latter making use of ideas from complex dynamics and holomorphic deformations of quasiconformal mappings. In the first straightforward case, let f be conformal off the set E, assumed to be compact, with $f(z) = z + O(\frac{1}{z})$ as $z \to \infty$. Then we have the much better bound

$$|f(E)| \le K|E|. \tag{12}$$

Indeed, from (5), (3) we have $\overline{\partial} f = \mu \partial f = \mu + \mu S(\mu) + \mu S(\mu S(\mu)) + \cdots$ with $\partial f = 1 + S(\overline{\partial} f)$. Thus,

$$\begin{aligned} |f(E)| &= \int_E J_f \, dm = \int_E (|\partial f|^2 - |\overline{\partial} f|^2) \\ &= |E| + \int_E 2 \operatorname{Re}(S(\overline{\partial} f)) + \int_E (|S(\overline{\partial} f)|^2 - |\overline{\partial} f|^2). \end{aligned}$$

Since S is an isometry on L^2 and $\overline{\partial} f$ vanishes outside E, the latter integral is non–positive. For the first integral represent $\overline{\partial} f$ with the above series and iterate the estimates

$$\int_E |S(\mu g)| \le \sqrt{|E|} \Big(\int_{\mathbf{C}} |S(\mu g)|^2 \Big)^{\frac{1}{2}} \le \sqrt{|E|} \parallel \mu \parallel_\infty \Big(\int_E |g|^2 \Big)^{\frac{1}{2}}$$

with $g = 1$, $g = S\mu$, $g = S(\mu S(\mu)), \ldots$ This yields

$$|f(E)| \le |E| + 2 \parallel \mu \parallel_\infty |E| + 2 \parallel \mu \parallel_\infty^2 |E| + \cdots = |E| \Big(1 + \frac{2 \parallel \mu \parallel_\infty}{1 - \parallel \mu \parallel_\infty} \Big) = K|E|.$$

Returning to the general case, we may assume that E is a finite union of disjoint disks all contained in the unit disk. Using the measurable Riemann mapping theorem any K-quasiconformal mapping f can be decomposed as $f = g \circ h$ where both g, h are K-quasiconformal, such that h is conformal in E and g in $\mathbf{C} \setminus h(E)$. By the above first step we may now concentrate on the second, complementary case and assume that the restriction $f|_E$ is conformal; in addition we may assume that f is conformal near ∞.

To reduce the problem further, consider the holomorphic deformations

$$\lambda \mapsto f_\lambda(z) \equiv f^{\lambda\mu}(z), \ \ \lambda \in \Delta, \ \ \|\mu\|_\infty = 1.$$

Here $f_0 = id$ and the coefficient μ is chosen so that $f_\lambda = f$ when $\lambda = \frac{K-1}{K+1}$. In particular,

$$K(f_\lambda) \equiv \frac{1 + |\lambda|}{1 - |\lambda|}.$$

Applying the classical Koebe's distortion theorem for $f|_E$ we have

$$B_i(\lambda) \equiv B\Big(f_\lambda(z_i), \frac{r_i}{4}|f'_\lambda(z_i)|\Big) \subset f_\lambda B(z_i, r_i),\ 1 \le i \le n,\ \text{where } E = \bigcup_{i=1}^{n} B(z_i, r_i). \tag{13}$$

Now the round disks $B_i(\lambda)$ do not intersect; in fact they produce a holomorphic family or a holomorphic motion of disjoint disks. Moreover, the original problem reduces to showing

$$\sum_{i=1}^{n} |B_i(\lambda)| \le C\Big(\sum_{i=1}^{n} |B_i(0)|\Big)^{\frac{1-|\lambda|}{1+|\lambda|}}. \tag{14}$$

This estimate has a natural interpretation in terms of complex dynamics: iterating the configuration of disks we obtain Cantor sets and can express (14) in this set up. Using the so-called thermodynamical formalism, c.f. [9], the bound (14) with $C = 1$ is equivalent to the statement that under a holomorphic deformation g_λ of a locally conformal repeller $g_0 = g$, the topological pressure [9] of the function $-2\log|g'_\lambda|$ changes according to

$$\frac{1+|\lambda|}{1-|\lambda|}P(-2\log|g'_0|) \le P(-2\log|g'_\lambda|) \le \frac{1-|\lambda|}{1+|\lambda|}P(-2\log|g'_0|).$$

Since this follows easily [3] we have a proof for Theorem 3.1. □

Later Eremenko and Hamilton [14] streamlined the argument along the corresponding lines, without using thermodynamical formalism. This also showed that if the quasiconformal mapping f is conformal outside the unit disk as well as in the set E and if we normalize f by $f(z) = z + O(\frac{1}{z})$ as $z \to \infty$, then one can take $M_K = 1$ in (11). That will be useful below, in obtaining G-closure bounds in homogenization problems, see Theorem 4.3.

Theorem 3.1 has a number of consequences on quasiconformal and related topics. We start with the optimal integrability of the derivatives of quasiconformal mappings.

COROLLARY 3.2 *If f is a K-quasiconformal mapping in $\Omega \subset \mathbf{C}$ then $f \in W^{1,p}_{loc}(\Omega)$ for all $p < \frac{2K}{K-1}$.*

Remark. The example (10) shows that one cannot take $p = \frac{2K}{K-1}$.

Proof of Corollary. Since the result is local we may assume that $\Omega = f(\Omega) = \Delta$ and then reflect f to a global K-quasiconformal mapping. Also as $|\partial_\alpha f|^2 \le KJ_f$ a.e, it is enough to estimate the area of $E_s = \{z \in \Delta : J_f(z) \ge s\}$. By Theorem 3.1

$$s|E_s| \le \int_{E_s} J_f \, dm = |fE_s| \le M_K|E_s|^{\frac{1}{K}}$$

so that $|E_s| \le Cs^{-K/(K-1)}$. Expressing the integrals of $J_f^{p/2}$ in terms of the distribution functions $|E_s|$ proves the corollary. □

As is well-known, the higher integrability estimates are related to the removability results of bounded quasiregular functions [19]. A refinement of the above argument gives the following counterpart of the classical Painleve-theorem

COROLLARY 3.3 [4] *If $E \subset \mathbf{C}$ has Hausdorff $\frac{2}{K+1}$-measure zero, then the set E is removable for all bounded K-quasiregular functions.*

Moreover, for all $t > \frac{2}{K+1}$ there are sets E of dimension $\dim_H(E) = t$ not removable for some bounded K-quasiregular functions.

Returning to the topic of the previous section, Theorem 3.1 also controls the distortion of Hausdorff dimension under general holomorphic motions.

COROLLARY 3.4 . [3] *Given a holomorphic motion $\Phi : \Delta \times E \to \overline{\mathbf{C}}$ of a subset $E \subset \overline{\mathbf{C}}$ write $E_\lambda = \Phi_\lambda(E)$. Then*

$$\frac{1-|\lambda|}{1+|\lambda|}\Big(\frac{1}{\dim_H(E)} - \frac{1}{2}\Big) \le \frac{1}{\dim_H(E_\lambda)} - \frac{1}{2} \le \frac{1+|\lambda|}{1-|\lambda|}\Big(\frac{1}{\dim_H(E)} - \frac{1}{2}\Big). \tag{15}$$

Moreover, for some sets E and motions Φ we have the equality (in one of the estimates).

Sketch of proof. By Theorems 2.2, 2.3 we may assume that $E_\lambda = f(E)$, where $f : \mathbf{C} \to \mathbf{C}$ is K-quasiconformal and $K = \frac{1+|\lambda|}{1-|\lambda|}$. If E is covered by round disks B_i, $1 \le i \le n$, then for any $p < \frac{K}{K+1}$

$$\begin{aligned}\sum_i |fB_i|^s &\le \sum_i \Big(\int_{B_i} {J_f}^p \, dm\Big)^{\frac{s}{p}} |B_i|^{s(1-\frac{1}{p})} \\ &\le \Big(\sum_i \int_{B_i} {J_f}^p \, dm\Big)^{\frac{s}{p}} \Big(\sum_i |B_i|^{\frac{p}{p-s}s(1-\frac{1}{p})}\Big)^{1-\frac{s}{p}}\end{aligned}$$

The estimate shows that $\dim_H(fE) < 2s$ whenever $\dim_H(E) < 2s\frac{p-1}{p-s}$ and this proves the claim.

Examples where one has the equality are described in [3]. □

An interesting particular case of the above result arises when $\dim_H(E) = \dim_H(E_0) = 1$; then Corollary 3.4 says that

$$\dim_H(E_\lambda) \le 1 + |\lambda|, \quad \lambda \in \Delta,$$

and that there are motions E_λ for which this holds as an equality for all $\lambda \in \Delta$. However, in the proof of the equality in [3] the holomorphically moving sets were Cantor sets, and thus totally disconnected. Hence it is interesting to know how much the holomorphic deformations can distort the dimension of *curves.* In his study [27] Ruelle showed that the Julia sets in the main cardioid of the Mandelbrot set, c.f. Example 2.4, give an example of a motion of $\mathbf{S}^1$ with

$$\dim_H(E_\lambda) \le 1 + .09|\lambda|^2 + O(|\lambda|^3).$$

These and other related results suggest the following

CONJECTURE 3.5 *We have*

$$\dim_H(E_\lambda) \leq 1 + |\lambda|^2 , \quad \lambda \in \Delta .$$

for all holomorphic motions E_λ of $\mathbf{S}^1$.

4. Applications

We conclude with examples, three topics of a quite different nature, where the results of the previous section can be applied. As the first let us consider

4.1. Singular integrals. Since in the plane quasiconformal mappings are solutions to the linear first order elliptic equation (3), the Beltrami differential equation, it is evident that they have close connections to singular integrals and especially to the Beurling operator S, c.f. (5). In particular, as shown by Gehring and Reich [15] Theorem 3.1 is equivalent to the following estimate.

COROLLARY 4.1 *There is a constant $\alpha \geq 1$ such that for any measurable set $E \subset \Delta$,*

$$\int_\Delta |S\chi_E| \, dm \leq |E| \log \frac{\alpha}{|E|}. \tag{16}$$

Note that in case $E = B(0,r)$ and $r < 1$, $\int_\Delta |S\chi_E| \, dm = |E| \log \frac{\pi}{|E|}$.

The corollary also yields sharp bounds on the $L^1 - LlogL$ norm of the Beurling operator. Namely

$$\int_\Delta |Sv| \, dm \leq \int_\Delta |v| \log\left(1 + \frac{|v|}{|v|_\Delta}\right) dm + c \log\left(\int_\Delta |v| \log\left(1 + \frac{|v|}{|v|_\Delta}\right) dm\right) \tag{17}$$

for all $v \in L\log L(\Delta)$; here $|v|_\Delta = \frac{1}{\pi}\int_\Delta |v| \, dm$ is the integral mean of $|v|$. This bound is optimal, up to the correct value of c.

In addition for bounded functions, normalized so that $|\omega(z)| \leq \chi_\Delta(z)$ *a.e.*, similar estimates show [17], [3]

$$|\{z \in \Delta : |\operatorname{Re} S\omega(z)| > t\}| \leq 2\alpha\pi e^{-t}. \tag{18}$$

Consequently, we now know the behaviour of the Beurling transform at L^1, at L^2 (where it is an isometry) and at L^∞. However, the important question of the precise value of the L^p-norm remains open. A large amount of evidence supports the following conjecture of Iwaniec.

CONJECTURE 4.2 *For all $1 < p < \infty$, $\|S\|_p = \max\{p-1, 1/(p-1)\}$.*

The best upper bound known so far, $4 \max\{p-1, 1/(p-1)\}$, is due to Banuelos and Wang [6]. They achieve this bound by using probabilistic methods and the inequalities of Burkholder [11].

4.2. Elliptic equations. Consider the partial differential equation

$$\sum_{i,j=1}^{2} \partial_i \sigma_{ij}(x) \partial_j u(x) = 0 \quad \text{a.e. in } \Omega, \tag{19}$$

or briefly $\mathrm{Div}(\sigma(x)\nabla u) = 0$, where σ is measurable and for a.e. x symmetric and uniformly elliptic,

$$\alpha|\xi|^2 \leq \big(\sigma(x)\xi, \xi\big) \leq \beta|\xi|^2, \qquad \xi \in \mathbf{R}^2. \tag{20}$$

By results of Lavrentiev, Bers and others the solutions can be interpreted in terms of quasiregular functions. This leads to sharp regularity results for the solutions, in terms of the smallest Sobolev spaces containing them, or sharp criteria for sets $E \subset \mathbf{R}^2$ removable for every solution.

In fact, suppose first that $\det\sigma(x) \equiv 1$ a.e. Then the measurable Riemann mapping theorem can be used to factor the solutions of (19) as $u = w \circ f$ where w is harmonic, $\triangle w = 0$, and f is K-quasiconformal, $K = \sqrt{\frac{\alpha}{\beta}}$. In this case the results of the previous section can be used directly.

If the ellipticity matrix does not have a constant determinant, quasiconformal methods are still applicable: If J denotes the standard complex structure of $\mathbf{R}^2$,

$$J = \begin{pmatrix} 0 & -1 \\ 1 & 0 \end{pmatrix}$$

then given any solution u of (19), $\mathrm{Curl}(J\sigma(x)\nabla u) = \mathrm{Div}(\sigma(x)\nabla u) = 0$ and hence we have a local conjugate v with $J^T\nabla v = \sigma\nabla u$. Moreover, the function $f = (u, v)$ is quasiregular. Note however, that the dilatation $K(f)$ now depends in a complicated manner on all entries of σ rather than just on the ellipticity coefficients α and β.

4.3. Homogenization and optimal G-closure bounds. Composite materials consist of mixtures of two or more phases of materials having some fine scale structure. Since the differential equations describing the behaviour of composites have highly oscillating parameter fields, the solution of the macroscopic response of the material is a very complicated problem. In the homogenization one idealizes this situation by letting the scale parameter of the microstucture tend to zero; in the limit one gets a homogenous medium with corresponding effective material properties approximating the original mixture. The question in the G–closure problem is to determine the set of all possible effective material properties made of mixtures of two or more phases with prescribed volume fractions.

Mathematically this can be formulated by the Y-periodic matrix fields σ, $Y = (0,1)^2$, which are uniformly elliptic as in (20). If Ω is a simply connected bounded open set in $\mathbf{R}^2$ with Lipschitz boundary, denote by $u_\varepsilon \in \mathrm{W}_0^{1,2}(\Omega)$ the solution of the conductivity problem

$$-\mathrm{Div}(\sigma(\frac{x}{\varepsilon})\nabla u_\varepsilon) = f \qquad \text{in } \Omega.$$

Then the following hold (see [7, 31])

$$u_\varepsilon \rightharpoonup u_0 \quad \text{in } W_0^{1,2}(\Omega), \tag{21}$$

$$\sigma(\frac{x}{\varepsilon})\nabla u_\varepsilon \rightharpoonup \sigma_{eff}\nabla u_0 \quad \text{in } (L^2(\Omega))^2, \quad \varepsilon \to 0, \tag{22}$$

where the constant symmetric matrix σ_{eff}, the effective conductivity, is independent of $f \in W^{-1,2}(\Omega)$. It is well-known (see [7, 31]) that σ_{eff} can be obtained from the Dirichlet variational principle

$$(\sigma_{eff}\xi, \xi) = \inf_{v \in W^{1,2}_{per}(Y)} \int_Y \big(\sigma(x)(\nabla v(x) + \xi), \nabla v(x) + \xi\big)\, dx\ . \tag{23}$$

We assume now that the composite consists of k materials and model each material by a constant, possibly anisotropic conductivity matrix $D^i = \mathrm{diag}(d_1^i, d_2^i)$, $1 \le i \le k$, satisfying (20). If in the composite we denote by Y^i the part of Y occupied by the i'th material, let us assume that the volume $|Y^i| = p_i$ is fixed, but otherwise Y^i can be arbitrary. Moreover, we do not fix the pointwise orientation of the materials; hence the class of microstructures considered in (21), (22) is the set of all σ's having the representation

$$\sigma(x) = \sum_{i=1}^{k} \chi^i(x) R(x) D^i R^T(x), \tag{24}$$

where χ^i is the characteristic function of the set Y^i and R is a measurable field of rotations. Let $G(D) \equiv G(D^1, \dots, D^k, p_1, \dots, p_k)$ denote the set of all effective conductivities σ_{eff} in (21),(22) arising from σ's of the form (24); intuitively this corresponds to the set of all possible macroscopic conductivities of the mixtures of the materials D^i, with given volume fractions p_i. The problem is then to characterize the sets $G(D)$ or to find the optimal bounds for σ_{eff} depending only on the D^i and p_i, $1 \le i \le k$.

Since the problem is rotationally invariant, it is enough to find bounds for the eigenvalues of the symmetric matrixes σ_{eff}. Moreover, as proposed by Nesi [26] and Kohn [18] we can study the so–called "upper overall conductivity"

$$(\sigma^*\xi, \xi) = \inf_{v \in W_0^{1,2}(\Omega)} \int_\Omega (\sigma(x)(\nabla v(x) + \xi), \nabla v(x) + \xi)\, dx \tag{25}$$

whose bounds obtained below turn out to be optimal also for σ_{eff}.

The basic approach to these questions is the translation method due to Lurie & Cherkaev [21], Murat & Tartar [25] and Tartar [30]. In particular, in the case of the mixtures of *isotropic* materials these authors have completely characterized the set $G(D)$, i.e. found optimal bounds on the corresponding eigenvalues. However, in case of anisotropic materials one needs refinements of the method to retain the optimality. Nesi [26] made the important discovery that quasiconformality is now the essential tool.

To explain this briefly, it can be shown [5], [26] that applying arguments from the translation method, one can prove for any microstructure σ as in (24) and for any $|t| < \text{essinf}\sqrt{\det\sigma(x)}$ that

$$\frac{1}{\text{Tr}(C\sigma^* C^T) + 2t\text{det}C} = \inf \int_\Omega \left\{ \frac{\text{Tr}(D\Psi\sigma D\Psi^T) - 2t\det(D\Psi)}{\det\sigma - t^2} \right\} dx, \tag{26}$$

where the infimum is taken over those $\Psi \in W^{1,2}(\Omega;\mathbf{R}^2)$ for which $L_C(\Psi) \equiv \int_\Omega \text{Tr}(D\Psi(\text{adj}C)^T)\,dx = 1$. The identity (26) holds for each constant matrix $C \in \mathbf{R}^{2x2}$ separately.

As Nesi realized the correct test functions in (26) are quasiconformal mappings; c.f [5] for further considerations and explanations why this should be the case. In particular, by the measurable Riemann mapping theorem there exits a (unique) quasiconformal mapping Ψ_σ, conformal off Ω, normalized by $\Psi_\sigma(z) = z + O(\frac{1}{z})$ and satisfying

$$2\big(\det\sigma(x)\big)^{\frac{1}{2}} J_{\Psi_\sigma}(x) = \text{Tr}\big(D\Psi_\sigma(x)\sigma(x)\big(D\Psi_\sigma(x)\big)^T\big). \tag{27}$$

Applying this to (26) yields

$$\frac{1}{\text{Tr}(C\sigma^* C^T) + 2t\det C} \leq \frac{1}{\big(L_C(\Psi_\sigma)\big)^2} \int_\Omega \frac{2\big((\det\sigma(x))^{\frac{1}{2}} - t\big) J_{\Psi_\sigma}(x)}{\det\sigma(x) - t^2}\, dx \tag{28}$$

$$= \frac{1}{\big(L_C(\Psi_\sigma)\big)^2} \int_\Omega \frac{2J_{\Psi_\sigma}(x)}{(\det\sigma(x))^{\frac{1}{2}} + t}\, dx \tag{29}$$

for all $|t| < \text{essinf}\sqrt{\det\sigma(x)}$. Note in particular that it was precisely the change of coordinates by (27) that removed the above singularity.

To proceed further optimal quasiconformal control is required. Since our knowledge is still limited here, let us concentrate on the case where one of the phases, the "more conductive", is isotropic. That is, our materials are

$$D^1 = \text{diag}(d,d), \quad D^2 = \text{diag}(\delta_1,\delta_2), \quad \delta_1 \geq \delta_2, \quad \text{and} \quad d \geq \delta \equiv \sqrt{\delta_1\delta_2}. \tag{30}$$

Therefore the inequality (28) reduces to

$$\frac{1}{\text{Tr}(C\sigma^* C^T) + 2\delta\det C} \leq \frac{1}{\big(L_C(\Psi_\sigma)\big)^2} \Big(\frac{2}{d+\delta} \frac{|\Psi_\sigma(E)|}{|\Omega|} + \frac{1}{\delta} \frac{|\Psi_\sigma(F)|}{|\Omega|} \Big), \tag{31}$$

where E and F are the sets containing the phases D^1 and D^2, respectively, with the fixed volume ratios, say, $p = |E|$ and $1 - p = |F|$. In particular, the bounds for the global conductivity now depend on finding the correct quasiconformal estimates for this weighted area-expression. Using the results of [3], [14] Nesi was hence able to deduce the following.

THEOREM 4.3 [26] *Under the assumption* (30) *on the materials, the eigenvalues of* σ^* *satisfy*

$$\sigma_1^* + \sigma_2^* \geq -2\delta + \frac{4\delta}{1 + \frac{\delta-d}{\delta+d}p^K}. \tag{32}$$

Further, we have the equality when $(\det\sigma)^{-1/2}\sigma$ *is the matrix dilatation of the radial stretch (10).*

It turns out that the matrix C in (26) determines a tangent line of the boundary of $G(D)$. However, to use this information the precise value of the variational integral must be determined and due to different normalizations each C requires a different quasiconformal estimate. Using such an approach the recent work [5] of M. Miettinen and the author found the tangent lines at the "cornerpoint" $(m(D), h(D)) \in \partial G(D)$ as well as reduced the determination of the whole lower curve of $G(D)$ to (conjectural) quasiconformal bounds. Here the corner points arise from classical bounds using the Dirichlet variational principle (23),

$$h(D) \equiv \left(\frac{p}{d} + \frac{1-p}{\delta_2}\right)^{-1} \leq \sigma_1^*, \sigma_2^* \leq m(D) \equiv pd + (1-p)\delta_1. \tag{33}$$

THEOREM 4.4 [5] *Under the assumption* (30),

$$\sigma_1^* + \left(p + (1-p)K\frac{d}{\delta}\right)^2 \sigma_2^* \geq m(D) + \left(p + (1-p)K\frac{d}{\delta}\right)^2 h(D). \tag{34}$$

The equality holds for the rank-one laminates σ determining $(m(D), h(D))$.

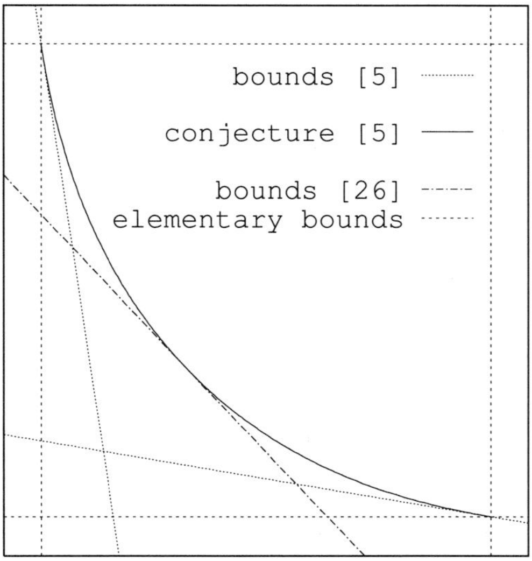

Added in proof. The precise expression for the lower boundary curve of $G(D)$, conjectured in [5], was recently proved by Milton and Nesi.

References

[1] Ahlfors, L., On quasiconformal mappings. *J. Analyse Math.*, 3 (1954), 1–58.
[2] Ahlfors, L. & Bers L., Riemann's mapping theorem for variable metrics. *Ann. Math.*, 72 (1960), 385–404.
[3] Astala, K., Area distortion of quasiconformal mappings. *Acta Math.*,173 (1994), 37–60.
[4] Astala, K. Painleve's theorem and removability properties of planar quasiregular mappings. *Preprint*
[5] Astala, K & Miettinen M., On quasiconformal mappings and 2-d G-closure problems. *Arch. Rational Mech. Anal.* (To appear)
[6] Banuelos, R. & Wang, G., Sharp inequalities for martingales with applications to the Beurling-Ahlfors and Riesz transforms, *Duke Math. J.* 80 (1995) 575–600.
[7] Bensoussan, A., Lions, J.L. & Papanicolaou, G. *Asymptotic Analysis for Periodic Structures* North Holland, 1978.
[8] Bojarski, B., Generalized solutions of a system of differential equations of first order and elliptic type with discontinuous coefficients. *Math. Sb.*, 85 (1957), 451–503.
[9] Bowen, R., *Equilibrium States and the Ergodic Theory of Anosov Diffeomorphisms.* Lecture Notes in Math., 470. Springer-Verlag, New York-Heidelberg, 1975.
[10] Branner, B. & Fagella, N., Homeomorphisms between Limbs of the Mandelbrot set. Journal of Geometric Analysis (To appear)
[11] Burkholder, D., Sharp inequalities for martingales and stochastic integrals. *Asterisque* 157–158 (1988), 75–94.
[12] Carleson, L. & Gamelin T., *Complex dynamics.* Universitext: Tracts in Mathematics. Springer-Verlag, New York, 1993.
[13] Douady, A. & Hubbard, J. On the dynamics of polynomial-like mappings. *Ann. Sci. École Norm. Sup.* (4) 18 (1985), no. 2, 287–343.
[14] Eremenko, A. & Hamilton, D. On the area distortion by quasiconformal mappings. *Proc. Amer. Math. Soc.* 123 (1995), no. 9, 2793–2797.
[15] Gehring, F.W. & Reich E., Area distortion under quasiconformal mappings. *Ann. Acad. Sci. Fenn. Ser. A. I.*,388 (1966), 1–14.
[16] Hempel, J.A. Precise bounds in the theorems of Schottky and Picard. *J.London Math. Soc* 21 (1980), 279–286.
[17] Iwaniec, T. & Kosecki R., Sharp estimates for complex potentials and quasiconformal mappings. Preprint of Syracuse University.
[18] Kohn, R.V., The relaxation of a double energy. *Continuum Mech. Thermodyn.* 3 (1991) 193–236.
[19] Lehto O. & Virtanen K., *Quasiconformal mappings in the plane.* Second edition. Springer-Verlag, New York-Heidelberg, 1973.
[20] Lehto O., *Univalent functions and Teichmüller spaces* Springer-Verlag, New York-Heidelberg, 1987.
[21] Lurie, K.A. & Cherkaev, A.V., Exact estimates of conductivity of composites formed by two isotropically conducting media taken in prescribed proportion. *Proc. Roy. Soc. Edinburgh* 99A (1984) 71–87.
[22] Mañé R., Sad P. & Sullivan D., On the dynamics of rational maps. *Ann. Sci. École Norm. Sup.*, 16 (1983), 193–217.
[23] Martin, G. The distortion theorem for quasiconformal mappings, Schottky's theorem & holomorphic motions. *Proc. Amer. Math. Soc.* (To appear)
[24] Mori, A., On an absolute constant in the theory of quasiconformal mappings. *J. Math. Soc. Japan*, 8 (1956), 156–166.

[25] Murat, F. & Tartar, L., Calcul des variations et homogénéisation., pp. 319–369 *in* Les Méthodes de l'Homogénéisation: Théorie et Applications en Physique. Eyrolles, 1985

[26] Nesi, V. Quasiconformal mappings as a tool to study certain two-dimensional G-closure problems. *Arch. Rational Mech. Anal.* 134 (1996), no. 1, 17–51

[27] Ruelle, D. Repellers for real analytic maps. *Erg. Th. and Dynam. Sys.*, 2 (1982), 99–107.

[28] Slodkowski, Z., Holomorphic motions and polynomial hulls. *Proc. Amer. Math. Soc.*, 111 (1991), 347–355.

[29] Sullivan, D. Quasiconformal homeomorphisms and dynamics. I. Solution of the Fatou-Julia problem on wandering domains. *Ann. of Math.* (2) 122 (1985), no. 3, 401–418.

[30] Tartar, L., Estimation fines des coefficient homogénéisés. pp. 168–187 *in* Pitman Research Notes in Mathematics no.s 125, 1985

[31] Zhikov, V.V., Kozlov, S.M. & Oleinik, O.A. *Homogenization of Differential Operators and Integral Functionals* Springer-Verlag, New York-Heidelberg, 1994

Progress in Mathematics, Vol. 168,

A Combinatorial Approach to Combings and Framings of 3-manifolds

Riccardo Benedetti

Dipartimento di Matematica
Via F. Buonarroti, 2
I-56127 Pisa, Italia
benedett@dm.unipi.it

1. Introduction

The results presented in this talk form a proper part of joint work with Carlo Petronio (see [3]). On the one hand, this work was intended to be a contribution to a *constructive* approach to the geometry and topology of 3-manifolds; on the other hand, it should provide some evidence of the fact that the theory of (eventually *branched*) *standard spines* could represent a natural unified framework for several different facets of 3-dimensional topology. It is in this very limited way that my talk interprets the *motto* of the Congress on the Unity of Mathematics.

The main authors who contributed to developing the classical theory of standard spines of 3-manifolds are Casler [6], S.V.Matveev [14] and Piergallini [16]. Mostly in the oriented case (see also [1]), standard spines nicely support a *combinatorial realization* (see later for a definition of this notion) of the class of 3-manifolds up to diffeomorphism, that, in a sense, is the most efficient one as the related *calculus* is generated by only one local move.

A beautiful example of realized constructive 3-dimensional geometry and topology is Weeks's *SnapPea* package that implements many achievements of Thurston's *geometrization program*; standard spines (in the *dual* but equivalent form of *topological ideal triangulations*) are widely employed in this context, for example in the treatment of the Thurston Hyperbolic Dehn Surgery theorem (see also [4] for a discussion of this point).

The interest in effective combinatorial presentations of 3-dimensional topological objects increased in recent years with the development of the theory of *Quantum Invariants* (see Turaev's book [18]) and their relatives (see [12] and [13]). They are ultimately expressed by explicit formulas, invariant with respect to the calculus of suitable combinatorial presentations. The combinatorial realization based on (classical) standard spines is appropriate for the definition and analysis of, for instance, the so-called *Turaev-Viro invariants* (see also [2]). On the other hand, refined versions of these invariants actually work for *structured* (such as *Combed*, *Framed* or *Spun*) manifolds. It would be interesting to have one framework in which to compute all the invariants, and the first step in this direction would be a unified combinatorial realization of such structured manifolds.

It is well known that every homotopy class of co-oriented distributions of tangent 2-planes in an oriented closed 3-manifold can be realized by foliations [19] or contact structures (see e.g.[7]). Recently, deep investigations have been devoted to the classification of contact structures and the relationships between *taut* foliations and *tight* contact structures. It would be interesting to develop a constructive approach to such theories and, again, a combinatorial realization of combed manifolds (that is equivalent to a realization of 3-manifolds with a homotopy class of cooriented distribution of planes) should be a first step.

We now come to the real content of the talk, and introduce a few slightly more formal notions. Let M be a compact, closed, connected and oriented 3-manifold; a *combing* c of M is a non-vanishing tangent vector field on M; a *framing* f of M consists of a triple of linearly independent vector fields on M inducing the orientation. Combings and framings are considered up to homotopy. Consider the class $\mathcal{C}$ (resp. $\mathcal{F}$) of pairs (M, c) (resp. (M, f)) up to the equivalence relation generated by orientation-preserving diffeomorphisms of manifolds. My aim is to present, in a unified setting, a *combinatorial realization* of $\mathcal{C}$ and $\mathcal{F}$; namely for $\mathcal{M} \in \{\mathcal{C}, \mathcal{F}\}$ I will define a set of combinatorial objects $\Theta(\mathcal{M})$ (in fact, a certain class of decorated finite graphs) and an effective *reconstruction map*

$$\Phi : \Theta(\mathcal{M}) \rightarrow \mathcal{M}$$

such that:

1. Φ is onto;

2. The equivalence relation on $\Theta(\mathcal{M})$ induced by Φ is generated by a *finite* set of *local* moves; these generate the *calculus* of the realization.

In the same framework, a realization for *Spun* 3-manifolds can also be obtained.

Note that the requirements of *finiteness* and *locality* of the calculus make our notion of combinatorial realization somewhat demanding: for example, no presentation via surgery along framed links and any known version of the *Kirby calculus* is a realization in the present sense.

The central geometric object of our construction is the notion of *branched standard spine*, obtained by combining the classical theory of standard spines, recalled above, with that of (oriented) branched surfaces, and suitably encoded in terms of a *finite graphic calculus.*

The notion of branched surface was originally introduced in [20] for the study of hyperbolic attractors (see also [5]). Later on, in a number of papers (see e.g. [8],[9] and the references quoted therein) branched surfaces have been viewed mostly as codimension one objects capable of supporting and generalizing classical notions like incompressibility.

Branched spines (not necessarily standard - I refer to [3] for a discussion of the importance of the *cellularity condition*, i.e. the standardness, in order to get a local calculus as required) have been tacitly considered, even if not explicitely defined and studied, in Gillman-Rolfsen's papers about the Zeeman conjecture (see [10]) and in Ishi's papers on *flow-spines* (see [11]).

I would also cite [17] in which Turaev first addresses the difficulty of a combinatorial treatment of non-singular vector fields up to homotopy; the difficulty being mostly due to their nature of affine spaces.

I will refer to [3] for the details of definitions and proofs, for some first applications, a more complete discussion about related problems and perspectives and more complete references to the literature. I will confine myself here to the statement of some ready-to-use results, with a few comments.

2. Normal o-graphs

By a *normal o-graph* we understand any object Γ with the properties that we will now describe:

1. The *support* of Γ is a compact planar curve with only simple normal crossings as singularities (exactly like the images of usual generic planar projections of spatial links) ; some crossings are *marked* and called the *vertices* of Γ. We assume that there is at least one vertex. Therefore the support of Γ can be viewed as the image of a planar immersion of an abstract graph Λ with quadrivalent vertices: the four edge-branches at each vertex are coupled two-by-two and the image of a neighbourhood of each vertex is normalized to be a simple normal crossing, so that the coupled branches go "straight"; these are the marked crossings; elsewhere the immersion is generic eventually with other unmarked simple normal crossings. We assume that Λ is *connected* and we call *edges* of Γ the images of the edges of Λ.

2. At each vertex of Γ an *under-or-over* specification (as in usual planar link diagrams) is given.

3. Every edge of Γ is oriented in such a way that the orientations match across the vertices.

Note that the edge-orientation determines on Γ a well-defined set of *oriented circuits* (oriented cycles with normal crossings).

We regard the set of normal o-graphs up to the relation generated by planar isotopy and the local moves of *Reidemeister type* of Figure 1, where the orientations are implicit. Denote by Θ the quotient set. We will often confuse an o-graph with its equivalence class.

3. Decoding normal o-graphs

(A). From Γ to an oriented 3-manifold with boundary $W(\Gamma)$ endowed with an oriented branched spine. Consider the plane containing Γ as the (xy)-plane of the (xyz)-space. Specify at each unmarked crossing of Γ an under-over branch, in an arbitrary way, so that Λ is now actually embedded into the space. Each edge of Γ can be regarded as a simple arc that we represent abstractly as an oriented segment, with endpoints labeled by $+$ or $-$ according to whether the corresponding branch at the vertex is over or under. The result of the application of the rules suggested by Figures 2 to 4 is the following: we have realized a spatial embedding of

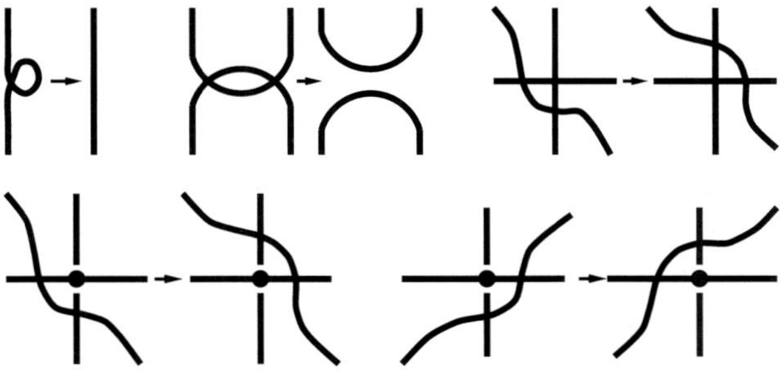

Figure 1: Reidemeister-type moves on normal o-graphs

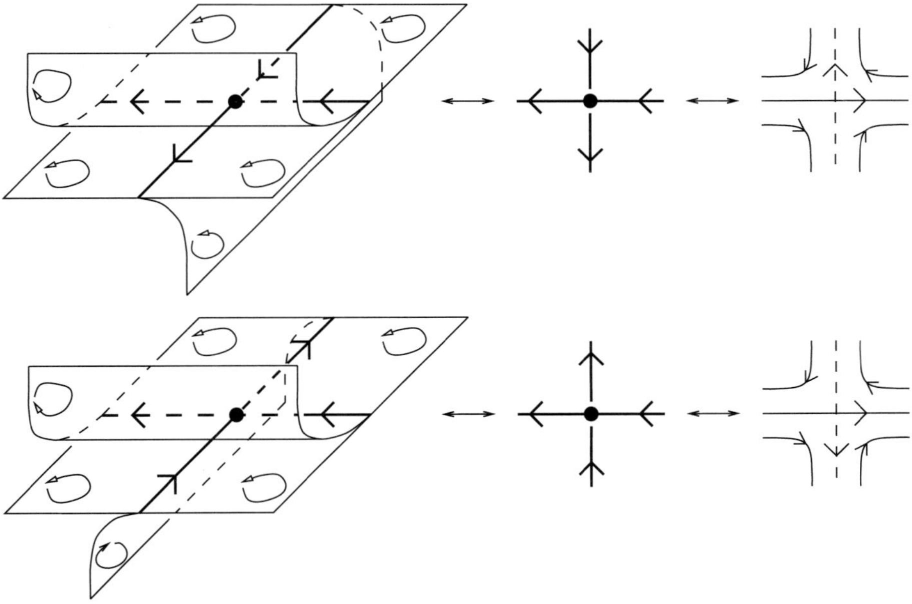

Figure 2: Decoding a normal o-graph: local picture at a vertex

a regular neighbourhood $N(\Gamma)$ in $P = P(\Gamma)$ of the singular set $S(P)$ of an oriented branched surface P. In the figures one can see only $N(\Gamma)$, the whole of P being obtained by gluing a 2-disc to each boundary component. One can prove that P is a standard spine of an oriented 3-manifold with boundary $W = W(\Gamma)$: W is a regular neighbourhood of P, and it is constructed from a regular neighbourhood U of $S(P)$ in the 3-space by attaching a 2-handle to the boundary of U along each boundary component of $N(\Gamma)$; note that $N(\Gamma)$ is properly embedded in U and transverse to its boundary. W is oriented according to the 3-space orientation. P is a transversely oriented branched spine of W (as P and W are oriented).

(B). How to reconstruct the boundary of W. The result of the application of the rules suggested by Figure 5 is a surface with boundary (actually, a regular neighbourhood of a graph with trivalent vertices). One can prove (see also [15]) that

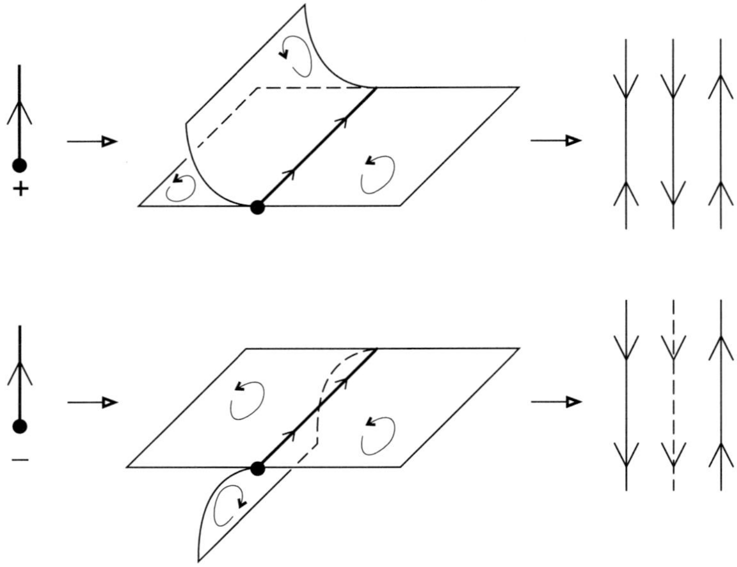

Figure 3: Decoding a normal o-graph: local pictures at the ends of an edge

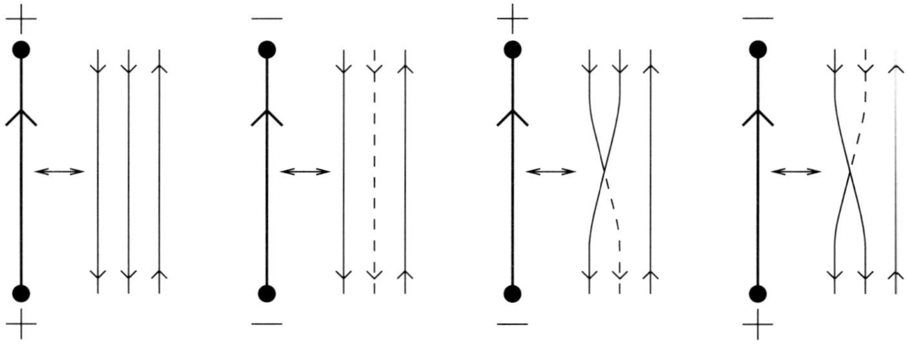

Figure 4: Decoding a normal o-graph: local picture along an edge

the closed surface obtained by gluing a 2-disc to each boundary component is isomorphic to the boundary of W. Note also that the figure suggests a *bicoloration* of the boundary of W, the meaning of which will be explained later.

4. The reconstruction map $\Phi : \Theta(\mathcal{C}) \to \mathcal{C}$

$\Theta(\mathcal{C})$ is defined as the subset of Θ which consists of the normal o-graphs Γ such that:

1. the boundary of $W(\Gamma)$ is a 2-sphere;
2. Γ has only one circuit.

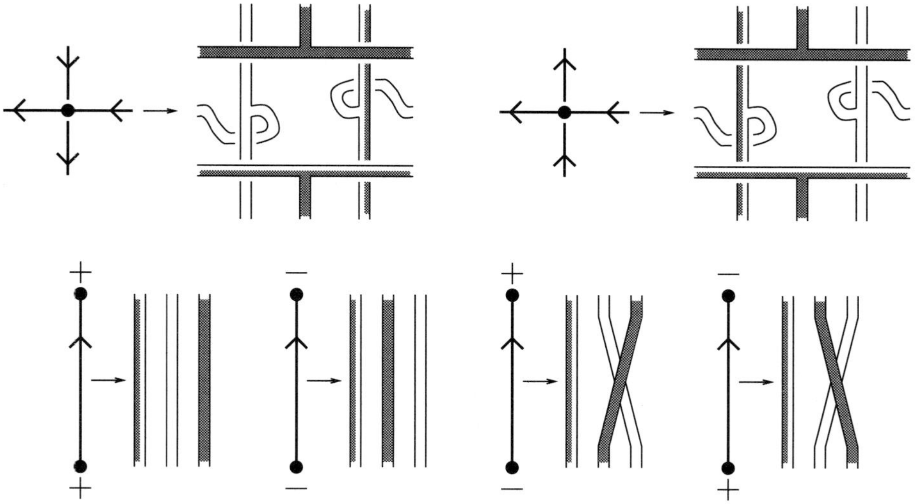

Figure 5: Reconstruction of the boundary

We define now

$$\Phi(\Gamma) = (M, c) = (M(\Gamma), c(\Gamma))$$

as follows: M is obtained by gluing a 3-disc to the boundary of W. There is a natural non-vanishing vector field on W normal to the branched spine $P = P(\Gamma)$, as suggested by the cross-section of Figure 6; the above condition (2) implies that this field is tangent to the boundary of W only along a single simple curve γ, which is *concave* with respect to the field. Therefore it is possible to extend this field to a combing c of M, which is uniquely defined up to homotopy. Note that the curve γ separates the boundary of W into two regions (both homeomorphic to discs in this case) which can be colored in a different way according to whether the vector field points outwards or inwards on them. This explains the bicoloration mentioned during the reconstruction of the boundary from an o-graph.

A first non-trivial result is the following:

THEOREM 1 *The map Φ is onto.*

5. The combing calculus

Consider the local moves of Figures 7 and 8 (when an orientation is not specified it is arbitrary and is compatible before and after the move). They generate a calculus on $\Theta(\mathcal{C})$ called the *combing calculus.* We have:

THEOREM 2 *The equivalence relation on $\Theta(\mathcal{C})$ induced by Φ coincides with the relation induced by the combing calculus.*

We note that forgetting the orientation (i.e the branching) and regarding the effect at the level of bare spines, each of the moves of Figure 8 is in fact the classical

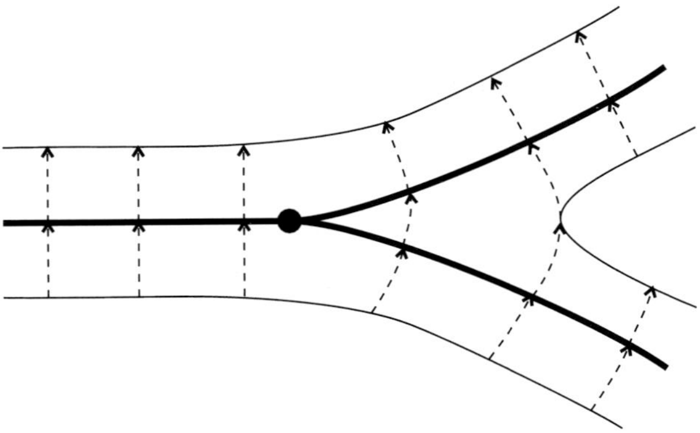

Figure 6: Cross-section of the combing

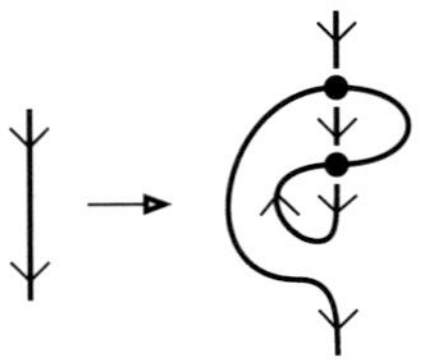

Figure 7: First move of the combing calculus

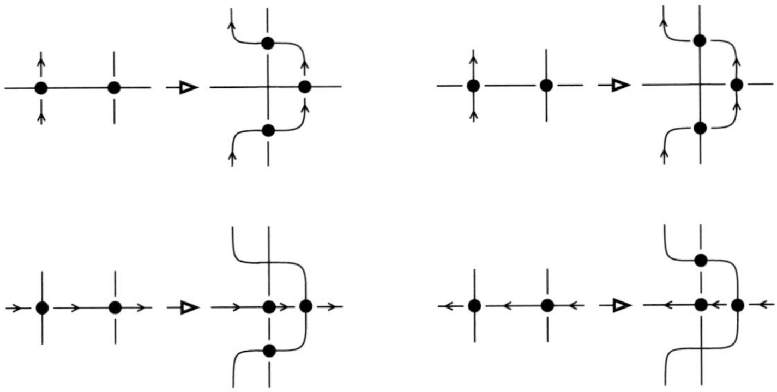

Figure 8: More moves of the combing calculus

Matveev-Piergallini move; actually, these are exactly all the branched versions of the MP-move which preserve condition (2) in the definition of $\Theta(\mathcal{C})$. On the other hand, the proof of our theorem concerning the combing calculus does not depend on the completeness result of the MP-calculus for classical (unbranched) standard spines.

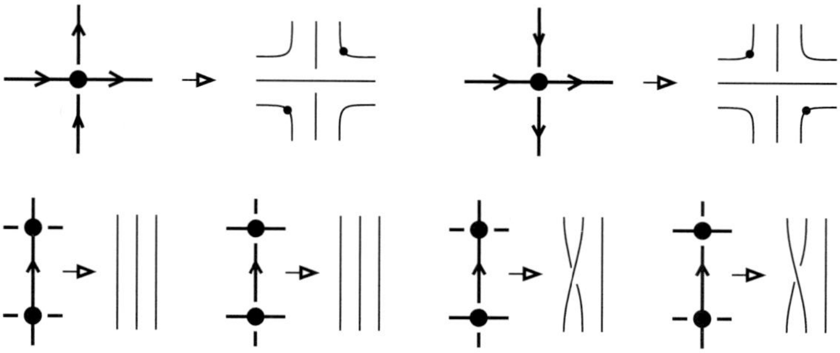

Figure 9: Computation of the Euler cochain

6. The Euler cochain

Complete the rules of Figures 2 and 3 by adding two dots on the boundary of $N(\Gamma)$ corresponding to each vertex, as illustrated in Figure 9. It is possible to prove that each boundary component of $N(\Gamma)$ contains an even number of dots. Associate to each 2-disc of $P \setminus N(\Gamma)$ the number $h = 1 - k$, where $2k$ is the number of dots on its boundary. In this way we obtain an integral 2-cochain $e(\Gamma)$, cellular with respect to the natural cellular decomposition of P, called the *Euler cochain* of the branched spine P. The name is justified by the following proposition.

PROPOSITION 3 *The Euler cochain $e(\Gamma)$ of $P(\Gamma)$ represents the Euler class of the field of 2-planes normal to the combing associated (up to homotopy) to the branched spine $P(\Gamma)$.*

Note that the sum of the integers h also gives the usual Euler-Poincaré characteristic of P, i.e. 1 in the present case.

7. Framings

The following facts hold true:

1. A combing of M defined by a graph $\Gamma \in \Theta(\mathcal{C})$ extends to a framing if and only if the normal Euler class is zero, hence if and only if there exists an integral cellular 1-cochain g having coboundary $\delta g = e(\Gamma)$. To each such 1-cochain one can constructively associate a framing of M which extends the combing. Moreover all framings whose first vector is homotopic to the given combing are associated to some 1-cochain g with $\delta g = e(\Gamma)$.

2. Two integral 1-cochains g and h such that $\delta g = \delta h = e(\Gamma)$ define homotopic framings if and only if $((g - h) \bmod 2)$ is a $\mathbb{Z}_2$-coboundary.

3. For every g as before, and for every 0-cochain s with coefficients in $\mathbb{Z}_2$, there exists an integral 1-cochain h such that $\delta h = e(\Gamma)$ and

$$((h - g) \bmod 2) = \delta s.$$

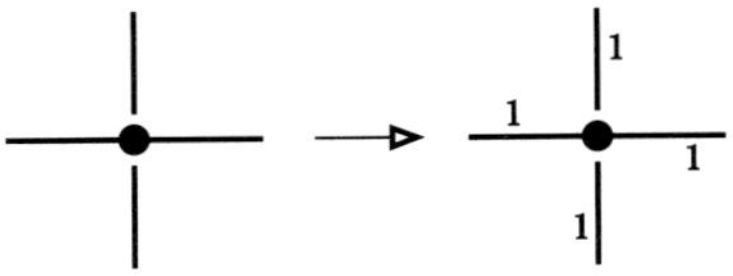

Figure 10: First move of the framing calculus

4. Assume that Γ and Γ' are elements of $\Theta(\mathcal{C})$ equivalent by the combing calculus and such that the associated combing extends to a framing. Let g be an integral 1-cochain which represents such a framing f, as stated in point 1, with respect to Γ; then there exists an integral 1-cochain g' on $P(\Gamma')$ which represents a framing f' on M homotopic to f. If Γ and Γ' are related by one move only, in the various instances the cochain g' can be computed effectively.

So the *framing calculus* is the following: $\Theta(\mathcal{F})$ consists of pairs (Γ, r) where Γ belongs to $\Theta(\mathcal{C})$ and r is a cellular 1-cochain in $P(\Gamma)$ with coefficients in $\mathbb{Z}_2$, with the property that there exists an integral 1-cochain g such that $\delta g = e(\Gamma)$ and $r = (g \bmod 2)$. Of course we can represent such a pair by a normal o-graph of $\Theta(\mathcal{C})$ having the edges suitably colored by 0 or 1. The elementary moves of the framing calculus are the following:

1. Replace (Γ, r) by $(\Gamma, r + \delta\hat{v})$ where $\hat{v}$ is the dual of a vertex. This move is graphically described in Figure 10.

2. For each move of the combing calculus enhance it to a move on $\Theta(\mathcal{F})$ following the rules of Figure 11. Note that by means of the move of Figure 10 we can always put an edge with different ends in a situation where it has a certain $\mathbb{Z}_2$ coefficient (e.g. 0).

We can summarize our results concerning framed 3-manifolds as follows:

THEOREM 4 *The reconstruction map $\Theta(\mathcal{F}) \to \mathcal{F}$ is surjective, and the equivalence relation it generates coincides with the relation generated by the framing calculus.*

8. Spin manifolds

For *any* Γ in $\Theta(\mathcal{C})$, the reduction mod 2 of the Euler cochain $e(\Gamma)$ is the second *Stiefel – Whitney* class of the 3-manifold $M = M(\Gamma)$ that is *zero* when M is parallelizable. Consider the set $\Theta(\mathcal{S})$ consisting of pairs (Γ, g), where Γ belongs to $\Theta(\mathcal{C})$ and g is a (mod2)-cellular 1-cochain on $P = P(\Gamma)$ such that $\delta g = e(\Gamma)$ (mod2) and these 1-cochains are regarded up to 0-coboundaries (mod 2). Note that this set of 1-cochains is naturally an affine space over the cellular 1-cohomology of P with $\mathbb{Z}_2$-coefficients. Again we can represent such a pair by a Γ in $\Theta(\mathcal{C})$ with suitable $\mathbb{Z}_2$ coefficients on the edges. In fact $\Theta(\mathcal{S})$ can be used as the domain of the reconstruction map of a combinatorial realization of the class of spun 3-manifolds. Note that there is a natural forgetting map from $\Theta(\mathcal{F})$ to $\Theta(\mathcal{S})$ that corresponds

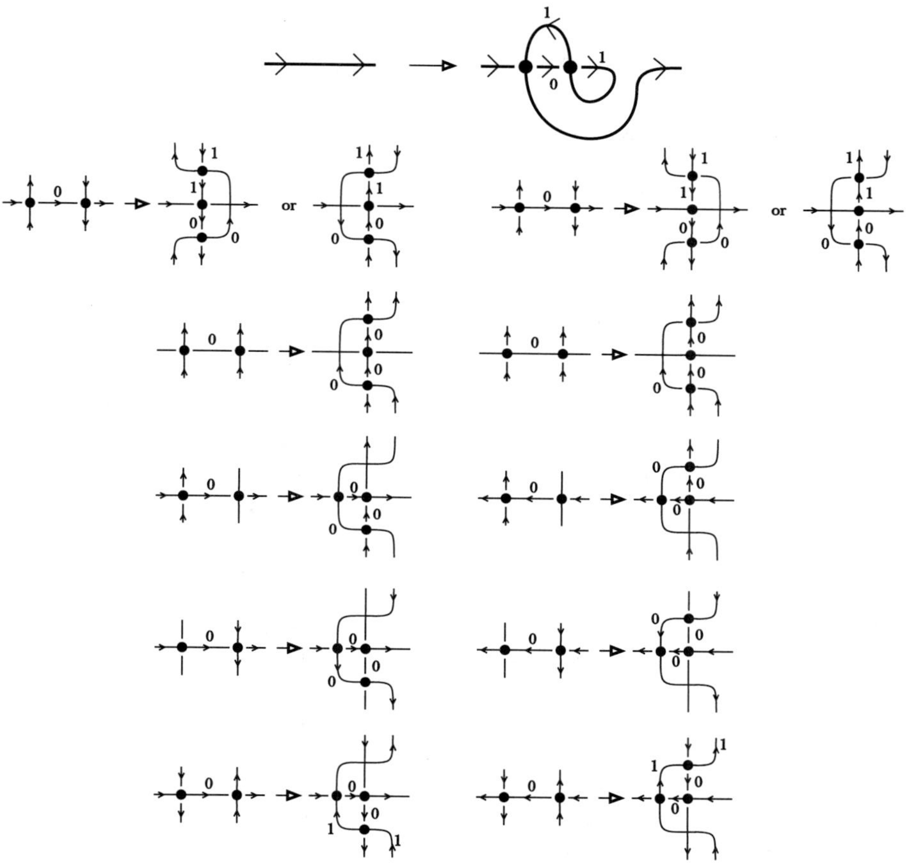

Figure 11: More moves of the framing calculus

to the spin structures carried by framings. In order to build the *spin calculus* of this realization, one proceeds as follows:

1. One adds to the combing calculus one more local move called the *Combinatorial Pontrjagin move*, as it is a combinatorial counterpart of the classical Pontrjagin construction. The meaning of this move is that any two Γ_1 and Γ_2 in $\Theta(\mathcal{C})$ can be connected by a finite sequence of moves of this *extended calculus*. Note that, by the way, we have obtained another combinatorial relization of the class of oriented closed 3-manifolds, based on branched standard spines and independent of the classical approach via bare standard spines recalled in the introduction.

2. One enhances the moves of the extended calculus in order to take into account the (local) modifications of the 1-cochains encoding the spin structures. Thus the spin calculus consists of exactly the same moves of the framing calculus completed by the enhanced combinatorial Pontrjagin move that one can see

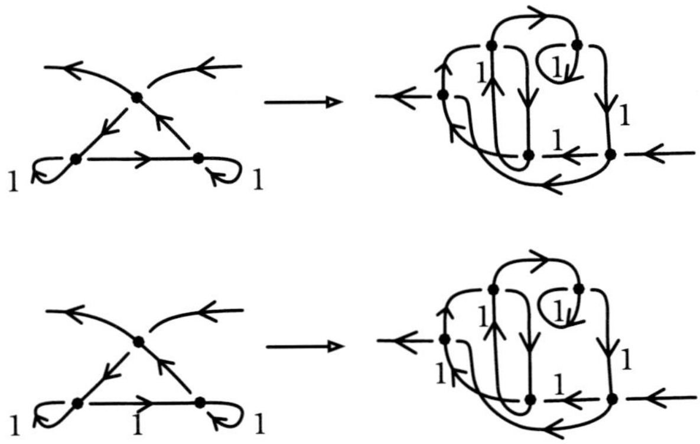

Figure 12: Combinatorial Pontrjagin move

in Figure 12 (by forgetting the $\mathbb{Z}_2$ coefficients on the edges one gets the pure Pontrjagin move).

References

[1] R. Benedetti – C. Petronio, *A finite graphic calculus for 3-manifolds*, Manuscripta Math. **88** (1995), 291–310.

[2] R. Benedetti – C. Petronio, *On Roberts' proof of the Turaev-Walker theorem*, Journal of Knot Theory and Its Ramifications, Vol. 5, No. 4 (1996), 427–439.

[3] R. Benedetti – C.Petronio, *Branched standard spines of 3-manifolds*, LNM 1653, Springer-Verlag, Berlin, Heidelberg, 1997.

[4] R.Benedetti – C.Petronio, *Lectures on Hyperbolic Geometry*, Universitext Springer-Verlag, 1992.

[5] J. Christy, *Branched surfaces and attractors I*, Trans. Amer. Math. Soc. **336** (1993), 759–784.

[6] B.G. Casler, *An embedding theorem for connected 3-manifolds with boundary*, Proc. Amer. Math. Soc. **16** (1965), 559–566.

[7] Y.A. Eliashberg, *Contact 3-manifolds twenty years after J. Martinet's work*, Ann. Inst. Fourier (Grenoble) **42** (1992), 165–192.

[8] Floyd – Oertel, *Incompressible surfaces via branched surfaces*, Topology (1984), 117–125.

[9] D. Gabai – U. Oertel, *Essential laminations in 3-manifolds*, Ann. of Math. **130** (1989), 41–73.

[10] D. Gillman – D. Rolfsen, *The Zeeman conjecture for standard spines is equivalent to the Poincarè conjecture*, Topology **22** (1983), 315–323.

[11] I. Ishii, *Moves for flow-spines and topological invariants of 3-manifolds*, Tokyo J. Math. **15** (1992), 297–312.

[12] L. Kauffman – D.E. Radford, *Invariants of 3-manifolds derived from finite dimensional Hopf algebras*, J. Knot. Theory Ramifications **4** (1995), 131–162.

[13] G. Kupenberg, *Non-involutory Hopf algebras and 3-manifold invariants*, Preprint 1995.

[14] S.V. Matveev, *Transformation of special spines and the Zeeman conjecture*, Math. USSR-Izv. **31** (1988), 423–434.

[15] C. Petronio, *Standard spines and 3-manifolds*, Tesi di Perfezionamento, Scuola Normale Superiore, Pisa, 1995.

[16] R. Piergallini, *Standard moves for standard polyhedra and spines*, Rendiconti Circ. Mat. Palermo **37**, suppl. 18 (1988), 391–414.

[17] V.G. Turaev, *Euler structures, nonsingular vector fields, and torsion of Reidemeister type*, Math. USSR-Izv. **34** (1990), 627–662.

[18] V.G. Turaev, *Quantum Invariants of Knots and 3-manifolds*, Studies in Math. De Gruyter, Berlin 1994.

[19] J.Wood, *Foliations on 3-manifolds*, Ann.Math. **89** (1969), 336–358.

[20] R.F. Williams, *Expanding attractors*, Publ. Math. IHES **43** (1974), 169–203.

Progress in Mathematics, Vol. 168,

Algebra and Combinatorics

Christine Bessenrodt

Institut für Algebra und Geometrie, Universität Magdeburg
D-39016 Magdeburg, Germany

1. Introduction

A very fruitful area of interaction between algebra and combinatorics has always centered around the symmetric groups and related groups. There are many aspects of such interactions; for example, group-theoretic investigations dealing with statistics on permutations, studies on the Bruhat order on symmetric groups, or invariant-theoretic investigations on symmetric polynomials.

The focus of this article are the representations of the symmetric and related groups and their associated combinatorics of partitions. There are many connections from this area both to other algebraic as well as to other combinatorial topics. Some groups closely related to the symmetric groups are the alternating groups A_n and the double covers $\widetilde{S}_n$ and $\widetilde{A}_n$ of the symmetric and alternating groups; the representation theory of these families of groups is highly interrelated as we will illustrate.

As the symmetric groups are Weyl groups, it is natural to try to generalize some of the results on these groups to Weyl groups or even Coxeter groups. This has been pursued successfully in several directions, although we shall not go into this here.

Another close relative is the family of general linear groups GL_m; this provides a further strong link to invariant theory. We will digress into an application of the representation theory of the symmetric groups to a classical problem of invariant theory later on.

The study of representations of symmetric groups and general linear groups is strongly knitted together via Schur algebras, Hecke algebras and their quantum analogues. All these fast developing topics would require surveys of their own and are not treated here; the interested reader is referred to [J90], [Mar] and the literature cited there.

From the very beginning of the complex representation theory of the symmetric groups, the connections with symmetric functions have played an important rôle. For an excellent treatment of this subject we refer the reader to the recent extended edition of Macdonald's monograph [Mac].

Besides permutations, the main combinatorial objects that we encounter in this area are partitions (and their generalisations). This provides a strong link to number theory. It has turned out that it is extremely fruitful to represent partitions graphically by Young diagrams (or: Ferrers diagrams) respectively shifted diagrams. Filling the nodes or boxes of these diagrams leads to Young tableaux and shifted tableaux which are of great importance both in combinatorics as well as in representation theory. Studying p-modular representations motivates the introduction of a parameter p into the combinatorial investigations (which usually does not have to be a prime on the combinatorial side); this has been a rich source of inspiration for combinatorial notions and problems. As the irreducible representations of the symmetric groups (and related families of groups) are naturaly labelled by combinatorial objects a recurring theme is to find combinatorial answers (in terms of the labelling objects) for algebraic questions. Conversely, algebraic methods lead to combinatorial theorems that often have no direct combinatorial proof (so far).

A particularly striking example of a new combinatorial notion coming from representation theory is the notion of a normal resp. good box of a partition introduced by Kleshchev in the course of his work on modular branching problems. This same notion has turned out to occur prominently also in other contexts; for the connections with Hecke algebras at roots of unity and crystal bases of quantum affine algebras the reader is referred to the article by Lascoux, Leclerc and Thibon [LLT96] and the literature cited there. In our survey of Kleshchev's work we will restrict our attention to the original representation theoretical problems.

In the following we will briefly describe some of the general results and problems of the representation theory of finite groups to give an idea of how to put the developments on the representation theory of the symmetric groups into context, but the main focus will be the study of the representations of the symmetric groups and their covering groups and the combinatorial themes important in this area.

2. Ordinary representation theory of finite groups

We consider a finite group G, and we let A be a commutative ring (with 1).
Then a *(linear) representation* of G is a homomorphism $G \to GL(V)$ resp. $G \to GL_m(A)$ from G to the group of invertible transformations on V, where V is a finitely generated free A-module (of rank m). Taking traces gives the associated character χ_V. With respect to this G-action V is a module for the *group algebra* AG, which is the algebra of formal sums $\sum_{g\in G} a_g g$ with coefficients in A and componentwise addition and multiplication induced from the multiplication in G (linearly extended).

The AG-module V (resp. the corresponding representation) is *irreducible* if it contains no non-trivial G-invariant A-submodule; the corresponding character χ_V is then also called irreducible. These are the basic building blocks for all representations in the way described in more detail below.

We first consider the situation of *ordinary representation theory.*
Here, we take A to be a field K of characteristic 0, which is "sufficiently large" for G (e.g. $K = \mathbb{C}$); indeed, for $G = S_n$ the field $K = \mathbb{Q}$ is already large enough. Some of the most important basic properties of ordinary representations of G are collected in the following (see [CR, F]):

THEOREM 2.1

(a) *(Maschke) KG is semisimple, i.e. any K-representation of G splits into a direct sum of irreducible representations of G.*

(b) *The number of irreducible K-representations of G equals the number $k(G)$ of conjugacy classes of G.*

(c) *The K-representations of G are determined (up to isomorphism) by their characters.*

(d) *Let $Irr(G)$ denote the set of irreducible characters of G over K; then*

$$|G| = \sum_{\chi \in \mathrm{Irr}(G)} \chi(1)^2 .$$

Thus, for a given finite group G, the basic problem of ordinary representation theory is to determine the irreducible K-characters and K-representations of G.

3. Ordinary representations of the symmetric groups and their combinatorics

For the symmetric groups S_n this classification of irreducible representations was achieved early in the history of representation theory by Frobenius. Important at all stages in the development of the representation theory of S_n was to find the right combinatorial notions. In the case of the ordinary representation theory, the fundamental associated combinatorial objects are partitions and tableaux.

A *partition* $\lambda = (\lambda_1, \ldots, \lambda_l)$ of $n \in \mathbb{N}$ is a non-increasing sequence $\lambda_1 \geq \ldots \geq \lambda_l > 0$ of integers with $\sum_{i=1}^{l} \lambda_i = n$, for short we write: $\lambda \vdash n$. The integer $l = l(\lambda)$ is the *length* of λ, the numbers λ_i are the *parts* of λ. We also write the partition exponentially as $\lambda = (l_1^{a_1}, \ldots, l_m^{a_m})$, $l_1 > \ldots > l_m > 0$, $a_1, \ldots, a_m \in \mathbb{N}$. Counting the partitions of a fixed number $n \in \mathbb{N}$ gives the partition function

$$p(n) = |\{\lambda \mid \lambda \vdash n\}| ;$$

this has been studied in depth since Euler in combinatorics as well as in number theory [A76].

In particular, Ramanujan obtained deep results on congruence properties of $p(n)$ for which combinatorial proofs were given only much later; a number of new results on congruence properties were proved in recent years by combinatorialists as well as by number theorists. From the representation-theoretic point of view the partition function is important for the symmetric groups since the conjugacy classes of S_n are naturally labelled by partitions of n, and thus the number of irreducible ordinary representations of S_n equals $p(n)$ by Theorem 2.1.

Indeed more is true (for detailed information we refer the reader to [JK, Sag] or [Mac]):

THEOREM 3.1 (Frobenius (1900)) *The irreducible complex characters of S_n are* naturally *labelled by partitions of n; we abbreviate this set of characters by $Irr(S_n) = \{[\lambda] \mid \lambda \vdash n\}$. The character values of the character $[\lambda]$ are given via the* Schur functions s_λ.

It has turned out to be extremely fruitful to represent a partition graphically as follows. For $\lambda = (\lambda_1, \ldots, \lambda_l) \vdash n$, its *Young diagram* $Y(\lambda)$ has λ_i boxes in row i.

Example. For $\lambda = (4^2, 2, 1)$, its Young diagram $Y(\lambda)$ looks like this:

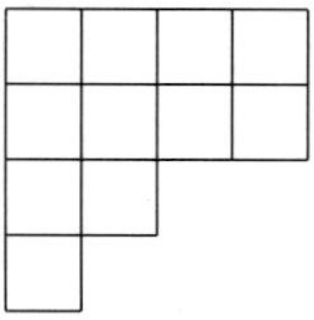

A particular rôle e.g. for induction arguments is played by the so-called hooks. The (i,j)-hook $H_{i,j}$ in λ consists of the box at position (i,j) together with all boxes in $Y(\lambda)$ to the right and below. The hooklength $h_{i,j}$ counts the number of boxes in $H_{i,j}$. An *l-hook* of λ is a hook of length l in λ. The *leg length* $L(H_{i,j})$ is the number of boxes below the(i,j)-box in λ.

Example. For λ as above, here is its Young diagram with the (1,2)-hook $H_{1,2}$ indicated; here, $h_{1,2} = 5$ and $L(H_{1,2}) = 2$.

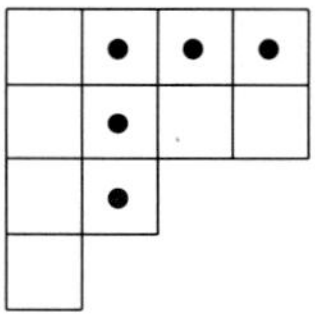

Corresponding to an(i,j)-hook $H_{i,j}$, λ contains an(i,j)-rim hook $R_{i,j}$, which connects the end box of the ith row with the end box of the jth column along the rim of the diagram. Removal of $H_{i,j}$ from λ then means the removal of $R_{i,j}$ from λ; the resulting partition is denoted $\lambda \setminus H_{i,j}$.

Example. We take again $\lambda = (4^2, 2, 1)$; below the rim hook $R_{1,2}$ is indicated, then the removal process to obtain $\lambda \setminus H_{1,2} = (3, 1^3)$.

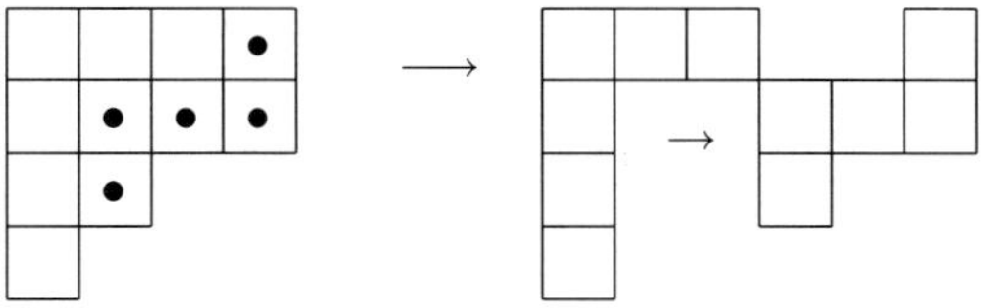

The following formula allows the character values to be computed recursively (see [JK, Sag]):

THEOREM 3.2 (Murnaghan-Nakayama formula) *Let $\lambda \vdash n$, $\sigma_\alpha \in S_n$ of cycle type $\alpha \vdash n$, e a part of α. Then*

$$[\lambda](\sigma_\alpha) = \sum_{h\ e-\text{hook in } \lambda} (-1)^{L(h)} [\lambda \setminus h](\sigma_{\alpha \setminus e})$$

An important special case is the restriction to the subgroup S_{n-1}; a removable 1-hook is called a *removable box* in λ.

Theorem 3.3 (Branching Theorem)

$$[\lambda]|_{S_{n-1}} = \sum_{A \text{ removable box in } \lambda} [\lambda \setminus A]$$

For studying the representations themselves rather than only their characters we have to introduce the notion of tableaux, which also occurs in other contexts.

For a partition $\lambda \vdash n$, a *λ-tableau* t is a filling of the boxes of the Young diagram $Y(\lambda)$ with the numbers $1, \ldots, n$. A λ-tableau is *standard* if its entries increase along rows to the right and down the columns.

Example. Here are two $(4, 4, 2, 1)$-tableaux of which only the second is standard:

$$\begin{array}{cccc} 10 & 2 & 5 & 8 \\ 3 & 1 & 9 & 6 \\ 7 & 11 & & \\ 4 & & & \end{array} \qquad \begin{array}{cccc} 1 & 2 & 4 & 7 \\ 3 & 6 & 10 & 11 \\ 5 & 9 & & \\ 8 & & & \end{array}$$

A standard tableau corresponds to a path in the *Young graph* which is the infinite graph having all partitions as its vertices, and where two vertices are joined if the corresponding partitions $\lambda \vdash n$ and $\mu \vdash n+1$ differ only by adjoining one box to λ to obtain μ.

The character value at 1 is the dimension of the corresponding representation. There are several ways of computing this dimension for an ordinary irreducible representation of S_n of which we recall the following [JK]:

Theorem 3.4 (Dimension formulae) *Let $\lambda \vdash n$ be a partition. Then*

(a) *(Hook formula)* $[\lambda](1) = \dfrac{n!}{\prod \textit{hooklengths in } \lambda}$

(b) $[\lambda](1) = f^\lambda = |\{\textit{standard } \lambda\textit{-tableaux}\}|$

Note that the equality

$$f^\lambda = \frac{n!}{\prod \text{hooklengths in } \lambda}$$

is a purely combinatorial statement; for a nice 'probabilistic' proof of this due to Greene, Nijenhuis and Wilf see [Sag].

From Theorem 2.1 we also obtain the following combinatorial result:

$$n! = \sum_{\lambda \vdash n} (f^\lambda)^2$$

A 'bijective proof' of this is given by the Robinson-Schensted-Knuth algorithm which has generalisations and variations to a number of other problems; for an account of this algorithm we also refer to [Sag].

Explicit irreducible matrix representations for S_n have been given by Young; in fact, he constructed the so-called seminormal, orthogonal and natural representations (see [JK, J, Sag]).

4. Spin representations of the symmetric groups and the combinatorics of strict partitions

Already early on in representation theory representations of groups were studied not only as linear transformations of affine spaces but also as transformations on projective spaces. Equivalently, a *projective representation* of a group G on a K-vector space V is given by a map

$$T : G \to GL(V)$$

satisfying $T(1_G) = \mathrm{id}_V$ and there is a 2-cocycle $\alpha : G \times G \to K^*$ such that $T(x)T(y) = \alpha(x,y)T(xy)$ for $x, y \in G$; for a trivial cocycle α the corresponding representation is a linear representation. The map T induces a homomorphism $\hat{T} : G \to PGL(V)$, which is also called a projective representation of G.

In the following we restrict our attention to the case $K = \mathbb{C}$.
There are close connections between projective and linear representations; e.g., it is well-known that to study the relation between linear representations of a group and its normal subgroups projective representations come into play (see [CR]). The fundamental work on projective representations was done by Schur [S04, S07] who showed that for complex representations the problem of classifying the corresponding projective representations can always be linearized in the sense that projective representations of the original group G correspond to linear representations of a representation group $\widetilde{G}$, which is a central extension of G by its Schur multiplier $M(G)$. In the case of the symmetric groups S_n, for $n \leq 3$ the Schur multiplier is trivial and all projective representations can be linearized. For $n \geq 4$, the Schur multiplier is of order 2, and there are two such representation groups $\widetilde{S}_n$ (isomorphic only for $n = 6$). Since their representation theory is very closely related it suffices to consider one of these; we choose the one which is described via generators and relations as

$$\begin{aligned}\widetilde{S}_n \;=\; <t_1, \ldots, t_{n-1}, z \mid\; & z^2 = 1, t_i^2 = z, 1 \leq i \leq n-1; \\ & t_{i+1}t_it_{i+1} = t_it_{i+1}t_i, 1 \leq i \leq n-2; \\ & t_it_j = zt_jt_i \text{ for } |i-j| > 1, 1 \leq i, j \leq n-1>\end{aligned}$$

for $n \in \mathbb{N}$. Note that modulo $<z>$ the generators t_i correspond to the generators $s_i = (i\; i+1)$ of S_n. This group $\widetilde{S}_n$ is a central non-split extension of S_n by $<z>$, a double cover of S_n, and it is a representation group for S_n, for $n \geq 4$.

So we then have a correspondence:

$$\begin{array}{ccc} \text{projective representations} & S_n & \to & PGL_m(\mathbb{C}) \\ \updownarrow & & & \\ \text{linear representations} & \widetilde{S}_n & \to & GL_m(\mathbb{C}) \end{array}$$

The proper projective representations of S_n (i.e. those which are not equivalent to a linear representation of S_n) then correspond to linear representations of $\widetilde{S}_n$ on which z acts as -1; these representations are called *spin representations*.

The classification of the irreducible proper projective S_n-representations over the complex numbers now corresponds to the classification of the irreducible complex spin $\widetilde{S}_n$-representations. This classification was achieved by Schur in his 1911 paper, where he gave the answer in terms of characters, similar to the classification given by Frobenius in the linear case. For this, he first determined the conjugacy classes of these groups; for the spin representations only those classes are relevant which do not contain an element y as well as the element yz, these classes are called *split classes.* There are two types of such classes which are canonically labelled by the set $\mathcal{O}(n)$ of partitions of n with odd parts only on the one hand, and the set of partitions of n into distinct parts with an odd number of even parts on the other hand.

As we will see, partitions into distinct parts (also called: *strict partitions*) play a special rôle for spin representations similar to the rôle of partitions in the linear case. We let $\mathcal{D}(n)$ be the set of partitions of n into distinct parts, and we let $\mathcal{D}^-(n)$ resp. $\mathcal{D}^+(n)$ denote the sets of those partitions in $\mathcal{D}(n)$ with an odd resp. an even number of even parts; the corresponding partitions are called *odd* resp. *even partitions.* Note that by a well-known Theorem of Euler, the set $\mathcal{O}(n)$ is equinumerous with the set $\mathcal{D}(n)$.

The ordinary spin representations of $\widetilde{S}_n$ are now given via their characters as follows (we refer the reader to [HH, Jo89] or [St90] for more details):

THEOREM 4.1 (Schur 1911) *A complete list of irreducible complex spin characters of $\widetilde{S}_n$ is given as follows:*
For each $\lambda \in \mathcal{D}^+(n)$ there is a self-associate *spin character $\langle\lambda\rangle = sgn \cdot \langle\lambda\rangle$ (where sgn is the sign character induced from S_n); for each $\lambda \in \mathcal{D}^-(n)$ there is an* associate pair *of spin characters $\langle\lambda\rangle, \langle\lambda\rangle' = sgn \cdot \langle\lambda\rangle$.*
The values of the spin characters on the $\mathcal{O}(n)$ classes are determined by the Schur Q-functions Q_λ, while the values on the $\mathcal{D}^-(n)$ classes are given explicitly.

The main part of the character values is determined by the coefficients of the expansion of the Schur Q-function into power sum functions (similar as in the S_n-case), and it turned out to be quite difficult to extract the values from there.

In contrast to the theory of linear representations of the symmetric groups it took a long time before the right combinatorial tools for the investigation of the spin representations of their double covers were found that led to further progress in the area.

The right notions turned out to be the *shifted diagrams* which differ from the Young diagrams by an indentation along the diagonal. Corresponding to the (standard) tableaux in the linear case, *shifted (standard) tableaux* are defined to be the fillings of such shifted diagrams (increasing along rows and down columns).

Example. For $\lambda = (4, 3, 1)$ its shifted diagram and a shifted standard tableau look like

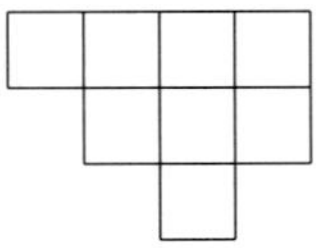

1	2	4	5
	3	6	8
		7	

For a connection with the ordinary case the *shift-symmetric diagram* of a partition is sometimes useful; this is obtained by gluing the shifted diagram to its reflection along the diagonal.

Example. For $\lambda = (4, 3, 1)$ its shift-symmetric diagram is depicted by:

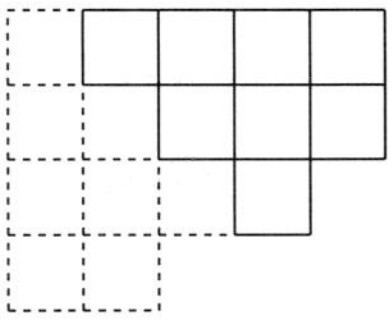

What is still missing at this point is the analogue of the hooks and the removal process of hooks from partitions in the ordinary linear case. Indeed, the subject lay dormant for about 50 years after Schur's paper. The required combinatorial notions of *bars*, their *bar lengths* and the process of *bar removal* were introduced 1965 by Morris ([Mo65], see also [MO88]). The bars correspond to boxes in the shifted diagram of λ, the length of a bar is given by the length of the corresponding hook in the shift symmetric diagram. Removing a p-bar is then achieved by either removing the last p boxes in a row if the resulting diagram (after reordering rows) is again a strict partition, or by removing two rows with a total of p boxes. The bar removal process then leads to the notion of $\bar{p}$-cores, corresponding to the p-cores in the previous situation. We denote by $\lambda_{(\bar{p})}$ the *p-bar-core* resp. $\bar{p}$-core of a strict partition λ.

Example. Take again $\lambda = (4, 3, 1)$. In the shift symmetric diagram below the bar lengths are filled into the corresponding boxes.

$$\begin{array}{ccccc} . & 7 & 5 & 4 & 2 \\ . & . & 4 & 3 & 1 \\ . & . & . & 1 & \\ . & . & & & \end{array}$$

Removing the bar of length 5 from λ is achieved by removing the parts 4 and 1 from λ, and thus gives the partition $(3) = \lambda_{(\bar{5})}$.

With these notions Morris derived a recursion formula for character values analogous to the Murnaghan-Nakayama formula (though in the spin case it is somewhat more complicated by the appearance of certain 2-powers); on the basis of this it was then possible to study the spin characters in more detail.

We only want to state the special case of the branching of spin characters from $\widetilde{S}_n$ to $\widetilde{S}_{n-1}$ explicitly. For $\lambda \in \mathcal{D}(n)$ we put

$$\langle\hat{\lambda}\rangle = \begin{cases} \langle\lambda\rangle & \text{if } \lambda \in \mathcal{D}^+(n) \\ \langle\lambda\rangle + \langle\lambda\rangle' & \text{if } \lambda \in \mathcal{D}^-(n) \end{cases}$$

and set

$$\begin{aligned} M(\lambda) &= \{\mu \in \mathcal{D}(n-1) \mid \mu \text{ is obtained from } \lambda \text{ by removing a 1-bar } \}\,. \\ M(\lambda)' &= \{\mu \in M(\lambda) \mid l(\mu) = l(\lambda)\}\,, \end{aligned}$$

THEOREM 4.2 (Branching Theorem for spin characters [Mo65])
If $\lambda \in \mathcal{D}^+(n)$, then

$$\langle\lambda\rangle|_{\widetilde{S}_{n-1}} = \sum_{\mu \in M(\lambda)} \langle\hat{\mu}\rangle\,.$$

If $\lambda = (\lambda_1, \ldots, \lambda_l) \in \mathcal{D}^-(n)$, *then*

$$\begin{array}{lll} \langle\lambda\rangle|_{\widetilde{S}_{n-1}} & = \langle\lambda\rangle'|_{\widetilde{S}_{n-1}} = \sum_{\mu \in M(\lambda)} \langle\mu\rangle & \textit{if } \lambda_l > 1 \\ \langle\lambda\rangle|_{\widetilde{S}_{n-1}} & = \sum_{\mu \in M(\lambda)'} \langle\mu\rangle + \langle\lambda_1, \ldots, \lambda_{l-1}\rangle & \textit{if } \lambda_l = 1 \end{array}$$

(and similarly for the associate character in the second case).

For the spin representations there are dimension formulae similar to the ones for linear representations; indeed, the second formula for the spin character degrees follows from the Branching Theorem above:

THEOREM 4.3 (Dimension formulae) *Let* $\lambda \in \mathcal{D}(n)$. *Then*

$$\langle\lambda\rangle(1) = 2^{[\frac{n-l(\lambda)}{2}]} \frac{n!}{\prod \textit{bar lengths}} = 2^{[\frac{n-l(\lambda)}{2}]} g^\lambda$$

where g^λ *counts the number of standard shifted* λ*-tableaux.*

A different route to the character values was taken in recent years by combinatorialists who identified the Schur Q-function as a tableaux-generating function (again this is an analogue to the ordinary Schur functions s_λ); for very readable accounts of these developments the reader is referred to the work by Stembridge [St89, St90] who has given e.g. also projective analogues of induction from Young subgroups and the important Littlewood-Richardson rule. These investigations in the intersection area of spin characters of S_n, shifted tableaux and symmetric functions have attracted much interest by combinatorialists in the past decade.

In comparison with the development of the linear representations of S_n where explicit matrix representations were given early on by Young, it was rather late before explicit projective matrix representations for S_n were given by Nazarov [N90].

5. Modular representation theory: local-global principles

In this section, we give a brief introduction to Brauer's theory of modular representations and blocks (see [CR, F]). Modular representation theory considers representations of finite groups G at positive characteristic $p > 0$, i.e.

$$G \to GL_F(V) \quad \text{resp.} \quad G \to GL_n(F)$$

with F a field of characteristic $p > 0$. As the representation theory at characteristic p not dividing the group order $|G|$ is very similar to the ordinary case, it is always tacitly assumed that p divides the group order.

While it is important to study modular representations of groups for its own sake, the interest was kindled by the applications of such results to ordinary representations. The link between ordinary and modular representations is given via *integral* representations which are defined over a suitable ring R. Typically, R is a (complete) discrete valuation ring with residue field $F = R/\text{rad } R$ of characteristic $p > 0$ and quotient field K of characteristic 0; for example, R could be taken to be the localization of $\mathbb{Z}$ at p or the p-adic integers. If F and K are sufficiently large for the group G under consideration (containing suitable roots of unity), the triple (F, R, K) is called a *p-modular splitting system for G*.

There are in this situation two types of local-global principles at work. One is of a number-theoretic type: representations over $\mathbb{Z}$ are studied by looking at the corresponding representations p-locally and p-modularly. The other is of a group-theoretic nature: once the prime p dividing the group order $|G|$ is involved, one considers p-local subgroups of G, i.e. subgroups of G having a normal p-subgroup.

Of course, in reducing modulo p, information is lost. Starting from an ordinary representation of G over K on a K-vector space V, one can always find a corresponding integral representation of G over R on an R-lattice $\widehat{V}$ (i.e. a finitely generated R-free R-module), so $V = K \otimes_R \widehat{V}$. This lattice may then be reduced modulo p (resp. modulo a prime ideal $\wp$ containing p) to give a modular representation of G on the F-vector space $\overline{V} = F \otimes_R \widehat{V}$.

So we have the following situation:

$$\begin{array}{ccccccc} & R & & & & \widehat{V} & \\ \swarrow & & \searrow & & \swarrow & & \searrow \ \text{modular reduction} \\ \mathrm{Quot}(R) = K & & R/\mathrm{rad}\ R = F & & V & & \overline{V} \end{array}$$

Now the choice of the lattice is not unique, and in fact, the modular reduction $\overline{V}$ is also not uniquely determined (up to isomorphism). But the fundamental fact is the following (see [CR]):

THEOREM 5.1 (Brauer-Nesbitt 1937) *The composition factors of the FG-module $\overline{V}$ are uniquely determined by the KG-module V (i.e. independent of the choice of $\widehat{V}$).*

Example. Let $G = S_3$, $R = \mathbb{Z}_{(3)}$, $K = \mathbb{Q}$, $F = GF(3)$. The group S_3 acts on $V = K^3$ in a natural way by

$$\sigma(a_1, a_2, a_3) = (a_{\sigma^{-1}(1)}, a_{\sigma^{-1}(2)}, a_{\sigma^{-1}(3)})$$

This induces an action of S_3 on $W = \{(a, b, c) \in V \mid a + b + c = 0\}$ in a natural way. Then W is an irreducible KG-module. We choose

$$\widehat{W} = \{(a, b, c) \in R^3 \mid a + b + c = 0\}\,.$$

Then $\overline{W} = \{(a, b, c) \in F^3 \mid a + b + c = 0\}$. The FG-module $\overline{W}$ has a unique G-invariant subspace $F(1, 1, 1) \leq \overline{W}$. Hence $\overline{W}$ is reducible, but it is at least indecomposable. By considering the corresponding matrix representation with respect to the basis $\{(1, 1, 1), (1, -1, 0)\}$ of $\overline{W}$ one easily sees that $\overline{W}$ has the composition factors F (the trivial module FG-module) and the sign representation (which are obviously irreducible as they are 1-dimensional).

From this example we see that modular group algebras are in general not semisimple (not every representation splits as a direct sum of irreducible representations). But the irreducible (also called *simple*) modules are still important building blocks for the representations.

We collect some of the main basic results from modular representation theory in the following theorem (the reader is referred to [CR] or [F] for more detailed information on modular representations):

THEOREM 5.2 *Let (F, R, K) be a p-modular splitting system for the finite group G. Then the following holds:*

(a) *(Maschke) If $p \mid |G|$, then FG is not semisimple.*

(b) *(Krull-Schmidt) Any representation of G has a unique decomposition into indecomposable direct summands.*

(c) *The number of irreducible F-representations of G equals the number of p-*regular *conjugacy classes of G (i.e. the conjugacy classes of elements whose order is not divisible by p).*

(d) *The composition factors (with multiplicities) of an F-representation of G are uniquely determined by its* Brauer character *(which has values in R and is defined on the p-regular classes only).*

To a certain extent at characteristic p the Brauer characters play the rôle of the ordinary irreducible characters at characteristic 0; both correspond to the irreducible representations.

One of the main problems in modular representation theory is then to determine the irreducible Brauer characters and the corresponding F-representations of G.

6. Connections between ordinary and modular representations: the decomposition matrix

We described above the connection between ordinary and modular representations given via integral representations. For every ordinary irreducible character χ resp. K-representation V with $\chi = \chi_V$ of G the composition factors of the associated FG-module $\overline{V}$ were found to be uniquely determined by χ (independent of the choice of V). This information is collected in the *p-decomposition matrix* of G. For a precise definition, let

$$\mathrm{Irr}(G) = \{\chi_1, \ldots, \chi_r\}$$

be the set of irreducible characters of G, and let $\widehat{V_i}$ be an RG-lattice affording χ_i, for $i = 1, \ldots r$ (so $r = k(G)$ is the number of conjugacy classes of G).
Let

$$\mathrm{IBr}(G) = \{\phi_1, \ldots, \phi_s\}$$

be the set of irreducible Brauer characters of G, corresponding to the simple FG-modules $T_1, \ldots, T_s$ (so $s = l(G)$ is the number of p-regular classes of G).
Then the decomposition matrix $D = (d_{ij})_{i,j}$ at characteristic p of G is defined by

$$d_{ij} = \text{multiplicity of the simple module } T_j \text{ as a composition factor in } \overline{V_i}$$

With respect to a suitable ordering of the irreducible representations at characteristic p and 0 respectively, the decomposition matrix has the form

$$D = \begin{pmatrix} \boxed{*} & & & & 0 \\ & \boxed{*} & & & \\ & & \ddots & & \\ & 0 & & \ddots & \\ & & & & \boxed{*} \end{pmatrix}$$

and no further such decomposition is possible, i.e. we have indecomposable matrix blocks along the 'diagonal'.

Corresponding to these indecomposable submatrices of the decomposition matrix, the irreducible representations are partitioned into the *p-blocks* of G. For a p-block B of G, its main arithmetical invariants are:

$$k(B) = |\mathrm{Irr}(B)| = |\{\text{irreducible } K\text{-representations } V_i \text{ belonging to } B\}|$$

$$l(B) = |\mathrm{IBr}(B)| = |\{\text{irreducible } F\text{-representations } T_j \text{ belonging to } B\}|$$

In algebra terms, the blocks may also be defined by decomposing the group algebra RG resp. FG into indecomposable two-sided ideals; these indecomposable ideals are then also called the p-blocks of G. They are generated by primitive central idempotents. An indecomposable (and in particular, an irreducible) representation is then sorted into the unique block whose idempotent operates as the identity on the representation (all other blocks annihilate the representation).

The main structural invariant of a p-block B is its *defect group* $\delta(B)$, which is a p-subgroup of G, unique up to conjugation in G. The defect group of a block roughly corresponds to a p-Sylow subgroup of a group G. The *defect* $d(B)$ is defined by $|\delta(B)| = p^{d(B)}$; it is a further important arithmetical invariant of the block. For example, if p^a is the maximal p-power dividing $|G|$, then $p^{a-d(B)}$ divides the dimensions of all representations belonging to B. A block is semisimple if and only if its defect is zero; in this case it has exactly one ordinary and modular irreducible character and the decomposition matrix is $D = (1)$.

The main representation-theoretical problems in the block theory of finite groups are then the following:

Given a finite group G and a p-modular splitting system (F, R, K) for G, determine

(i) The dimensions of the irreducible representations of G at characteristic 0 and p (at least the p-parts of the dimension);

(ii) $k(B)$, $l(B)$ for each p-block B of G;

(iii) the defect $d(B)$, and the defect group $\delta(B)$ for each p-block B of G;

(iv) the distribution of the irreducible representations into p-blocks;

(v) the decomposition matrix at characteristic p.

There are a number of long-standing conjectures relating these representation theoretical invariants, some of the most important going back to Brauer. We mention the following two (see [O93]):

CONJECTURES 6.1 (Brauer)
Let B be a p-block of the finite group G. Then:

(a) $k(B) \leq |\delta(B)|$.

(b) (Height 0 conjecture) *A defect group $\delta(B)$ is abelian if and only if for all $\chi \in Irr(B)$ we have $|\chi(1)|_p = |G : \delta(B)|_p$ (here n_p denotes the exact p-power dividing $n \in N$).*

These conjectures have been affirmed in many cases and there has been a lot of deep work on these questions by many representation theorists but they have still not been solved.

There are some other conjectures of a similar nature (see [O93]), and results about the relationship between these conjectures ([Kü87, KR96]). There are also a number of deep conjectures that relate global to p-local data, i.e. they provide relations between representations of G resp. of a block B of G and representations of a p-local subgroup resp. blocks of p-local subgroups. Some of the most prominent of these are due to Alperin (Alperin's Weight Conjecture: counting p-modular irreducible representations p-locally), Alperin-McKay (counting characters of height 0 p-locally), Broué (conjecture on isotypies and perfect isometries) and Dade (counting characters p-locally). Stating these important conjectures precisely would require going into more notations and definitions; see the references mentioned above for more details.

7. Invariants of pairs of non-commuting $n \times n$ matrices

We would now like to describe an interesting application of representation theory to a classical problem in invariant theory; we refer the reader to [BLB91] for the literature quoted.

Sylvester investigated the question of whether the field of rational matrix invariants of pairs of non-commuting complex $n \times n$ matrices is rational over $\mathbb{C}$, i.e. whether it is purely transcendental over the ground field. The invariance here means the invariance with respect to the usual action of $GL(n, \mathbb{C})$ by simultaneous conjugation, or equivalently, the corresponding action of $PGL(n, \mathbb{C})$. In 1883, Sylvester showed that this is true for $n = 2$, and the corresponding five algebraically independent invariants are the traces of the two matrices, their squares and their product.

For many decades there was no progress until Formanek proved rationality for n=3 and 4 in 1979 and 1980. The main ingredient was a translation of the problem into a lattice context, due to Procesi and Formanek. We let $K_{2,n} = \mathbb{C}(M_n \times M_n)^{PGL_n}$ be the field of rational matrix invariants. Then

$$K_{2,n} \simeq \mathbb{C}(G_n)^{S_n}$$

where G_n is a specific $\mathbb{Z}S_n$ lattice. Formanek was then able to find the S_n-invariants for the lattice G_n.

Whilst there was only slow progress on the rationality question itself a number of weaker properties were investigated, e.g. by Saltman. From a geometric point of view it was reasonable to study a property still very close to rationality; in our context the question is then whether $K_{2,n}$ is stably rational over the ground field, i.e. whether a suitable rational extension of $K_{2,n}$ is rational over the ground field.

In [BLB91] we could settle stable rationality for n=5 and 7 by investigating modular S_n-representations and gluing p-local solutions to a global integral solution over $\mathbb{Z}$; we used the computer algebra system CAYLEY to obtain the so-called table of marks for the symmetric groups S_5 and S_7. In subsequent work of Schofield resp. Katsylo, a Reduction Theorem was proved which then implies (on the basis of the before mentioned results) stable rationality of the field of matrix invariants for all divisors of $420 = 2^2 \cdot 3 \cdot 5 \cdot 7$.

In recent work by Beneish a new route to such stable rationality results is investigated which avoids the use of a computer and might also lead to results for larger integers n.

8. Modular representations of the symmetric groups and p-combinatorics

In this section we describe the progress on some of the main representation-theoretical problems for the representations of the symmetric groups.

By Theorem 5.2 we know that the number of p-modular irreducible S_n-representations is the number of p-regular conjugacy classes, i.e.

$$l(S_n) = |\{\lambda = (\lambda_1, \ldots, \lambda_l) \mid p \not| \lambda_i \text{ for all } i\}|$$

By a well-known partition identity (see [A76]) this also satisfies

$$l(S_n) = |\{\lambda = (l_1^{a_1}, \ldots, l_m^{a_m}) \mid l_1 > \ldots > l_m > 0, a_i < p \text{ for all } i\}|$$

These latter partitions are called *p-regular*, and they will serve as the labels of the modular irreducible S_n-representations.

We have to define some more combinatorial notions based on the parameter p. Given $\lambda \vdash n$, successively remove p-hooks as long as possible. This gives a (uniquely determined!) partition $\lambda_{(p)}$, the *p-core* of λ. The number of p-hooks removed is the *weight* $w(\lambda)$ of λ.

Example. Let $\lambda = (6, 4, 3)$, $p = 3$. We can then remove three 3-hooks in the way indicated below (the i-th 3-hook corresponds to the three boxes marked with i):

			2	2	2
	3	1	1		
3	3	1			

Thus $\lambda_{(3)} = (3, 1)$, $w(\lambda) = 3$.

For the symmetric groups we now have a very nice combinatorial algorithm for determining the p-blocks; this Theorem is still called the Nakayama Conjecture though it was proved by Brauer and Robinson 1947 (see [JK]):

THEOREM 8.1 (Nakayama Conjecture 1940, Brauer-Robinson 1947)
Let $\lambda, \mu \vdash n$. Then $[\lambda], [\mu]$ belong to the same p-block if and only if $\lambda_{(p)} = \mu_{(p)}$. Furthermore, p-blocks of defect 0 correspond to p-core partitions.

The combinatorial definitions above make sense for any $t \in \mathbb{N}$, not necessarily a prime; let

$$c_t(n) := |\{\lambda \vdash n \mid \lambda \ t\text{-core partition}\}|$$

So by the Theorem above $c_p(n)$ counts the p-blocks of defect 0 for S_n, if p is a prime. For the connection of t-cores with other combinatorial questions we refer the reader to [GKS90].

For $t \leq 3$, not every $n \in \mathbb{N}$ has a t-core. For $t = 2$ a 2-core of n exists if and only if n is a triangular number; also for $t = 3$, the natural numbers n with $c_t(n) > 0$ are classified.

It was an open question whether for $t \geq 4$ t-cores exist for all n; Erdmann and Michler dealt with the cases $t = 5$ and $t = 7$ [EM96] and recently the problem has been completely solved by Granville and Ono [GO96].

THEOREM 8.2 (t-core Conjecture, [GO96])
For all $t \geq 4$ and $n \in \mathbb{N}$, $c_t(n) > 0$.

Their proof uses deep number-theoretic methods; these arguments have then been reduced to a shorter and more elementary number-theoretic proof by Kiming [Ki96].

The Theorem by Granville and Ono also has as its consequence the last missing piece in the classification of finite simple groups with a p-block of defect 0 (which had been Problem 19 on Brauer's famous list of problems [Br63]), we refer the reader to [GO96] for more details on this.

Based on the results about the block distribution of characters, the block invariants have also been determined (see [O93]). Note that by Theorem 8.1 a p-block B of a symmetric group is characterized by a p-core and the common weight $w = w(B)$ of the partitions λ with $[\lambda]$ belonging to B.

THEOREM 8.3 *Let B be a p-block of S_n, μ its p-core, $w = w(B)$ its weight. Then the following holds.*
(a) *Its defect group $\delta(B)$ is isomorphic to a Sylow p-subgroup of S_{pw}.*
(b) *$k(B) = k(p, w)$, where $k(r, s)$ counts all r-tuples of partitions with total sum s.*
(c) *$l(B) = k(p-1, w)$.*

With these invariants the sizes of the indecomposable submatrices of the decomposition matrix of S_n are known; while the p-decomposition numbers for S_n are only partially known at least a general result on the 'shape' of the decomposition matrix is available. We recall that the partitions are partially ordered by the *dominance order* $\trianglerighteq$ defined by

$$\mu = (\mu_1, \ldots, \mu_k) \trianglerighteq \lambda = (\lambda_1, \ldots, \lambda_l) \text{ if and only if } \sum_{i=1}^{j} \mu_i \geq \sum_{i=1}^{j} \lambda_i \text{ for all } j\,.$$

THEOREM 8.4 (Farahat-Müller-Peel [FMP76], James [J]) *Order the p-regular partitions lexicographically (decreasing). Then the p-decomposition matrix for S_n has the form*

$$
\begin{array}{c} \\ \textit{p-regular} \\ \textit{partitions} \\ \\ \textit{p-singular} \\ \textit{partitions} \end{array}
\overset{\textit{p-regular partitions}}{\left(\begin{array}{ccccc}
1 & & & & \\
 & 1 & & 0 & \\
 & & \ddots & & \\
 & * & & \ddots & \\
 & & & & 1 \\
\hline
 & & * & & \\
\end{array}\right)}
$$

Furthermore, the right-most non-zero entry in the row labelled by a partition α is a 1 in the column labelled by the p-regular partition α^R, where α^R is obtained by a combinatorial 'regularization' from α. The only non-zero entries in this row are in columns labelled by p-regular partitions $\beta \trianglerighteq \alpha^R$.

The decomposition matrices have been computed for small n; see [JK] for the decomposition numbers up to $n \leq 13$, $p = 2$ and $p = 3$, and [Ben87] for $p = 2$ and $n = 14$ and 15 (here there is one ambiguity left). Further information on specific decomposition numbers is also available; we refer the reader to the survey by James [J87] for the status up to 1987.

More recently, Kleshchev [K97] has obtained further specific decomposition numbers in the course of his work on branching results which we will describe below.

But the main problem is still open at this time:

Give a combinatorial algorithm for determining all decomposition numbers $d_{\lambda,\mu}$!

So far, we have said little about the representations of S_n themselves. Perhaps the most important representations for the symmetric groups are the *Specht modules* (for an explicit description see [J, JK]). The Specht module S^λ associated with $\lambda \vdash n$ is a combinatorially defined $\mathbb{Z}S_n$-module such that S^λ has a $\mathbb{Z}$-basis naturally indexed by standard λ-tableaux, and such that $S^\lambda_{\mathbb{Q}} = \mathbb{Q} \otimes_{\mathbb{Z}} S^\lambda$ is an (absolutely) irreducible $\mathbb{Q}S_n$-module affording the character $[\lambda]$.

Let F be a field of characteristic $p > 0$. Then for $\lambda \vdash n$ a p-regular partition, the Specht module $S^\lambda_F = F \otimes S^\lambda$ has a unique irreducible quotient D^λ. The set

$$\{D^\lambda \mid \lambda \vdash n \text{ } p\text{-regular}\}$$

is then a complete set of non-isomorphic irreducible F-representations of S_n; thus the modular irreducible S_n-representations are naturally indexed by p-regular partitions of n in a way compatible with the labelling of the S_n-representations at characteristic 0.

9. Modular branching of S_n-representations and normal and good boxes of partitions

In characteristic 0 we have seen that there is a very nice description of the restriction of the Specht module $S_{\mathbb{Q}}^{\lambda}|_{S_{n-1}}$ resp. its character $[\lambda]|_{S_{n-1}}$ for $\lambda \vdash n$, namely just those representations resp. characters of S_{n-1} occur in the restriction that are labelled by the partitions $\mu \vdash n-1$ obtained from λ by removing a box from its Young diagram.

In characteristic p, the corresponding problem is much harder, in particular since the modular group algebra is no longer semisimple. Starting out with the question of when the restriction $D^{\lambda}|_{S_{n-1}}$ of a modular irreducible representation D^{λ} is again irreducible, Jantzen and Seitz [JS92] obtained a number of results on the modular branching problem and put forward a conjecture describing those p-regular partitions $\lambda \vdash n$ for which the restriction $D^{\lambda}|_{S_{n-1}}$ is irreducible; this generalised an earlier conjecture for the case $p = 2$ proposed by Benson [Ben87].

This conjecture and a number of surprising branching results were then proved in a series of papers by Kleshchev ([K95a, K95b, K96a, K97]). The strategy used had appeared already in the work of Jantzen and Seitz, and even earlier in work of Carter and Payne [CP80]: translate the problems into problems about SL_n-modules resp. GL_n-modules and work in the corresponding Lie algebra situation. Proving S_n-results via the GL_n-route had also been a successful strategy in other work, notably for results about the decomposition matrix by Donkin [D85], and work on special dimension formulae by Erdmann [E95] and Mathieu [M96].

In the afore-mentioned series of papers, Kleshchev introduced several new combinatorial concepts which have already turned out to play an important rôle in other contexts (see [LLT96])

First we have to define the *p-residue diagram* which introduces the prime p into the Young diagram. For this, the (i,j)-box is filled with the residue $j-i$ (mod p), for each box in the Young diagram of λ. The *p-content* $c(\lambda) = (c_0, \dots, c_{p-1})$ of λ is then defined by counting the multiplicities c_i of each p-residue i in the p-residue diagram of λ. It is well-known that the p-block of $[\lambda]$ is not only determined by the p-core of λ as described before, but also by the p-content $c(\lambda)$.

Example. Let $p = 3$, $\lambda = (7, 4^2, 2, 1^2)$ has 3-content $c(\lambda) = (6, 7, 6)$ as can be seen from the 3-residue diagram:

0	1	2	0	1	2	0
2	0	1	2			
1	2	0	1			
0	1					
2						
1						

Considering the Young diagram of $\lambda \vdash n$ we want to pay particular attention to the residues of the removable boxes at the corners of $Y(\lambda)$, and to the residues of the boxes that could be added to $Y(\lambda)$ to give a partition of $n+1$; we call these latter boxes *indent* boxes.

We attach a sign + to each removable box, a sign − to each indent box of λ, and associate a *signature sequence* to λ by going down from the top row to just below the bottom row (and from left to right in a row) in the p-residue diagram of λ, writing down the residues of all removable and indent boxes with the corresponding sign.

Example. For $p = 3$, λ as above, we obtain the signature sequence

$$0+\ 1-\ 0-\ 1+\ 1+\ 2-\ 0-\ 1+\ 0-$$

Now for a given p-residue r consider the sign subsequence $s_1, \ldots, s_m$ of the signature sequence consisting of the signs attached to boxes of residue r only. Count the signs as $+1$ and -1 respectively, and set

$$h(0) = 0 \quad \text{and} \quad h(i) = \sum_{j=1}^{i} s_j \,.$$

Then a removable box of residue r corresponding to the sign $s_i = +1$ is a *normal* box of λ if

$$h(i) > h(k) \quad \text{for } k = 0, \ldots, i-1\,.$$

A normal box is called a *good* box of λ if

$$h(i) = \max\{h(j) \mid s_j \text{ corresponds to a normal box}\}$$

If A is the removable box of λ corresponding to s_i, then we also write $h(A) = h(i)$, and we call $h(A)$ the *height* of A.

In the **example** above, the first (and only) removable box of 3-residue $r = 0$ is normal and good, the second and third removable boxes of residue $r = 1$ are normal and the third is good, and there is no normal box for the 3-residue $r = 2$.

Before we state Kleshchev's results on modular branching we recall that the *socle* $\mathrm{soc}(M)$ of a module M is the largest completely reducible submodule of M; it is the direct sum of the simple submodules of M. For a removable box A in $Y(\lambda)$, $\lambda \vdash n$, we denote by $\lambda \setminus A$ the corresponding partition of $n-1$.

THEOREM 9.1 (Kleshchev) *Let λ be a p-regular partition of n, $n \in \mathbb{N}$, $n \geq 2$. Then the following holds:*

(i) *soc* $(D^\lambda|_{S_{n-1}}) \simeq \bigoplus_{A \text{ good}} D^{\lambda\setminus A}$.

(ii) $D^\lambda|_{S_{n-1}}$ *is completely reducible if and only if all normal boxes in λ are good.*

(iii) *Let A be a removable box of λ such that $\lambda\setminus A$ is p-regular. Then the multiplicity of $D^{\lambda\setminus A}$ in $D^\lambda|_{S_{n-1}}$ is given by*

$$\left[D^\lambda|_{S_{n-1}} : D^{\lambda\setminus A}\right] = \begin{cases} h(A) & \text{if } A \text{ is normal in } \lambda \\ 0 & \text{else} \end{cases}$$

As a consequence of these results, one easily deduces that the restriction $D^\lambda|_{S_{n-1}}$ has at most p indecomposable summands, and all these summands belong to different p-blocks.

Clearly, the branching results also give good bounds for the dimension of the irreducible module D^λ: consider those paths from the empty partition $\emptyset$ to λ in the Young graph which are constructed along p-regular partitions by only adding normal boxes and take into account the heights of the added boxes along the path. Unfortunately, not all composition factors of $D^\lambda|S_{n-1}$ are of the form $D^{\lambda\setminus A}$, so in general we do not obtain the exact dimension by the procedure above. But below we will see that at least some more composition factors can be "explained"; so the dimension bound can be improved by taking these into account.

It is still a major open problem to determine the dimensions of the modular irreducible representations D^λ; for some special partitions λ answers are available by the work of Erdmann [E95], Mathieu [M96] and Kleshchev [K96b].

10. The Mullineux Conjecture and p-conjugation of p-regular partitions

Another application of the branching results was the proof of the long-standing Mullineux Conjecture. This describes the result of tensoring an irreducible S_n-representation with the sign representation; of course, this tensor product is again an irreducible representation, but the question is: what is the labelling partition?

In characteristic 0 it is well-known and not hard to show that

$$\mathrm{sgn} \otimes S^\lambda_{\mathbb{Q}} \simeq S^{\lambda'}_{\mathbb{Q}}$$

where λ' is the partition conjugate to λ (obtained by reflecting the Young diagram $Y(\lambda)$ along the main diagonal).

In characteristic $p > 0$ we have to describe the p-regular partition λ^P with

$$\mathrm{sgn} \otimes D^\lambda \simeq D^{\lambda^P} .$$

From all available data it was clear that the answer must be quite complicated. In 1979, Mullineux ([Mu79a], [Mu79b]) defined a combinatorial involution $\lambda \mapsto \lambda^M$ on p-regular partitions; he then conjectured for this p-analogue of conjugation:

$$\lambda^M = \lambda^P \quad \text{for all } p\text{-regular partitions } \lambda .$$

A lot of evidence was collected for this conjecture, both of algebraic and combinatorial nature. Investigating the number of fixed points of this map led to a very general partition identity including known identities of a similar type as special cases, and implying a number of classical and new identities ([AO91, Bes91, Bes95a]). The breakthrough finally came with Kleshchev's work who used his modular Branching Theorem for reducing the Mullineux Conjecture to the following purely combinatorial statement [K96a] which was subsequently proved:

THEOREM 10.1 ([FK97, BO98a]) *If A is a good box in the p-regular partition λ, then there exists a good box B in λ^M such that*

$$(\lambda \setminus A)^M = \lambda^M \setminus B .$$

A very long and complicated proof of this was found by Ford and Kleshchev [FK97]; soon afterwards a shorter proof with further insights into the behaviour of the good boxes of a partition was provided by [BO98a], where also some new combinatorial notions have been introduced that have already proved useful for other problems.

As mentioned before, application of the Mullineux map explains further composition factors of $D^\lambda|_{S_{n-1}}$; it also produces further decomposition numbers as

$$d_{\lambda,\mu} = d_{\lambda',\mu^M} \quad \text{for all } \lambda, \mu \vdash n,\ \mu\ p\text{-regular}$$

A further important consequence of the combinatorial answer to the tensor product question is that it opens up the road to investigating the modular representations of the alternating groups A_n in more detail [BO98b]. For the knowledge of the classification of the p-modular irreducible A_n-representations it is crucial to know which irreducible S_n-representations split on restriction to A_n. For $p = 2$, the answer was given by Benson [Ben87]; for odd characteristic p, exactly those representations split which are labelled by fixed points under the Mullineux map.

11. Modular spin representations of $\widetilde{S}_n$ and $\bar{p}$-combinatorics

For the ordinary characters of S_n, a nice combinatorial answer for their p-block distribution was given with the Nakayama Conjecture Theorem 8.1. In the case of the irreducible spin characters of $\widetilde{S}_n$ it took much longer before even a conjecture on the block distribution was put forward. Based on the information obtained by his recursion formula, Morris formulated such a conjecture for odd characteristic p [Mo65]. Then 20 years passed before it was proved by Humphreys and with different methods by Cabanes. For odd primes p, the p-blocks of $\widetilde{S}_n$ either contain only linear or only spin representations. The distribution of the linear characters is determined by the p-cores as in the S_n case; the p-blocks of spin characters are determined by the $\bar{p}$-cores:

THEOREM 11.1 (Morris Conjecture [Mo65]; Humphreys [H86], Cabanes [C88]) *Let $p \neq 2$ be a prime. For $\lambda \neq \lambda_{(\bar{p})}$ the $\bar{p}$-core determines the p-block of $\langle\lambda\rangle$. If $\lambda = \lambda_{(\bar{p})}$, then $\langle\lambda\rangle$ (as well as $\langle\lambda\rangle'$ if λ is non-self-associate) forms a p-block of defect 0.*

For the case $p = 2$ a corresponding conjecture was formulated only in 1986 by Knörr and Olsson [O87] on the basis of calculations of Benson. This conjecture was then proved in 1993 [BO97a] by completely different methods as in the odd characteristic case. In contrast to the case of odd p, the 2-blocks are 'mixed' in the sense that they contain linear as well as spin characters. The 2-block invariants are 2-cores as in the linear S_n-case.

THEOREM 11.2 (Knörr-Olsson Conjecture, [BO97a]) *Let $p = 2$. For a partition $\lambda = (\lambda_1, \dots, \lambda_l) \in \mathcal{D}(n)$, let*

$$\text{dbl}\,(\lambda) = \left(\left[\frac{\lambda_1+1}{2}\right], \left[\frac{\lambda_1}{2}\right], \left[\frac{\lambda_2+1}{2}\right], \left[\frac{\lambda_2}{2}\right], \dots, \left[\frac{\lambda_l+1}{2}\right], \left[\frac{\lambda_l}{2}\right] \right)$$

be the doubling *of λ. Then $\langle\lambda\rangle$ and* [dbl (λ)] *belong to the same 2-block of $\widetilde{S}_n$. Thus the 2-core of the doubling of λ determines the 2-block of $\langle\lambda\rangle$.*

Similarly as t-cores have been investigated before, we are now interested in counting $\bar{p}$-cores for $p \neq 2$. Erdmann and Michler [EM96] have studied the cases $p = 5$ and 7; for odd primes $p \leq 7$ the number $c_{\bar{p}}(n)$ of $\bar{p}$-cores is known for all $n \in \mathbb{N}$. Also, Kiming [Ki97] has shown that $c_{\bar{p}}(n) > 0$ for all primes $p \geq 7$, $n \in \mathbb{N}$.

What is known about the p-block invariants of modular spin representations?

As in the case of linear representations of the symmetric groups S_n, combinatorial formulae for $k(B)$ and $l(B)$ are known for the p-blocks of the covering groups $\widetilde{S}_n$, due to the work of Olsson [O90, O92] for $p \neq 2$, and [BO97a] for $p = 2$. The structure of the defect groups for $p \neq 2$ was determined by Cabanes [C88]; for $p = 2$ this is a consequence of the result in the S_n case. We also know that (up to associates for odd partitions) the number of p-modular spin representations equals the number of partitions in

$$\mathcal{D}_p(n) = \{\lambda = (\lambda_1, \ldots, \lambda_l) \in \mathcal{D}(n) \mid \lambda_i \not\equiv 0 \,(\mathrm{mod}\ p),\ i = 1, \ldots, l\}$$

In comparison to the S_n case an analogue to the Specht modules is missing in the case of the double covers $\widetilde{S}_n$, and there are no naturally defined p-modular irreducible spin representations for $p \neq 2$.

To obtain general results on the decomposition matrix a suitable set of labels for the modular irreducible $\widetilde{S}_n$-representations is needed.

In the case $p = 2$ these are just the 2-regular partitions as any 2-modular irreducible $\widetilde{S}_n$-representation has the central subgroup $<z>$ in its kernel and hence is in fact an irreducible S_n-representation. In this situation, the problem is to determine the rows in the decomposition matrix corresponding to the spin characters. Generalizing the information for special spin characters given by Benson [Ben87] we have obtained in [BO97a] as much information on the spin rows as is known in the ordinary case by Theorem 8.4:

THEOREM 11.3 ([BO97a]) *Let $\lambda \in \mathcal{D}(n)$. Let $\mathrm{dbl}^2(\lambda)$ denote the 2-regularization of $\mathrm{dbl}(\lambda)$, and let $m_0(\lambda)$ be the number of even parts of λ. Then the 2-modular composition factors of the spin representation labelled by λ are given by:*

$$\overline{\langle\lambda\rangle} \sim 2^{[m_0(\lambda)/2]} D^{\mathrm{dbl}^2(\lambda)} + \sum_{\substack{\mu \rhd \mathrm{dbl}^2(\lambda) \\ \mu \text{ 2-regular}}} c_\mu D^\mu .$$

For $p \neq 2$ and 'small' n, decomposition matrices for the spin representations have been computed by Morris and Yaseen ([MY88], [Y87]). Their calculations indicated that a shape result similar to the one given in Theorem 8.4 might hold in the spin case, though complicated by the phenomenon of associate spin representations. The first proofs of Theorem 8.4 did not make use of the Specht modules but used character induction, with control on the blocks by the p-residue diagram. The idea in the spin case was to generate suitable partition labels in trying to apply a similar induction process for spin characters. The crucial combinatorial notion that is needed for the analogue of the p-combinatorics in the spin case is the $\bar{p}$-residue diagram for $p \neq 2$.

For a partition $\lambda \in \mathcal{D}(n)$ its *$\bar{p}$-residue diagram* is the λ-part of the *shifted $\bar{p}$-residue diagram*

$$\begin{array}{ccccccccccccc}
1 & 2 & \dots & \frac{p-1}{2} & \frac{p+1}{2} & \frac{p-1}{2} & \dots & 2 & 1 & 1 & 2 & \dots & \\
 & 1 & 2 & \dots & \frac{p-1}{2} & \frac{p+1}{2} & \frac{p-1}{2} & \dots & 2 & 1 & \dots & & \\
 & & 1 & 2 & \dots & \dots & & & & & & & \\
 & & & \ddots & & & & & & & & &
\end{array}$$

Counting the multiplicity c_i of i in the $\bar{p}$-residue diagram of λ gives the *$\bar{p}$-content* $\bar{c}(\lambda) = (1^{c_1}2^{c_2}\dots \frac{p+1}{2}^{c_{(p+1)/2}})$ of λ. Similar to the case of p-cores, the $\bar{p}$-content of λ determines the $\bar{p}$-core $\lambda_{(\bar{p})}$ [MY86].

Thus we can control the distribution of the constituents of $\langle\lambda\rangle \uparrow^{\tilde{S}_{n+1}}$ into p-blocks; by only adding boxes of a specified $\bar{p}$-residue to the shifted diagram of λ in each step we obtain the constituents in a single block. This is the principle of $(r,\bar{r})$-induction introduced by Morris and Yaseen [MY88]. The partition labels we want to construct as analogues of the p-regular partitions should be of 'high type' with respect to a suitable ordering so that the corresponding column of the decomposition matrix has its highest non-zero entry in a row corresponding to a spin character with this same partition label.

For achieving this, we construct partitions of n by the following *top node algorithm.* We set $\mathcal{C}_p(1) = \{(1)\}$. Assume that $\mathcal{C}_p(n-1)$ has already been constructed. Then the partition λ belongs to $\mathcal{C}_p(n)$ if it can be constructed from some $\mu \in \mathcal{C}_p(n-1)$ by adding a box to the $\bar{p}$-residue diagram of μ which is highest among the indent boxes of μ with the same $\bar{p}$-residue. For example, one easily checks that $\mathcal{C}_3(5) = \{(5),(4,1)\}$.

The problem now is to find an internal description of these sets and to compare the cardinality of $\mathcal{C}_p(n)$ with the cardinality of $\mathcal{D}_p(n)$.

These are purely combinatorial problems, which have been solved for $p=3$ and $p=5$, showing that in these cases the partitions in $\mathcal{C}_p(n)$ may serve as modular spin labels.

THEOREM 11.4 ([BMO94, ABO94, ABO96]) *Let $n \in \mathbb{N}$. Then*

$$\begin{aligned}\mathcal{C}_3(n) = \{\lambda = (\lambda_1,\dots,\lambda_l) \ \in \mathcal{D}(n) \mid \lambda_i - \lambda_{i+1} \geq 3\,, i = 1,\dots,l-1;\\ \lambda_i - \lambda_{i+1} > 3 \ \textit{if}\ \lambda_i \equiv 0\ (\textit{mod}\ 3)\,, i = 1,\dots,l-1\}\end{aligned}$$

and

$$\begin{aligned}\mathcal{C}_5(n) = \{\lambda = (\lambda_1,\lambda_2,\dots,\lambda_l) \in \mathcal{D}(n) \mid \lambda_i - \lambda_{i+2} \geq 5 \ \textit{for all}\ i \leq l-2;\\ \lambda_i - \lambda_{i+2} > 5 \ \textit{if}\ \lambda_i \equiv 0\ (\textit{mod}\ 5) \ \textit{or if}\ \lambda_i + \lambda_{i+1} \equiv 0\ (\textit{mod}\ 5)\,,\\ \textit{and there are no subsequences of the following types:}\\ (5j+3, 5j+2),\ (5j+11, 5j+9, 5j+5),\ (5j+10, 5j+6, 5j+4),\\ (5j+11, 5j+10, 5j+5, 5j+4),\ j \geq 0\}\end{aligned}$$

Furthermore,

$$|\mathcal{C}_p(n)| = |\mathcal{D}_p(n)|$$

for $p=3$ and $p=5$.

The enumerative identity for $p = 3$ is a special case of a partition identity due to Schur [S26].

A partition set $\mathcal{C}'(n)$ very similar to the set $\mathcal{C}_5(n)$ was defined by Andrews in the context of generalizing the Rogers-Ramanujan identities; there is an easy bijection between these two sets, and hence, the enumerative identity for $p = 5$ in the Theorem above is equivalent to the conjecture on the cardinality of $\mathcal{C}'(n)$ formulated by Andrews in 1974 ([A74a, A74b]).

Unfortunately, for $p \geq 7$, the sets $\mathcal{C}_p(n)$ are too small, and it is still an open problem how to find suitable partition labels in the general situation.

For any partition $\lambda \in \mathcal{C}_p(n)$, there are usually quite different construction paths in the top node algorithm; fortunately, for $p = 3$ and 5 there are such paths where we have reasonably good control over the constituents in the corresponding induced character of $\widetilde{S}_n$. With this we obtain an approximation to a column of the decomposition matrix which we label by λ, and as in the linear case we can then deduce a result on the shape of the decomposition matrix.

THEOREM 11.5 ([BMO94, Bes95b]) *Assume $p = 3$ or $p = 5$. Order the spin characters of $\widetilde{S}_n$ by first taking the ones with labels in $\mathcal{C}_p(n)$ in decreasing lexicographic order, and then the others (doubling rows if the labels belong to an associate pair); take as column labels for the decomposition matrix D the partitions in $\mathcal{C}_p(n)$ in decreasing lexicographic order (doubling columns if the corresponding irreducible modular spin representation is non-selfassociate; this depends only on the $\bar{p}$-core and the weight of the block to which the representation belongs). Then the spin part of the p-decomposition matrix for $\widetilde{S}_n$ has the form*

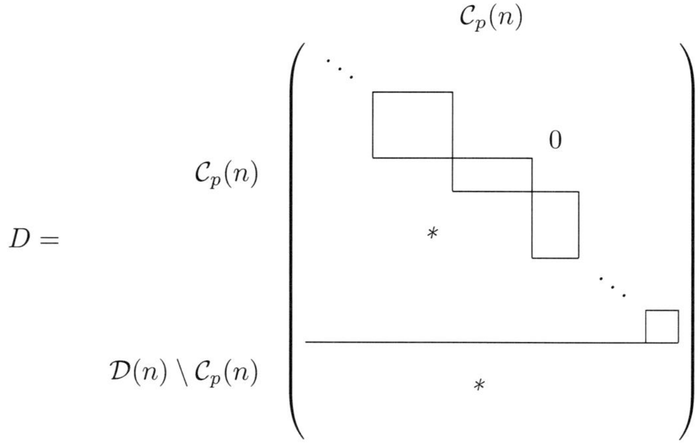

The block matrices appearing along the 'diagonal' in the upper part of the matrix are $s \times t$ matrix blocks for $s, t \in \{1, 2\}$ depending on the splitting of the corresponding rows and columns.

In the case $p = 3$ there is some more precise information available [BMO94]; similar to the linear S_n case, it is based on a regularisation process of partitions in the $\bar{p}$-residue diagram.

12. Open problems

In the previous sections we have already pointed out a number of open problems for S_n and $\widetilde{S}_n$-representations.

The main representation-theoretical invariants that have not yet been determined are the dimensions of the modular irreducible linear and spin representations. There are some partial answers available in the linear case and good bounds coming from Kleshchev's branching results, but a combinatorial formula similar to one of the dimension formulae in the ordinary case is still not at hand.

For the modular spin representations there is hardly anything known about this question. From the available data it seems that the modular spin branching might be governed by combinatorial notions resembling the good and normal boxes in the linear case; but even if such a behaviour holds the route to such results is not clear as the strategy in the linear case has so far no pendant in the spin case.

In comparison to general finite groups a surprisingly large amount is known about the blocks (in both the linear and the spin case) and about the decomposition numbers (in the linear case). The problem of determining the decomposition numbers by a combinatorial algorithm also has attracted much attention in related situations. The recent progress on such problems for Hecke algebras raises the hope of an answer at least in the case of the symmetric groups.

For the covering groups $\widetilde{S}_n$, there are still basic open problems on the modular representations. Analogues of Specht modules which could provide naturally defined modular irreducible representations are missing, and even more modestly, suitable labels of modular irreducible spin representations compatible with the labels of the ordinary spin representations are not yet available.

Many of the problems above have also been investigated for the alternating groups and their covering groups; while the block invariants have been determined to a large extent [O90, O92], a more detailed analysis of the modular representations is only now possible after the solution of the Mullineux problem [BO98b].

Finally, we would also like to mention that a number of main general conjectures on finite groups (see section 6) have been affirmed for the symmetric and alternating groups and their covering groups; for the latter groups in some cases only for primes $p \neq 2$. In particular, the following results have been achieved:

(i) Brauer's conjectures 6.1 and Olsson's conjecture (bounding the number of characters of height 0) hold for the p-blocks of S_n, A_n and their covering groups (for $\widetilde{A}_n$ at $p \neq 2$) ([O76, O84, O90, BO97a, BO97b]).

(ii) Alperin's weight conjecture has been affirmed for S_n at all primes [AF90], and for A_n, $\widetilde{A}_n$, $\widetilde{S}_n$ at primes $p \neq 2$ [MiO91].

(iii) Dade's conjecture has been affirmed for the symmetric groups at all primes p ([OU95] for $p \neq 2$, [An96] for $p = 2$).

(iv) Donovan's conjecture is that there are only finitely many Morita equivalence classes of blocks of group algebras with a prescribed defect group; this has been affirmed in the family of symmetric groups at all primes [Sco91], and for the family of their double covers at $p \neq 2$ [Ke95].

(v) Broué's Isotypy Conjecture has been affirmed for the symmetric and alternating groups [FH97].

References

[AF90] J. L. Alperin, P. Fong, Weights for symmetric and general linear groups, J. Algebra 131 (1990) 2–22

[An96] J. An, Dade's Conjecture for 2-blocks of symmetric groups, preprint 1996

[A74a] G. E. Andrews, *On the General Rogers-Ramanujan Theorem*, Memoirs of the Amer. Math. Soc., No. 152 (1974) ii+86 pp.

[A74b] G. E. Andrews, A general theory of identities of the Rogers-Ramanujan type, Bull. Amer. Math. Soc. 80 (1974) 1033–1052

[A76] G. E. Andrews, *The theory of partitions*, Encyclopedia of Mathematics and its Applications, Vol. 2, Addison-Wesley 1976

[AO91] G. E. Andrews, J. B. Olsson, Partition identities with an application to group representation theory, J. reine angew. Math. 413 (1991) 198–212

[ABO94] G. E. Andrews, C. Bessenrodt, J. B. Olsson, Partition identities and labels for some modular characters, Trans. Amer. Math. Soc. 344 (1994) 597–615

[ABO96] G. E. Andrews, C. Bessenrodt, J. B. Olsson, A refinement of a partition identity and blocks of some modular characters, Archiv d. Math. 66 (1996), 101–113

[Ben87] D. Benson, Some remarks on the decomposition numbers for the symmetric groups, In: Proc. Arcata Conf. on Representations of Finite Groups, Proc. Symp. Pure Math. 47, Part 1, Amer. Math. Soc., Providence (1987) 381–394

[Ben88] D. Benson, Spin modules for symmetric groups, J. London Math. Soc. (2) 38 (1988) 250–262

[Bes91] C. Bessenrodt, A combinatorial proof of a refinement of the Andrews-Olsson partition identity, Europ. J. Combinatorics 12 (1991), 271–276

[Bes95a] C. Bessenrodt, Generalisations of the Andrews-Olsson partition identity and applications, Discrete Math. 141 (1995), 11–22

[Bes95b] C. Bessenrodt, Representations of the covering groups of the symmetric groups and their combinatorics, Pub. I.R.M.A. Strasbourg, 1995, Actes 33e Sém. Loth., 1–25 (Electronic J. Sém. Loth. Combinatoire B33a, 29pp.)

[BLB91] C. Bessenrodt, L. Le Bruyn, Stable rationality of certain PGL_n quotients, Invent. math. 104 (1991), 179–199

[BMO94] C. Bessenrodt, A. O. Morris, J. B. Olsson, Decomposition matrices for spin characters of symmetric groups at characteristic 3, J. Algebra 164 (1994) 146–172

[BO94] C. Bessenrodt, J. B. Olsson, On Mullineux symbols, J. Comb. Theory (A) 68 (1994) 340–360

[BO97a] C. Bessenrodt, J. B. Olsson, The 2-blocks of the covering groups of the symmetric groups, Adv. Math. 129 (1997), 261–300

[BO97b] C. Bessenrodt, J. B. Olsson, Heights of spin characters in characteristic 2, in: Finite Reductive Groups, Related Structures and Representations, M. Cabanes (Ed.), Progress in Math. 141, Birkhäuser (1997), 51–71

[BO98a] C. Bessenrodt, J. B. Olsson, On residue symbols and the Mullineux Conjecture, J. Alg. Comb. (1998) (to appear)

[BO98b] C. Bessenrodt, J. B. Olsson, Branching of modular representations of the alternating groups, J. Algebra (to appear)

[Br63] R. Brauer, Representations of finite groups, Lect. on Modern Math. 1 (1963), Wiley, New York, 133–175

[C88] M. Cabanes, Local structure of the p-blocks of $\widetilde{S}_n$, Math. Z. 198 (1988) 519–543

[CP80] R. W. Carter, M. T. J. Payne, On homomorphisms between Weyl modules and Specht modules, Math. Proc. Cambrodge Philos. Soc. 87 (1980), 131–145

[CR] C. W. Curtis, I. Reiner, *Methods of representation theory, I, II*, Wiley 1981 and 1987

[D85] S. Donkin, A note on decomposition numbers for general linear groups and symmetric groups, Math. Proc. Camb. Phil. Soc. 97 (1985) 57–62

[E95] K. Erdmann, Tensor products and dimensions of simple modules for symmteric groups, Manusc. Math. 88 (1995) 357–386

[EM96] K. Erdmann, G. O. Michler, Blocks for symmetric groups and their covering groups and quadratic forms, Beiträge zur Algebra und Geometrie 37 (1996) 103–118

[FMP76] H. K. Farahat, W. Müller and M. H. Peel, The modular characters of the symmetric groups, J. Algebra 40 (1976) 354–363

[F] W. Feit, *The representation theory of finite groups*, North-Holland 1982

[FH97] P. Fong, M. E. Harris, On perfect isometries and isotypies in alternating groups, Trans. Amer. Math. Soc. 349 (1997), 3469–3516

[FK97] B. Ford, A. Kleshchev, A proof of the Mullineux conjecture, Math. Z. 226 (1997), 267–308

[GKS90] F. Garvan, D. Kim, D. Stanton, Cranks and t-cores, Inv. Math. 101 (1990) 1–17

[GO96] A. Granville, K. Ono, Defect 0 p-blocks for finite simple groups, Trans. Amer. Math. Soc. 348 (1996), 331–347

[HH] P. N. Hoffman, J. F. Humphreys, *Projective representations of the symmetric groups*, Clarendon Press, Oxford 1992

[H86] J. F. Humphreys, Blocks of projective representations of the symmetric group, J. London Math. Soc. (2) 33 (1986) 441–452

[J] G. James, *The representation theory of the symmetric groups*, Springer Lect. Notes Math. 682 (1978)

[J87] G. James, The representation theory of the symmetric groups, Proc. Symposia in Pure Math. 47 (1987) 111–126

[J90] G. James, Representations of S_n and GL_n and the q-Schur algebras, in: Topics in Algebra, Banach Center Publ. 26, 303–316, Warsaw 1990

[JK] G. James, A. Kerber, *The representation theory of the symmetric group*, Addison-Wesley (1981)

[JS92] J. Jantzen, G. Seitz, On the representation theory of the symmetric group, Proc. London Math. Soc. (3) 65 (1992) 475–504

[Jo89] T. Józefiak, Characters of projective representations of symmetric groups, Exposition. Math. 7 (1989) 193–247

[Ke95] R. Kessar, Blocks and source algebras for the double covers of the symmetric groups, PhD Thesis, Ohio 1995

[Ki96] I. Kiming, A note on a theorem of A. Granville and K. Ono, J. Number Th. 60 (1996) 97–102

[Ki97] I. Kiming, On the existence of $\bar{p}$-core partitions of integers, Quart. J. Math. Oxford Ser. (2) 48 (1997) 59–65

[K95a] A. Kleshchev, Branching rules for modular representations of symmetric groups I, J. Algebra 178 (1995) 493–511

[K95b] A. Kleshchev, Branching rules for modular representations of symmetric groups II, J. reine angew. Math. 459 (1995) 163–212

[K96a] A. Kleshchev, Branching rules for modular representations of symmetric groups III, J. London Math. Soc. 54 (1996) 25–38

[K96b] A. Kleshchev, Completely splittable representations of symmetric groups, J. Algebra 181 (1996) 584–592

[K97] A. Kleshchev, On decomposition numbers and branching coefficients for symmetric and special linear groups, Proc. London Math. Soc. (3) 75 (1997) 497–558

[Kü87] B. Külshammer, A remark on conjectures in modular representation theory, Arch. Math. 49 (1987) 396–399

[KR96] B. Külshammer, G. R. Robinson, Alperin-McKay implies Brauer's problem 21, J. Algebra 180 (1996) 208–210

[LLT96] A. Lascoux, B. Leclerc, J.-Y. Thibon, Hecke algebras at roots of unity and crystal bases of quantum affine algebras, Commun. Math. Phys. 181 (1996) 205–263

[Mac] I. G. Macdonald, *Symmetric Functions and Hall Polynomials*, 2nd ed., Clarendon Press, Oxford 1995

[Mar] S. Martin, *Schur algebras and representation theory*, Cambridge University Press 1993

[M96] O. Mathieu, On the dimensions of some modular irreducible representations of the symmetric groups, Lett. Math. Phys. 38 (1996) 23–32

[MiO91] G. O. Michler, J. B. Olsson, Weights for covering groups of symmetric and alternating groups, $p \neq 2$, Canad. J. Math. 43 (1991) 792–813

[Mo62] A. O. Morris, The spin representation of the symmetric group, Proc. London Math. Soc. (3) 12 (1962) 55–76

[Mo65] A. O. Morris, The spin representation of the symmetric group, Canad. J. Math. 17 (1965) 543–549

[MO88] A. O. Morris, J. B. Olsson, On p-quotients for spin characters, J. Algebra 15 (1988) 51–82

[MY86] A. O. Morris and A. K. Yaseen, Some combinatorial results involving shifted Young diagrams, Math. Proc. Camb. Phil. Soc. 99 (1986) 23–31

[MY88] A. O. Morris, A. K. Yaseen, Decomposition matrices for spin characters of symmetric groups, Proc. Royal Soc. Edinburgh 108A (1988) 145–164

[Mu79a] G. Mullineux, Bijections of p-regular partitions and p-modular irreducibles of symmetric groups, J. London Math. Soc. (2) 20 (1979) 60–66

[Mu79b] G. Mullineux, On the p-cores of p-regular diagrams, J. London Math. Soc. (2) (1979) 222–226

[N90] M. L. Nazarov, Young's orthogonal form of irreducible projective representations of the symmetric group J. London Math. Soc. (2) 42 (1990) 437–451

[O76] J. B. Olsson, McKay numbers and heights of characters, Math. Scand. 38 (1976) 25–42

[O84] J. B. Olsson, On the number of characters in blocks of finite general linear, unitary and symmetric groups, Math. Z. 186 (1984) 41–47

[O87] J. B. Olsson, Frobenius symbols for partitions and degrees of spin characters, Math. Scand. 61 (1987), 223–247.

[O90] J. B. Olsson, On the p-blocks of symmetric and alternating groups and their covering groups, J. Alg. 128 (1990) 188–213

[O92] J. B. Olsson, The number of modular characters in certain blocks, Proc. London Math. Soc. 65 (1992) 245–264

[O93] J. B. Olsson, *Combinatorics and representations of finite groups*, Vorlesungen aus dem Fachbereich Mathematik der Universität GH Essen, Heft 20, 1993.

[OU95] J. B. Olsson, K. Uno, Dade's conjecture for symmetric groups, J. Algebra 176 (1995), 534–560

[Sag] B. Sagan, *The Symmetric Group. Representations, combinatorial algorithms, and symmetric functions*, Wadsworth & Brooks 1991

[S04] I. Schur, Über die Darstellung der endlichen Gruppen durch gebrochene lineare Substitutionen, J. Reine Ang. Math. 127 (1904) 20–50

[S07] I. Schur, Untersuchung über die Darstellung der endlichen Gruppen durch gebrochene lineare Substitutionen, J. Reine Ang. Math. 132 (1907) 85–137

[S11] I. Schur, Über die Darstellung der symmetrischen und der alternierende Gruppe durch gebrochene lineare Substitutionen, J. Reine Ang. Math. 139 (1911) 155–250

[S26] I. Schur, Zur additiven Zahlentheorie, Sitzungsberichte Preuss. Akad. Wiss. (1926) 488–495

[Sco91] J. C. Scopes, Cartan matrices and Morita equivalence for blocks of the symmetric groups, J. Algebra 142 (1991) 441–455

[St89] J. R. Stembridge, Shifted tableaux and the projective representations of the symmetric groups, Adv. in Math. 74 (1989) 87–134

[St90] J. R. Stembridge, On symmetric functions and the spin characters of S_n, In: Topics in Algebra, Banach Center Publ. 26, Part 2, Warsaw 1990

[Y87] A. K. Yaseen, Decomposition matrices for spin characters of symmetric groups II, preprint 1987

Progress in Mathematics, Vol. 168,

Some Recent Results for the Ginzburg-Landau Equation

F. Bethuel

Université Paris-Sud
Laboratoire Analyse Numérique et EDP
URA CNRS 760, Bâtiment 425
91405 Orsay, France

I. Introduction

In this talk, I will present some recent results obtained in the study of Ginzburg-Landau functionals. These functionals were first introduced by V. Ginzburg and L. Landau in 1950 [GL], in the context of superconductivity and were aimed to model the energy state of a superconducting sample, in the presence of an exterior magnetic field. They appeared thereafter in various contexts, and in different forms: for instance, one may mention the Abelian-Higgs model, the Poliakov-t'Hooft monopoles in particle physics, but also nonlinear optics, fluid mechanics... One common feature of the above models is the major role of topological defects, which will often be termed vortices, in our context.

To make things more precise, we will start with a very simple model situation, which was studied in particular in a joint book with H. Brezis and F. Helein [BBH]. Let Ω be a smooth bounded and starshaped domain in $\mathbb{R}^2$ (for instance, the unit disk). We will consider complex-valued functions on Ω, i.e. maps v from Ω to $\mathbb{R}^2$. The simplest possible Ginzburg-Landau functional for such maps takes the form

$$E_\varepsilon(v) = \frac{1}{2}\int_\Omega |\nabla v|^2 + \frac{1}{4\varepsilon^2}\int_\Omega (1-|v|^2)^2.$$

Here ε is a parameter, describing some characteristic length. In the sequel, we will always be interested in the case where ε is small, and we will study the asymptotic properties as ε tends to zero. The potential $V(v) = \varepsilon^{-2}(1-|v|^2)^2$ forces $|v|$, for critical maps of E_ε, to be close to 1, and hence, stationary (or low energy) maps will be almost S^1−valued. However, at some points $|v|$ may have to vanish, introducing defects of topological nature.

In order to get a well-posed mathematical problem, we will prescribe Dirichlet boundary conditions (although these may not describe any realistic physical situation...). For that purpose, let g be a smooth map from $\partial\Omega$ to S^1. We set v equal to g on $\partial\Omega$. Hence it is natural to introduce the Sobolev space

$$H^1_g(\Omega;\mathbb{R}^2) = \{v \in H^1(\Omega;\mathbb{R}^2), \quad v = g \text{ on } \partial\Omega\}.$$

The functional E_ε is indeed well defined, even C^∞ on H^1_g, and verifies the Palais-Smale condition, so that we may use tools of the calculus of variation to find critical points (see, for instance [Str1] for a definition of the Palais-Smale condition). Critical points v of E_ε verify the Ginzburg-Landau equation

$$\begin{cases} -\Delta v = \dfrac{1}{\varepsilon^2} v(1-|v|^2) & \text{on } \Omega \\ v = g & \text{in } \partial\Omega. \end{cases} \tag{1}$$

Using standard tools in nonlinear elliptic equations, one may prove the following (see [BBH])

Proposition 1. *Any solution v to (1) is smooth, and verifies*

$$|v| \leq 1 \quad \text{on } \Omega \quad \text{(maximum principle).} \tag{2}$$

Moreover, there is a constant C depending only on g such that

$$|\nabla v| \leq \frac{C}{\varepsilon}, \tag{3}$$

$$\frac{1}{4\varepsilon^2}\int_\Omega (1-|v|^2)^2 \leq C. \tag{4}$$

Remark: The proof of (4) uses crucially the fact that Ω is starshaped. This follows from Pohozaev's identity. It may fail if Ω is not simply connected, but it is an open problem to determine whether simple connectedness is a sufficient condition for (4).

Since E_ε is strictly positive, one easily verifies that

$$\kappa_\varepsilon = \operatorname{Inf}\{E_\varepsilon(v), v \in H^1_g\}$$

is achieved, and hence (1) possesses minimizing solutions (not necessarily unique). We will denote these solutions by u_ε.

II. Asymptotic analysis of minimizing solutions

The winding number d of g (from $\partial\Omega$ to S^1) plays a crucial role in that analysis, forcing, in the case $d \neq 0$, the appearance of vortices and the divergence of κ_ε. Let us first look at the easy case $d = 0$.

II.1. The case $d = 0$. Here no vortices are necessary and there are smooth maps u_0 in H^1_g, such that $|u_0| = 1$. Hence using u_0 as a comparison map, we see that

$$\kappa_\varepsilon \leq \int_\Omega |\nabla u_0|^2 .$$

In particular κ_ε remains bounded, independently of ε, so that u_ε remains also bounded in H^1_g, and

$$\int_\Omega (1-|u_\varepsilon|^2)^2 \leq 4\kappa_\varepsilon \varepsilon^2 \to 0.$$

Extracting a subsequence, if necessary, we see that u_ε converges weakly in H^1 to some map u_* which is S^1-valued. With a little more work, we may assert that u_* is the solution of the minimization problem

$$\mathrm{Inf}\left\{\int_\Omega |\nabla v|^2,\quad v\in H^1_g(\Omega;S^1)\right\},$$

hence $u_* = \exp\ i\varphi_*$, where φ_* solves

$$\begin{aligned}\Delta\varphi_* &= 0 \quad \text{in } \Omega\\ \exp\ i\varphi_* &= g \quad \text{on } \partial\Omega.\end{aligned}$$

Finally, in [BBH2], we carried out more refined asymptotics, as for instance, the estimates

$$\|u_\varepsilon - u_*\|_{L^\infty} \le C\ \varepsilon^2.$$

II.2. The case $d\neq 0$. Assume for instance that $d>0$. This case is more involved since now $H^1_g(\Omega;S^1)=\emptyset$ (the fact that $C^0_g(\Omega;S^1)$ is empty reduces to standard degree theory: to the former assertion one has to work a little more). In particular $\kappa_\varepsilon\to+\infty$, and the problem becomes a problem of singular limit. Since $u_\varepsilon\in C^\infty$, the topology of the boundary data forces u_ε to vanish somewhere in Ω. The points where u_ε vanishes play an important role: the Dirichlet energy will concentrate in there neighborhood, accounting for the divergence of κ_ε. In [BBH] we proved the following

Theorem 1. *There exists a constant $C>0$, depending only on g such that*

$$|\kappa_\varepsilon - \pi d|\log\ \varepsilon|| \le C,\quad \forall\, 0<\varepsilon<1. \tag{5}$$

- *u_ε has exactly d zeroes, provided ε is sufficiently small (this relies on a work by P. Baumann, N. Carlson and D. Philips [BPC]).*
- *There exists exactly d points $a_1,\dots,a_d$ in Ω such that, up to a subsequence ε_n,*

  $$u_{\varepsilon_n}\longrightarrow u_*\ \text{ on any compact subset of } \Omega\setminus\bigcup_{i=1}^d\{a_i\}$$

  where

  $$u_* = \prod_{i=1}^d \frac{z-a_i}{|z-a_i|}\exp\ i\varphi,\ (\varphi \text{ being a harmonic function}).$$

  (in particular the winding number around each singularity is equal to $+1$).
- *The configuration a_i is not arbitrary, but minimizes on $\Omega^d\setminus\Delta$ (where Δ denotes the diagonal) a renormalized energy, which has (roughly) the form*

  $$W_g(a_1,\dots,a_d) = \pi\sum_{i\neq j}\log|a_i-a_j| + \text{boundary contributions}.$$

- *We have the expansion*

  $$\kappa_\varepsilon = \pi d|\log\ \varepsilon| + W_g(a_1,\dots,a_d) + d\gamma_0 + o(1),$$

  where γ_0 is some absolute constant.

Remark. The restriction Ω being starshaped in Theorem 1 can be removed, using a result by M. Struwe [Str2] (see also Del Pino-Felmer [DF]).

III. Asymptotics for non minimizing solutions

A similar analysis can be carried out for non-minimizing solutions. Let v_ε be solutions to (I). Then we have, if Ω is starshaped (see [BBH, Section X]):

Theorem 2.

- *There exists some constant $C > 0$ such that*

$$E_\varepsilon(v_\varepsilon) \leq C(|\log\ \varepsilon| + 1) \quad 0 < \varepsilon < 1. \tag{6}$$

- *There exists a subsequence ε_n, ℓ points $a_1, \ldots, a_\ell$, and ℓ* integers *$d_1, \ldots, d_\ell$ such that*

$$v_{\varepsilon_n} \to v_* \prod_{i=1}^{\ell} \left(\frac{z - a_i}{|(z - a_i)|} \right)^{d_i} \exp\ i\varphi, \quad \text{where } \varphi \text{ is harmonic.}$$

- *The configuration (a_i, d_i) is critical for the renormalized energy.*

Note that main difference between Theorem 1 and Theorem 2 is that vortices may have a multiplicity d_i, different from $+1$ (it might even be negative). Actually, the proof of Theorem 2 is much simpler than the proof of Theorem 1, the most difficult part in the proof of Theorem 1 being to prove that the multiplicity of vortices is exactly $+1$. Let us next give some indications about the proof of Theorem 2. The main point is to prove that solutions remain bounded in $W^{1,p}(\Omega)$, for some $p < 2$ (this is of course not true for $p = 2$). To that aim, the main ingredients in the proof are:

Step 1 (localizing the vortices). There exist constants N and λ depending only on g, ℓ points $x_1, \ldots, x_\ell$, in Ω, ($\ell \leq N$) such that

$$|v_\varepsilon| \geq \frac{1}{2} \quad \text{on } \Omega_\varepsilon = \Omega \backslash \bigcup_{i=1}^{\ell} B(x_i, \lambda_\varepsilon).$$

The proof of Step 1 combines Proposition 1 together with a standard covering argument.

Step 2 (rewriting the equation in form of an elliptic system). On Ω_ε we have

$$\frac{\partial}{\partial x}(v \times v_x) + \frac{\partial}{\partial y}(v \times v_y) = 0, \tag{7}$$

$$\frac{\partial}{\partial x}\left(-\frac{1}{\rho^2} v \times v_y\right) + \frac{\partial}{\partial y}\left(\frac{1}{\rho^2} v \times v_x\right) = 0, \tag{8}$$

where $\rho = |v|$.

In order to prove (7) it suffices to take the wedge product of the equation with the solution itself: the nonlinear term disappears (an idea first used by Shatah and Keller, Sternberg and Rubinstein (see [Sh] and [KRS])). The second equation expresses the fact that $v/|v|$ is S^1−valued. Equations (7) and (8) together from an elliptic system for the 1-form $\alpha = v \times v_x dx + v \times v_y dy$. We may then use ideas from Hodge de Rham theory as well as linear elliptic estimates due to Stampacchia to assert that

$$\int_\Omega |\alpha|^p \leq C_p \quad \text{for any } p < 2.$$

The final step is then

Step 3: For any $p < 2$, $\int_\Omega |\nabla \rho|^p \leq C_p$.

IV. The existence problem for non-minimizing solutions

In view of Section III, a natural question is to know whether non-minimizing solutions do exist, and if one is able to prescribe the multiplicity of the vortices. We begin this section with an example, which should give some insight.

IV.1. An example. Take $\Omega = D^2$ and the boundary value g of the form $g(\theta) = \exp\ id\theta$, (for $d \geq 1$). In view of the symmetry, one can find solutions $v(r,\theta)$ of the Ginzburg-Landau equation of the form (in polar coordinates)

$$v(r,\theta) = f_d(r) \exp\ id\theta,$$

where f_d verifies the ODE

$$r^2 f'' + r f' - d^2 f + \frac{1}{\varepsilon^2} r^2 f(1 - f^2) = 0, \tag{7}$$

$$f(0) = 0, \quad f(1) = 1. \tag{8}$$

Computing the energy of these radially symmetric solutions one sees that they are of order $\pi |d|^2 |\log\ \varepsilon|$; hence, if $d \geq 2$, and ε is sufficiently small they are non-minimizing (in the case $d = 1$, it is conjectured that the radially symmetric solution is minimizing: important progresses to establish that conjecture have been made by P. Mironescu [M1,M2]).

There is also a natural action of the group S^1, which leaves both the space and the functional invariant, namely

$$T_\alpha v(z) = \exp\ (-id\alpha) \cdot v(\exp\ i\alpha z)$$

for any $z \in D^2$, $\alpha \in [0, 2\pi]$ and $v \in H^1_g$. The functions left invariant by that group action are precisely the radially symmetric ones. Hence, by this action, if we have a solution which is not radially symmetric, we get a whole orbit of solutions. Using the Index theory developed by Faddell and Rabinowitz (some sort of Lyusternik-Schnirelmann theory in the presence of compact group actions), the computation of the Morse Index of the radially symmetric solution yields

Theorem 3. *There exists some universal constant $\mu_0 > 0$, such that if ε is sufficiently small, then, if $\Omega = D^2$, and $g = \exp\ id\theta$, (1) has at least $\mu_0 |d|^2$ orbit of solutions.*

The proof is given in Almeida-Bethuel [AB1]. Note however that we have no clear idea about the type of vortices obtained. To that aim, and also in view of extending Theorem 3 to more general situation, we need some better insight and to use more general variational methods.

IV.2. Variational methods. A satisfactory More theory for (1) has yet to be constructed. In [AB2], we established a first result in that direction. We proved

Theorem 4. *Let* $d \geq 2$. *If* ε *is sufficiently small, then equation* (1) *has at least three distinct solutions.*

Note also that other non-minimizing solutions have been produced by Lin [Li1] (local minimizers with vortices of opposite signs, or solutions corresponding to critical points of the renormalized energy). However, in view of Theorem 3, one might expect a much more complete theory in the near future.

Let us briefly describe some ideas of the proof of Theorem 4. As a standard argument in Morse theory, we consider the level sets

$$E_\varepsilon^a = \{v \in H_g^1(\Omega; \mathbb{R}^2), \quad E_\varepsilon(v) \leq a\}.$$

If E^a and E^b have different topologies, then there is a critical value between a and b, and hence a solution. The main ingredient in the proof of Theorem 4 is then

Proposition 2. *There exists a constant* $\kappa_0 > 0$, *such that for* $a = x_\varepsilon + \kappa_0$, *and* ε *sufficiently small, there exists a loop in* E^a *which is not contractible, i.e. a continuous map* $\gamma_\varepsilon : S^1 \to E^a$, *which is not extendable to* D^2 *in a continuous way.*

Since $E^\infty = H_g^1$ is a contractible space, this yields a non-minimizing solution. A third one can be obtained by a slightly different argument.

The proof of Propositon 2 rest on considering a similar problem, for vortices. As already noticed, vortices (for minimizers at least) are defined on the space $\Sigma = \Omega^d \backslash \Delta$, a space with a non trivial Poincaré group.

If Σ is the right space for the proof of Theorem 4, our analysis in [AB2] has shown that a more involved analysis should take into account vortices of opposite charges, and the fact that they might annihilate: these spaces have been studied from the topological point of view by D. Mc Duff [MD], and have a very rich topology, which in turn should have some impact on the existence problem of non-minimizing solutions.

V. Evolution equations

V.1. Heat-flow. For the heat-flow equation

$$\begin{cases} \frac{\partial u}{\partial t} - \Delta u = \frac{1}{\varepsilon^2} u(1 - |u|^2)^2 & \text{on } \Omega \times [0, [\infty[\\ u = g & \text{on } \partial\Omega, \end{cases}$$

F.H. Lin has proved that, in the right time scaling, vortices move according to the gradient of the renormalized energy [Li2].

V.2. Schrödinger equation. The Schrödinger equation

$$i\frac{\partial u}{\partial t} = \Delta u + \varepsilon^{-2} u(1 - |u|^2) \quad \text{on } \mathbb{R}^2 \times [0, +\infty[\tag{9}$$

is often termed the Gross-Pitaevsky equation. It plays an important role in non-linear optics and fluid mechanics. The parameter ε is sometimes interpreted as the inverse of the velocity of sound. In a joint work with J.C. Saut [BS], we established the existence of travelling waves for equation (9). These solutions exhibit two vortices of opposite topological charges, and have been previously computed numerically by physicist, or on a more formal level, by the method of asymptotic matched expansions (for references, see for instance Jones, Putterman and Roberts (JPR), or Pismen and Nepomnyashchy [PN]).

VI. Superconductivity

As mentioned in the introduction, Ginzburg-Landau functionals have been first introduced in the frame of superconductivity. In order to account for the electromagnetic effects, one introduces a vector potential A (a connection), and for stationary problems the functional writes

$$F_\varepsilon(u,A) = \frac{1}{2}\int_\Omega |\nabla_A u|^2 + |dA|^2 + \frac{1}{4\varepsilon^2}\int_\Omega (1-|u|^2)^2.$$

The asymptotic analysis of [BBH] has been extended to the former functional in [BR]. In an other direction the existence of permanent currents in non homotopically trivial superconducting samples has prompted many interesting mathematical works: Jimbo, Morita and Zhai [JMZ], Rubinstein and Sternberg [RS], Almeida [A]. In [A], it is proved that they correspond to configuration minimizing the energy $F(u,A)$ in a topological "sector", and that the "threshold energy" is of order $|\log\,\varepsilon|$ (i.e. the energy necessary to jump from one sector to the other).

References

[A] L. Almeida, Thesis.

[AB1] L. Almeida and F. Bethuel, Multiplicity results for the Ginzburg-Landau equation in presence of symmetries, to appear in *Houston J. of Math.*

[AB2] L. Almeida and F. Bethuel, Topological methods for the Ginzburg-Landau equation, to appear in J. Math. Pures et Appliquées (1998).

[BBH1] F. Bethuel, H. Brezis and F. Helein, Ginzburg-Landau vortices, Birkhäuser, 1994.

[BBH2] F. Bethuel, H. Brezis and F. Helein, Asymptotics for the minimization of a Ginzburg-Landau functional, *Calc. Var and PDE, 1,* 1993, 123–148.

[BCP] P. Bauman, N. Carlson and D. Philipps, On the zeroes of solutions to Ginzburg-Landau type systems, to appear.

[BR] F. Bethuel and T. Rivière, A minimization problem related to superconductivity, *Annales IHP, Analyse Non Linéaire,* 1995, 243–303.

[BS] F. Bethuel and J.C. Saut, Travelling waves for the Gross-Putaevskii equation, preprint.

[DF] M. Del Pino and P. Felmer, preprint.

[GL] V. Ginzburg and L. Laudau, On the theory of superconductivity, *Zh Eksper. Teoret. Fiz, 20,* 1950, 1064–1082.

[JMZ] S. Jimbo, Y. Morita and J. Zhai, Ginzburg-Landau equation and stable steady state solutions in a non-trivial domain, preprint.

[JPS] C. Jones, S.J. Putterman, and P.H. Roberts, Motion in Bose condensate V, *J. Phys. A.*, *19*, 1986, 2991–3011.

[KRS] J. Keller, P. Sternberg and J. Rubinstein, Reaction-diffusion processes and evolution to harmonic maps, *SIAM J. Appl. Math.*, *49*, 1989, 1722–1733.

[Li1] F.H. Lin, Solutions of Ginzburg-Landau equations and critical points of the renormalized energy, *Annales IHP, Analyse Non Linéaire, 12,* 1995, 599–622.

[Li2] F.H. Lin, Some dynamical properties of Ginzburg-Landau vortices, to appear in *CPAM.*

[McD] D. Mac Duff, Configuration spaces of positive and negative particles, *Topology, 14,* 1974, 91–107.

[Mi1] P. Mironescu, On the stability of radial solutions of the Ginzburg-Landau equation, *J. Funct. Anal., 130,* 1995, 334–344.

[Mi2] P. Mironescu, Les minimiseurs locaux pour l'équation de Ginzburg-Landau sont à symétrie variable, to appear in *C.R. Acad. Sci. Paris.*

[PN] L. Pismen and A. Nepomnyashchy, Stability of vortex rings in a model of superflow, *Physica D,* 1993, 163–171.

[RS] J. Rubinstein and P. Sternberg, Homotopy classification of minimizers of the Ginzburg-Landau energy and the existence of permanent currents, to appear.

[Sh] J. Shatah, Weak solutions and developments of singularities of the SU(2) σ-model, *Comm. Pure and Appl. Math., 41,* 1988, 456–469.

[Str1] M. Struwe, *Variation Methods: Applications to Nonlinear PDE and Hamiltonian Systems,* Springer, 1990.

[Str2] M. Struwe, On the asymptotic behavior of the Ginzburg-Landau model in 2 dimensions, *J. Diff. Integr. Equ., 7,* 1994, 1613–1624, Erratum *8,* 1995, 224.

Progress in Mathematics, Vol. 168, © 1998 Birkhäuser Verlag Basel/Switzerland

Mathematics, Parallel Computing and Reservoir Simulation

PETTER E. BJØRSTAD

Institutt for Informatikk
Høyteknologisenteret, University of Bergen
N-5020 Bergen, Norway
http://www.ii.uib.no/~petter, e-mail: petter@ii.uib.no

ABSTRACT. This paper will highlight, by way of examples, a few seemingly very different mathematical problems and show how they have direct relevance to the construction of efficient computational procedures for the simulation of oil reservoirs on parallel computers.

1. Partitioning of graphs

Consider a graph $G(V, E)$, having a set of vertices connected by edges. We associate work with each vertex and communication (or dependencies) with each edge. Our problem is to partition the graph into two parts such that the two sets are similar in size. We further want the dependencies (i.e., the number of edges) between the two sets to be as small as possible.

Mathematically this can be expressed by having a variable x_i at vertex V_i be one or minus one depending on which of the two sets V_i belongs to. A little reflection shows that our problem can be formulated

$$\min \sum_{E_{ij}} (x_i - x_j)^2 \quad \text{subject to} \quad \sum_i x_i = 0, \ x_i = \pm 1.$$

Define the degree d_i of vertex V_i as the number of edges connected to that vertex. We can formulate the minimization problem as an equivalent matrix problem by defining the **Laplacian** matrix of the graph G by

$$L_{ij} = \begin{cases} d_i & \text{if } \ i = j \\ -1 & \text{if } \ i, j \in E_{ij} \\ 0 & \text{otherwise} \end{cases}$$

It is straightforward to show that our problem is equivalent with

$$\min \ x^T L x \quad \text{subject to} \quad x^T e = 0 \ \text{ and } \ x_i = \pm 1,$$

where e is a vector of all ones. This problem is NP-hard, we therefore consider the problem without the discrete constraint on x, that is, we require instead that $x^T x = n$, where n is the size of our problem. Clearly, this minimum must be a

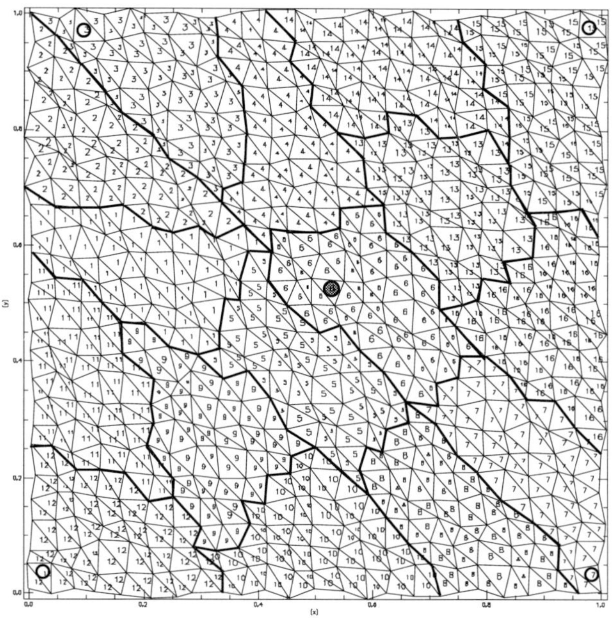

Figure 1: A simple oil reservoir model with four injection wells and one production well. The discrete model has been partitioned into 16 subdomains using the method in Section 1.

lower bound for our original problem since we minimize over a larger space. In order to solve the continuous problem we note that the matrix L is symmetric, semi-definite and that e is an eigenvector corresponding to $\lambda_1 = 0$. Since $x^T e = 0$, we can expand the solution $x = \sum_{i=2}^{n} a_i v_i$ in the set of orthogonal eigenvectors of L, and find by direct computation that

$$x^T L x = \sum_{i=2}^{n} a_i^2 \lambda_i \geq n\lambda_2,$$

since $x^T x = \sum_{i=2}^{n} a_i^2 = n$. Moreover, $x = \sqrt{n} v_2$ achieves this lower bound and therefore solves our problem.

The problem can therefore be solved by computing the eigenvector corresponding to the smallest positive eigenvalue of L. Very efficient algorithms for this problem are known.

It remains to find a good approximation to our original discrete problem. The closest discrete solution to our continuous solution can be easily constructed by computing the median value of the components x_i and then assigning -1 to all components having a lower value and the value $+1$ to the complement set.

Many refinements of the simple strategy outlined here are possible. The interested reader is referred to the recent journal article by Hendrickson and Leland [8] and the references to be found there for a more extensive study of these techniques. This exposition is a greatly simplified version of their more comprehensive and general theory. The software package Chaco [7] was used to produce the partioning of the grid in Figure 1.

2. Approximation and stability properties of the weighted L^2 projection

Our next problem is concerned with L^2 projections and their mathematical properties. Consider the model problem:

Find $u^* \in V(\Omega)$ such that

$$a(u^*, v) = f(v) \quad \forall v \in V, \tag{1}$$

with

$$a(u,v) = \sum_{i=1}^{N} \int_{\Omega_i} \rho_i \nabla u \cdot \nabla v \, dx$$

and

$$f(v) = \int_{\Omega} fv \, dx,$$

in an appropriate Sobolev space $V(\Omega)$. Note that our problem domain Ω is the union of subdomains Ω_i and that the parameter ρ_i can change across subdomains. Here ρ_i are positive constants and $\Omega = \cup_{i=1}^{N} \Omega_i$.

In practice we will compute an approximate solution to the above problem. Typically, we define a subspace $V_h \subset V$ and search for a solution $u \in V_h$ such that

$$a(u,v) = f(v) \quad \forall v \in V_h.$$

That is, we find $u \in V_h$ as an ortogonal projection (in the $a(\cdot,\cdot)$ inner product) of the solution $u^* \in V$ onto the subspace V_h. Similarly, corresponding to the subdomains Ω_i we define a coarse subspace V_H, where H is the diameter of the largest subdomain. In the regular case one can view $V_H \subset V_h$ as the space of functions that results from a coarse discretization of (1) using the subdomains as discretization elements of Ω.

Given $u \in V_h$, define the function $u_0 = Q_H^{\rho} u$ such that

$$(u_0, v) = (u, v) \quad \forall v \in V_H,$$

with

$$(u,v) = \sum_{i=1}^{N} \int_{\Omega_i} \rho_i uv \, dx,$$

a weighted L^2 inner product. In light of the definition above u_0 is the weighted L^2 projection of u on V_H.

Consider first the case when $\rho_i = 1$ corresponding to the normal L^2 inner product. The L^2 projection has a reasonable approximation property as can be seen from the following estimate (p. 165 in [9]):

$$||u - u_0||_{L^2(\Omega)} \leq CH|u|_{H^1(\Omega)},$$

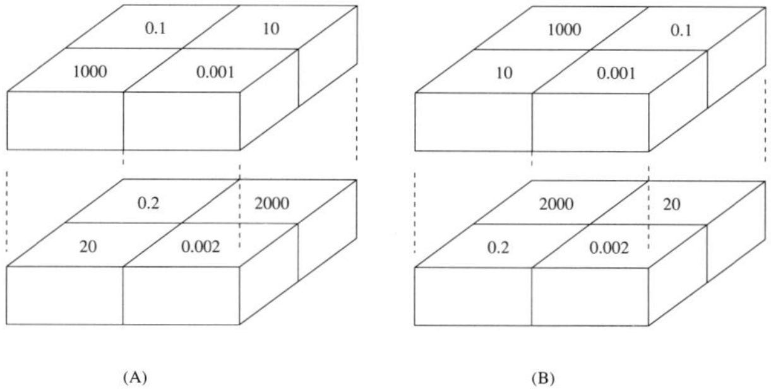

Figure 2: Non-quasimonotone (A) and quasimonotone (B) distribution of the coefficients ρ_i.

where the seminorm $|u|_{H^1(\Omega)} = a(u,u)^{1/2}$. From this estimate one can also prove H^1 stability of the projection, that is, $|u_0|_{H^1(\Omega)} \leq C|u|_{H^1(\Omega)}$. In other words, the energy of the projection $u_0 = Q_H u$ is bounded by the energy of the original function u.

When the coefficients ρ_i are allowed to have jumps from one subdomain to the next, the situation becomes more involved [4], [6], [10]. In fact, it has been shown that in the general case (see Figure 2A), the approximation property can only be estimated by:

$$\sum_{i=1}^{N} \rho_i ||u - u_0||^2_{L^2(\Omega_i)} \leq CH^2 \frac{H}{h} \sum_{i=1}^{N} \rho_i |u|^2_{H^1(\Omega_i)}, \tag{2}$$

while if the coefficients have a so-called quasimonotone distribution (see Figure 2B), we recover the estimate for the case $\rho_i = 1$,

$$\sum_{i=1}^{N} \rho_i ||u - u_0||^2_{L^2(\Omega_i)} \leq CH^2 \sum_{i=1}^{N} \rho_i |u|^2_{H^1(\Omega_i)}. \tag{3}$$

A quasimonotone distribution is verified by being able to traverse a path from any subdomain (across common faces) through neighboring subdomains (sharing a common vertex as in Figure 2) such that the value of the coefficient is monotone increasing to the subdomain with the largest value. We observe that the approximation property may deteriorate like $1/h$ in estimate (2), similarly we lose H^1 stability, the energy of the projection may exceed the energy of the original function u.

3. The oil reservoir problem

We now give a brief description of our application, the simulation of flow in porous media. The problem carries substantial economic interest in the petroleum industry. It is widely used for planning purposes, for reservoir management and for prediction of reservoir performance [2]. Another field of application is the study of ground water flow, in particular, the study of ground contamination by way of pollution simulation models [1].

A petroleum reservoir consists of hydrocarbons and other chemicals trapped in tiny pores in the rock. If the rock permits and if the fluid is sufficiently forced, it will flow from its current position in the rock to some other points in the reservoir. By injection of additional fluids and the release of pressure through the production of fluids at wells, the petroleum engineer can adjust the flow rate and modify the mixture of chemicals produced. In the associated reservoir flow problem, we are given the reservoir conditions (pressure and saturation) and the well flow conditions. The main problem is to model the fluid flow, especially to predict the fluid flow into production wells. The fluid flow problem can be expressed in terms of approximate mathematical equations. The physical laws that govern the flow can be derived from volume balance, phase equilibrium and conservation of mass, plus Darcy's Law stating that the fluid flow is proportional to pressure gradients and gravitational potential differences [1]. The constants of proportionality depend on permeability, viscosity and the density of the phases involved. The equations constituting a mathematical model for a reservoir are almost always too complex to be solved by analytical methods, even after many idealizations.

A simplified *prototype* black oil simulator [5] with incompressible flow and neglecting gravity can be reduced to an equation for the pressure

$$-\nabla \cdot \mathbf{C}(s, \mathbf{x}) \cdot \nabla p = q_1(\mathbf{x}, t), \tag{4}$$

and a saturation equation

$$\phi \frac{\partial s}{\partial t} + \mathbf{v}(\mathbf{x}) \cdot \nabla f(s) - \epsilon \nabla \cdot (\mathbf{D}(s, \mathbf{x}) \cdot \nabla s) = q_2(\mathbf{x}, t). \tag{5}$$

There are many different approaches to solving the equations. One can treat the complete nonlinear system using Newton's method [2] or decouple the system and iterate between the pressure equation and the saturation equation. The pressure equation is solved with the saturation fixed, and then the saturation equation is solved with the velocity (pressure) fixed. This splitting is not supported by a rigorous mathematical theory; rather it is heuristically motivated by the different behavior of the pressure equation and the saturation equation [1].

4. Domain decomposition

Next, we illustrate the importance and impact of the mathematical examples from the first two sections applied to the pressure equation (4). We note that this equation has a weak (variational) formulation similar to (1). We assume that the function $\mathbf{C}(s, \mathbf{x})$ is piecewise constant ($= \rho_i$) in subdomain Ω_i. This function reflects material properties, in particular, the permeability of the rock. The permeability in reservoirs may change by several orders of magnitude (across rock interfaces or layers) in a rather discontinuous manner. Our formulation in (1) is therefore often of interest.

The discrete version of (4) will produce a very large linear system of equations

$$Ax = b.$$

The matrix A typically inherits properties from the continuous problem and is symmetric and positive definite. Effective iterative algorithms for the solution of this equation are most often based on the idea of preconditioning, that is, the approximate construction of an inverse matrix B. One can then apply a basic iterative procedure of the form

$$x^{k+1} = x^k + B(b - Ax^k),$$

and accelerate this with a suitable Krylov subspace method [9].

The success of this approach depends on how well one can construct the matrix B. Domain decomposition methods try to construct B from exact or approximate inverses that are localized, that is, they approximate the inverse of A restricted to a subdomain.

Assume that we partition the original discrete problem using the techniques described in the first section. In this way we can define a set of subdomains Ω_i, with a balanced size and with a small number of interconnecting nodes. (The procedure in Section 1 defined two sets, but we can obviously apply this algorithm recursively; more refined techniques can further be used to define just the number of subdomains that we require.) Define the (rectangular) restriction matrix R_i applied to a vector x defined on all nodes, to return just the coefficients that belong to subdomain Ω_i. The local matrix associated with Ω_i is $A_i = R_i A R_i^T$, and a corresponding local inverse can be expressed as

$$B_i = R_i^T (R_i A R_i^T)^{-1} R_i.$$

In addition, these methods almost always need a representation that can capture the smooth or average behavior of the solution. This should be a matrix of much lower dimension than the original A, its role can equivalently be seen as a global coupling of all the localized approximations. If the subdomain decomposition forms a regular grid then this matrix is often just the stiffness matrix derived from a discretization where the subdomains are considered to be the basic discretization elements. We denote the inverse of such a low dimensional, coarse space approximation B_C.

A simple additive preconditioner can now be constructed by taking

$$B = B_C + \sum_i B_i.$$

In order for this to work well, we need to make an important modification to the approach outlined above. We must introduce a better coupling between the local spaces and the coarse space. This is often done by making the subdomains overlap each other. This modification produces the so-called Additive Schwarz preconditioners.

An alternative to overlap is to change or complement the coarse space matrix B_C. We briefly discuss two such alternatives where we avoid overlap. (Thus, we just keep the subdomains given by our graph partitioning algorithm) [3].

In the first method, called an Additive Diagonal Scaling method, we add yet another special matrix to our sum above. This matrix, which is diagonal,

Method	Additive Average			Additive Diagonal		
$\rho = 1$	39	(28.3)	204	28	(14.0)	574
Quasi-monotone	44	(31.0)	221	34	(15.1)	592
Not quasi-monotone	39	(20.9)	201	136	(363)	1028

Table 1: Comparison of two iterative methods. The number of iterations required to reduce the residual by six orders of magnitude is followed by the condition number of the preconditioned system in parenthesis. The last number given is the elapsed time in seconds when using a cluster of 8 Ultra Sparc processors. All subdomain and coarse problems are solved approximately by using only two symmetric Gauss-Seidel iterations.

arises from the restriction of the space V_h to the union of all interior subdomain interfaces and the use of an approximate bilinear form where we just sum products $u(x)v(x)$ for nodal values x on the interface. That is, we solve a diagonal problem for unknowns corresponding to all interior interface nodes.

In the second method, called an Additive Average method, we define a completely new B_C that includes all the interface variables, but in a way that leaves us with a system that is relatively easy to solve. This matrix corresponds to a new coarse space defined as the range of an interpolation-like operator that extends into the interior of each subdomain by just using the average value of the subdomain boundary nodal values (see [3] for details).

In Table 1 we compare the two methods for different distributions of the coefficient ρ_i. The computational example has 512 regular subdomains in three space dimensions, with a total of approximately a quarter of a million of unknowns. The time per iteration was about 4.2 second for all cases, however, the construction of the coarse part, B_C, of the preconditioner takes considerable time for the Additive Diagonal method. We immediately notice the poor behavior of the Additive Diagonal method in the non quasi-monotone case. It is, at this point, perhaps not surprising that this difference is directly related to the mathemathical properties of the L^2 projection discussed in Section 2. The analysis of the Additive Diagonal method depends on the approximation and H^1 stability properties of this projection. However, by constructing a completely different coarse space as we do in the Additive Average method, we are able to avoid this difficulty and indeed work with an interpolation operator which also has the desired approximation and stability properties in the general case where the ρ_i may have a non-quasimonotone distribution.

5. Concluding notes and remarks

We have seen how a discrete optimization problem can be approximated by a continuous problem which in turn reduces to the solution of a very special eigenvalue problem. Furthermore, this problem has important applications in the design of parallel computer algorithms, since it can be used to partition a problem among many processors in such a way that each processor receives a fair share of the work, with relatively little communication to other processors.

Also, we have outlined how quite abstract and subtle estimates of approximation and stability properties of the weighted L^2 projection shows up as the difference between 33 and 134 iterations in a numerical algorithm applied to an oil reservoir simulation problem. Furthermore, how such insights may help in the construction of numerical algorithms that are more robust with respect to large, discontinuous variations in, for example, the permeability of rock.

Mathematical techniques and knowledge covering a very broad area are needed in order to build a practical numerical algorithm that effectively can be used as a part in an even more complex computer simulation package for todays challenging engineering problems.

Acknowledgment. The author thanks Eero Vainikko for providing the computational results given in Table 1.

References

[1] M. B. Allen, G. A. Behie, and J. A. Trangenstein, *Multiphase Flow in Porous Media*, Springer-Verlag, 1988. Lecture Notes in Engineering.

[2] K. Aziz and A. Settari, *Petroleum reservoir simulation*, Elsevier Applied Science Publishers, 1979.

[3] P. Bjørstad, M. Dryja, and E. Vainikko, *Additive schwarz methods with no subdomain overlap and new coarse spaces*, in Domain Decomposition Methods in Sciences and Engineering, R. Glowinski, J. Périaux, Z. Shi, and O. B. Widlund, eds., John Wiley & Sons, 1996, pp. 141–157. Proceedings from the Eight International Conference on Domain Decomposition Metods, May 1995, Beijing.

[4] J. H. Bramble and J. Xu, *Some estimates for a weighted L^2 projection*, Math. Comp., 56 (1991), pp. 463–476.

[5] G. Chavent, G. Cohen, and J. Jaffra, *Discontinuous upwinding and mixed finite elements for two-phase flows in reservoir simulation*, Comp. Meth. in Appl. Mech and Eng., 47 (1984), pp. 93–118.

[6] M. Dryja, M. Sarkis, and O. B. Widlund, *Multilevel Schwarz methods for elliptic problems with discontinuous coefficients in three dimensions*, Tech. Report 662, Department of Computer Science, Courant Institute, March 1994.

[7] B. Hendrickson and R. Leland, *The Chaco user's guide, version 2.0*, Tech. Report SAND 94-2692, Sandia National Laboratories, July 1995.

[8] ——, *An improved spectral graph partitioning algorithm for mapping parallel computations*, SIAM J. Sci. Comput., 16 (1995).

[9] B. F. Smith, P. Bjørstad, and W. Gropp, *Domain Decomposition: Parallel Multilevel Methods for Elliptic Partial Differential Equations*, Cambridge University Press, 1996.

[10] J. Xu, *Counter examples concerning a weighted L^2 projection*, Math. Comp., 57 (1991), pp. 563–568.

Progress in Mathematics, Vol. 168, © 1998 Birkhäuser Verlag Basel/Switzerland

Large Deviations and Perturbations of Random Walks and Random Sufaces

ERWIN BOLTHAUSEN*

Institut für Mathematik
Universität Zürich
Winterthurerstrasse 190, CH-8057 Zürich

1. Introduction: Coin tossing and generalizations

The results we will discuss here are variations of the following simple observation: Consider a coin tossing sequence $X_1, X_2, \ldots$, the X_i taking values 0 or 1 with equal probabilities. By the law of large numbers,

$$\lim_{n\to\infty} \sum_{i=1}^{n} X_i/n = 1/2 \quad \text{a.s.}$$

As was known already to Jacob Bernoulli, for $\varrho > 1/2$, the probability of the event $\{\sum_{i=1}^{n} X_i/n \geq \varrho\}$ converges to 0 exponentially fast. What can one say about the conditional law of the sequence given this highly improbable event:

$$\mathcal{L}\left(X_i, \ldots, X_n \,\middle|\, \sum_{i=1}^{n} X_i/n \geq \varrho\right)?$$

The answer is obvious: For large n, this essentially is coin tossing with the shifted success probability ϱ. One has to be a bit careful with the formulation: It is not true that the conditional law of the full sequence of length n is close in total variation to the shifted coin tossing, but rather shorter pieces are: If $m(n) = o(n)$, then

$$\mathcal{L}\left(X_i, \ldots, X_{m(n)} \,\middle|\, \sum_{i=1}^{n} X_i/n \geq \varrho\right) \tag{1.1}$$

is in total variation close to shifted coin tossing. The situation is slightly more challenging for more general sequences $X_1, X_2, \ldots$ of i.i.d. random variables, having distribution μ, with exponential moments $\int e^{\lambda x}\mu(dx) < \infty$ for all $\lambda \in \mathbb{R}$, say. If $\varrho > EX_i$, then the answer is similar: (1.1) above converges in total variation to an i.i.d. sequence with a shifted distribution μ_ϱ. It is however no longer evident

*) partially supported by Swiss NF, contract no. 20-41925.94

what μ_ϱ should be. It is the unique probability distribution having mean ϱ and minimizing the entropy

$$h(\mu_\varrho \,|\, \mu) \triangleq \int \left(\log \frac{d\mu_\varrho}{d\mu}\right) d\mu_\varrho$$

among such measures ($h(\nu \,|\, \mu) \triangleq \infty$ if ν is not absolutely continuous with respect to μ), or equivalently, being the unique measure of the form $e^{\lambda x}\mu(dx)/z$ having mean ϱ, where z is the norming constant.

The result in this form is due to Csiszar [C] and Dembo & Zeitouni [DZ] (with more general types of conditionings). It fits in the general picture of equivalence of ensembles in statistical physics and has in this context been widely discussed and in much more general situations, for instance for Gibbs random fields (see Lanford [L], Bolthausen and Schmock [BS1], Georgii [G], and most recently Lewis, Pfister and Sullivan [LPS]).

From a large deviation viewpoint the above answer can be understood in the following way: Define the *empirical measure* by

$$L_n \triangleq \frac{1}{n}\sum_{j=1}^{n} \delta_{X_j}.$$

This is a random element in the space $\mathcal{M}_1^+(\mathbb{R})$ of probability measures on $\mathbb{R}$. A version of the law of large numbers states that L_n is (in weak topology) close to μ, with large probability. *Sanov's Theorem* then gives information on the small probabilities of deviations from this. Roughly (and not very precisely) stated, it says that

$$P(L_n \sim \nu) \sim \exp(-nh(\nu \,|\, \mu)).$$

One way to make it precise is in the form of a Laplace asymptotic: For any real valued, weakly continuous and bounded function F defined on $\mathcal{M}_1^+(\mathbb{R})$, one has

$$\lim_{n\to\infty} \frac{1}{n} \log E \exp(nF(L_n)) = \sup_\nu (F(\nu) - h(\nu \,|\, \mu)). \tag{1.1}$$

The function $h(.|\, \mu)$ is called the rate function of the large deviation principle. It should be noted that the expectation on the left-hand side of (1.1) is evaluated by such a statement only up to leading order (logarithmic) asymptotics.

From the Sanov theorem, it is easy to understand why the special μ_ϱ is appearing. Under the conditioning, the empirical measure settles near the unique distribution ν, which is the easiest to reach among the ones satisfying the constraint $\int x\nu(dx) \geq \varrho$, and this is just the μ_ϱ. Given that L_n settles there, it is not difficult to see that at least locally, the sequence is still close to an i.i.d sequence of random variables (in the above discussed sense). In fact, if one fixes L_n, the sequence of random variables is a sequence of drawings from a box of n values without replacement. It is well known that a subset of drawings, which is in size negligible compared with n, looks close to an independent sampling. This last property and variants of it are sometimes called "propagation of chaos".

Why are such conditionings interesting? They are special cases of a wide class of perturbations of a sequence of probability laws $\{P_n\}$ (in the above example the coin tossing laws), by so called "Hamiltonians" H_n, which are (extended) real valued functions defined on the appropriate configuration space. The new laws are (formally) given by

$$d\widehat{P}_n = \exp(-H_n)\, dP_n / z_n,$$

z_n being the norming, the so called partition function. Vaguely speaking, $\widehat{P}_n$ is the law of the "typical" configurations which contribute to this partition function. Conditionings on an event are obtained by setting $H_n = \infty$ on the complement of it, and zero otherwise.

We will discuss a number of situations where reasonings similar to the above type apply, but where the global picture is more delicate, and especially also the transition from a rough large deviation estimates of the type (1.1) to a precise description of the perturbed law. Often, there are also nontrivial problems around the variational problems, i.e. the problem how to characterize the elements which maximize expressions like the right-hand side of (1.1).

2. Self attracting random walks

We consider an ordinary, continuous time, symmetric random walk X_t, $t \geq 0$, on $\mathbb{Z}^d$, starting at 0. Such a walk stays for exponentially distributed holding times on points, and then jumps with equal probabilities to one of the $2d$ neighbors. We denote by P the law of this random walk on the space of right continuous pure jump paths. For $T > 0$, a self attraction on the set of paths of length T can be introduced by preferring paths where

$$\int_0^T \int_0^T 1_{X_t = X_s}\, dt\, ds = T^2 \sum_{x \in \mathbb{Z}^d} l_T(x)^2 = T^2 \|l_T\|_2^2 \tag{2.1}$$

is large. Here $l_T(x)$ is the mean sojourn time in x: $l_T(x) \triangleq |\{ t \leq T : X_t = x \}|/T$.

A natural choice for the law of a self attracting random walk would be to take minus the expression in (2.1) as a Hamiltonian. However, an easy rough estimation shows that this would make the self-attraction too strong to be interesting as the path would for large T collapse with large probability to the path identical zero. For various reason, the appropriate Hamiltonian is minus (2.1) divided by T. We therefore define the self-attracting path measure by

$$d\widehat{P}_{T,\beta} \triangleq \exp(\beta T \|l_T\|^2)\, dP / z_{T,\beta},$$

where $\beta > 0$ is a parameter, and $z_{T,\beta} \triangleq \int \exp(\beta T \|l_T\|^2)\, dP$ is the norming constant.

The rough (i.e. leading order) evaluation of this norming factor can be obtained from a large deviation principle of Donsker and Varadhan [DV] (see also [DS]) for l_T. Unfortunately, due to the lack of strong ergodicity properties of the random walk, there is only a "weak" large deviation principle, meaning that one gets a statement like (1.1) only for a very restricted class of functionals:

Theorem 2.1 (Donsker & Varadhan) *If* $F : \mathcal{M}_1^+(\mathbb{Z}^d) \to \mathbb{R}$ *is continuous, and* $\{\nu : F(\nu) \geq a\}$ *is compact in* $\mathcal{M}_1^+(\mathbb{Z}^d)$ *for all* $a \in \mathbb{R}$, *then*

$$\lim_{T\to\infty} \frac{1}{T} \log E \exp(TF(l_T)) = \sup_{\|g\|_2=1} \left\{ F(g^2) - \frac{1}{2}\|\nabla g\|_2^2 \right\},$$

here ∇g *is the discrete gradient of* g.

The conditions on F are rather restrictive, and it is known that one cannot give up the compact level condition in general. Nevertheless, in our case, although $\| \ \|_2$ does not satisfy this condition, one can show by an easy compactification argument, using also some special monotonicity property, that the conclusion is true:

$$\lim_{T\to\infty} \frac{1}{T} \log z_{T,\beta} = \sup_{\|g\|_2=1} \left(\beta \sum_x g(x)^4 - \frac{1}{2}\|\nabla g\|_2^2 \right) \triangleq b(\beta) \tag{2.2}$$

From the coin tossing example of section 1, one is tempted to conclude that with large $\widehat{P}_{T,\beta}$-probability, l_T is close to some g^2 that solves the above variational problem. Having the necessary information about the distribution of l_T, one can then try to get information about the distribution of the whole process, and especially about the distribution of X_T, like in the coin tossing example of section 1. There are, however, considerable difficulties which are partly connected with the shift degeneracy of the variational problem.

Remarks concerning the variational problem

(1) *There are not always solutions.* Solutions exist if and only if $b(\beta) > 0$. For $d = 1$, $b(\beta) > 0$ holds true for all $\beta > 0$. If $d \geq 2$, there exists $\beta_c(d) > 0$ such that $b(\beta) > 0$ for $\beta > \beta_c(d)$, and $b(\beta) = 0$ for $\beta < \beta_c(d)$. (see Brydges and Slade [BS]).

(2) If there are solutions at all, then all shifts are solutions, too. This *shift degeneracy* is the most delicate points for the probabilistic analysis.

(3) Let K be the set of probability densities g^2 where g maximizes the expression in (2.2). If $b(\beta) > 0$, then K is not empty by (1). By the observation (2), K cannot be compact, but it is *compact modulo shifts.* This is very important for the analysis of the problem.

(4) *Uniqueness modulo shifts?* It might be true that the solutions of the variational problem are always unique modulo shifts, but we don't know how to prove or to disprove this, even for $d = 1$. One can prove that it is correct for $\beta \geq 2d$ (see [BS2, section 5]). This problem appears to be very delicate, partially due to the lattice structure, which prohibits an application of symmetrization techniques. In the one dimensional case, numerical calculations suggest that the Euler equation has many stationary points, which are not shift equivalent. This is in sharp contrast to the corresponding variational problem on $\mathbb{R}$: $\beta \int g^4(x)dx - \frac{1}{2} \int \|\nabla g(x)\|^2 dx = \max$, where there is just one.

One can rule out all solutions of the Euler equation except the unimodal ones, as they cannot be maximizers, and the numerical calculations suggest that there remain just two candidates, which both are symmetric, but we have no proof that the solutions of the variational problem have to be symmetric. One of the symmetric solutions (of the Euler equation) is symmetric around some point in $\mathbb{Z}$, and the other one around a midpoint between two neighboring ones. For large β the "one summit" one is better, but for small β, the other one looks numerically better but the difference is extremely slight. This suggests that for at least one β, there are two solutions of the variational problem. However, we have no proof for this picture.

Despite the shift degeneracy of the variational problem, it turns out that the self interaction in case $b(\beta) > 0$ is strong enough to localize the distribution of the endpoint in a strong sense. This might be somewhat surprising, as the only ingredient which is not shift invariant is the starting point, whose influence is felt strongly on the whole time interval.

Theorem 2.2 (Bolthausen & Schmock, [BS2]).

a) *If $\beta > \beta_c(d)$, then there exists $\lambda_0(\beta) > 0$ such that*

$$\sup_{T>0} \widehat{E}_T\left(e^{\lambda|X_T|}\right) < \infty$$

when $\lambda < \lambda_0(\beta)$.

b)

$$\liminf_{T\to\infty} \widehat{E}_T(|X_T - \widehat{E}_T X_T|^2) > 0.$$

The theorem states that in a strong sense the distribution of the endpoint collapses and keeps just fluctuations of order 1, which however stay nontrivial.

This result complements the following one obtained by Brydges and Slade:

Theorem 2.3 (Brydges & Slade, [BS]).

There is a $\bar{\beta}_c(d) \leq \beta_c(d)$, which is positive for $d \geq 2$, such that X_T is diffusive for $\beta < \bar{\beta}_c(d)$, i.e. $X_T/\sqrt{T}$ under $\widehat{P}_T$ is asymptotically normally distributed with positive variance.

It is natural to conjecture that there is no intermediate regime, that is $\bar{\beta}_c(d) = \beta_c(d)$, but this is open at present. The transition from the diffusive behavior of Theorem 2.3 for small β to the one described in Theorem 2.2 for large β is called a *collapse transition.*

We do have more precise information on the distribution of X_T, and in fact on the whole path measure. It is in fact possible to prove a sort of a "propagation of chaos" result in the form of an asymptotic description of the law of the path and its endpoint conditioned on l_T. The main problem however is to determine the asymptotic distribution of l_T itself under $\widehat{P}_{T,\beta}$. Despite the shift degeneracy, we can prove tightness. A complete evaluation, however, we do have only if the solution of the variational problem is unique modulo shifts, i.e. when

$$K = \{ g_0^2(\cdot - x) : x \in \mathbb{Z}^d \} \tag{2.3}$$

holds for some g_0.

Theorem 2.4 (Bolthausen & Schmock [BS2]) *Assume that (2.3) is satisfied. Then* $\lim_{T\to\infty} \mathcal{L}_{\widehat{P}_T}(X_T) = g_0 \star g_0/z$, *where* z *is the norming constant*

The appearance of g_0 which is the square root of a probability density which is in K may be somewhat surprising. Very roughly, this is coming from the fact that the tightness is a "boundary effect" at the boundaries of the time interval. The starting point affects the distribution of l_T, and this affects the distribution of the endpoint, each by g_0 only, as from the boundary points, the relevant time interval fills only half the line. As the two boundary points are well separted, the influences are asymptotically independent, which explains the appearance of the convolution.

The main steps of the proof of the Theorems 2.2 and 2.4.

By point (3) above, K is compact modulo shifts. The discussion in Section 1 then suggests that under $\widehat{P}_T$, l_T is with large probability close to some element of K. This is true but rather delicate to prove, the difficulty coming from the "weakness" of the large deviation estimates, and the non compactness of K, which is due to the shift degeneracy.

Knowing that l_T is close to some element of K still leaves one with the most intricate problem, namely to compare probabilities of the type $\widehat{P}_T(l_T \sim \mu)$ with $\widehat{P}_T(l_T \sim \theta_x(\mu))$, where $\mu \in K$ and $\theta_x(\mu)$ is the shift of μ by $x \in \mathbb{Z}^d$. One has to show that, uniformly in T, the second is negligible compared with the first, provided $|x|$ is large. Such a comparison can not be made on the basis of rough (logarithmic) large deviation estimates, as one has to evaluate quotients for which asyptotic results like (1.1) are apparently useless. There do exist sharp expansions for large deviations, see e.g. [KT] and [BDT], but up to now, they do not solve the problems posed here. However, the evaluation of the quotients can be done by some combinatorial tricks.

Having the distribution of l_T under control, one can get the necessary information on the distribution of the path, and of X_T and its exponential moment. In the uniqueness case, i.e. when (2.3) is satisfied, one can completely determine the limiting distributions.

Possible extensions and related problems

1 The method is not restricted to quadratic functionals of l_T as $\|l_T\|_2^2$ is, and can be extended to other functionals which however have to possess some smoothness properties. An example is $\sum_x l_T(x)^3$. This has a collapse transition for $d = 1$, as was pointed out to the author by G. Slade, and is connected with the nonlinear focussing Schrödinger equation (see [LRS]).

2 A somewhat different type of self-attraction for the random walk has been investigated in [B2], and for a related model by Sznitman [S2]. Here the path measure P of the random walk is transformed by setting

$$d\widehat{P}_{T,\beta} = \exp(-\beta N_T) dP / z_{T,\beta}$$

where N_T is the number of different sites in $\mathbb{Z}^d$ which are visited by the random walk. The one dimensional case (in a slightly different setting) had been discussed by Schmock [S]. For $d = 2$ and all $\beta > 0$, the displacement of X_T under $\widehat{P}_{T,\beta}$ is of order $T^{1/4}$. Actually, $X_T/T^{1/4}$ is asymtotically stochastically bounded as $T \to \infty$. The set S_T of points visited by the random walk is with $\widehat{P}_{T,\beta}$-probability close to one (for large T) close to a tightly filled ball with radius proportional to $T^{1/4}$. The variational problem which is behind this model is closely related to the standard isoperimetric inequality.

Up to now, only the dimension 2 has been investigated in all details. However, the sharpenings of the isoperimetric inequality given in [H] (of which I had not been aware when writing the paper) should make it possible to show in arbitrary dimensions that the displacement is of order $T^{1/(d+2)}$ for all $\beta > 0$. (All the steps, except the one connected with the variational problem, and another one of probabilistic nature, which is easy for $d = 2$, are actually done in [B2] for arbitrary dimensions).

There is no collapse transition from small to large β in this case, but such a transition probably takes place when weakening the interaction in the following sense. We replace the "Hamiltonian" N_T by $T^{-\alpha} N_T$, $\alpha > 0$, and arrive in this way at a path measure $\widehat{P}_{T,\beta,\alpha}$ with a norming constant $z_{T,\beta,\alpha}$. The rough asymptotics for $z_{T,\beta,\alpha}$ has been investigated (for the slightly different Wiener sausage case) in [B1] and [S1]. There is a transition in the behavior of z depending on α, which should be reflected in the behavior of the path measure. This latter question has however not been investigated rigorously, and the following picture is conjectural and based up to now on non-rigorous considerations (joint investigation with M. van den Berg and F. den Hollander):

Assume $d \geq 3$ (a full discussion of the two dimensional case is probably more delicate). For $\alpha < 2/d$, the displacement of the endpoint is expected to be of order $T^{(1+\alpha)/(d+2)}$, and again, the set of visited points S_T is filling tightly a droplet of this radius. For $\alpha > 2/d$, the process is diffusive, and $X_T/T^{1/2}$ should just be asymptotically normally distributed. The critical case $\alpha = 2/d$ is particularly interesting for various reasons (It is probably connected with the tail behavior of the distribution of the size of the intersection set of two independent random walks of infinite length, which was recently investigated by Khanin, Mazel, Shlosman and Sinai [KMSS]). In that case, we expect that there is a transition from a diffusive behavior for small β to a collapsed phase, where the displacement of the endpoint is of order $T^{1/d}$. Even in the collapsed phase, S_T no longer is a tightly filled droplet, but should rather become porous.

3. A 2 + 1-dimensional Wulff-type droplet construction

I am presenting here a joint work with Dima Ioffe.
We consider a simple model of a two dimensional "random surface": the *lattice*

free field or harmonic crystal: If $N \in \mathbb{N}$ let $A_N \triangleq N[-1,1]^d \cap \mathbb{Z}^d$. The lattice free field $(X(i))_{i \in A_N}$ is the Gaussian random field defined by its distribution P_N on $\mathbb{R}^{A_N}$:

$$P_N(dx) \triangleq \frac{1}{Z_N} \exp\Bigg(-\frac{1}{2} \sum_{\substack{k,l \in A_N \cup \partial A_N \\ |k-l|=1}} (x_k - x_l)^2\Bigg) \prod_{k \in A_N} dx_k, \tag{3.1}$$

where ∂A_N is the outer boundary of A_N, and $x|_{\partial A_N} \equiv 0$.

We will mainly stick to the case $d = 2$. The 1 in 2+1 in the title of this section refers to the one-dimensionality of $\mathbb{R}$ which is the space in which the random field takes its values. It is easily calculated that for $d = 2$, $\mathrm{var}(X_0) \sim \log N$. Therefore, a limit of the measure as $N \to \infty$ does not exist, in contrast to the higher dimensional case $d \geq 3$.

The *anharmonic crystal* is the case where $(\ )^2$ in (3.1) is replaced by a more general function $V : \mathbb{R} \to [0, \infty)$ satisfying some growth condition. For $d = 1$, this is just an ordinary random walk which is tied down to 0 at the two boundary points of A_N. In higher dimensions, the anharmonic crystal is a much more complicated object. Up to now, our quantitatively precise results are restricted to the harmonic case.

We add now three ingredients to the harmonic crystal, all three being perturbations of a large deviation type, meaning that configurations are favoured which have small probabilities under the free field. These ingredients are: A hard wall condition, an attractive wall-surface interaction and a macroscopic volume constraint. The wall here is just the lattice $\mathbb{Z}^2$, interpreted as a two dimensional surface in our $2+1$-dimensional space. We will first introduce these ingredients and discuss their effect on the random surface separately. Our main object will be, however, to describe their joint effect.

The phenomenon we are investigating is known in the physics literature as wetting, by interpreting the part of the space between the random surface and the wall as a "liquid". The volume and positivity constraints have obvious interpretations, and the attractive surface to wall interaction is reflecting a tendency of the wall to stay dry. A $1+1$-dimensional case has first been investigated by de Coninck, Dunlop and Rivasseau [CDR]. Our work is an attempt to derive macroscopical interfaces (Wulff type shapes) from the microsopical model in higher than two dimensions. In two dimensions, there had been a lot of activities in recent years, see e.g. [DKS], [P] and [I1], [I2] for interfaces in the Ising model, and [ACC] for a precolation model. For the physical background see [Wi] and [Wu].

We now first describe the three interactions in greater details:

1. The hard wall condition. We condition the field upon staying positive, i.e. upon the event $\Omega^+ = \{X(i) \geq 0, i \in A_N\}$. The probability of this event is going to zero, as $N \to \infty$, for $d = 2$ it is of order $\exp(-c(\log N)^2)$ ([BDG]). The effect of conditioning on this event is quite interesting: it pushes the random surface to $+\infty$ in the limit as $N \to \infty$. This is coming from an interplay between large deviations for the global behavior of the random surface and the local fluctuations.The effect is called entropic repulsion (see Bricmont, El Mellouki, Fröhlich [BEF], Lebowitz

& Maes [LM]). For the harmonic case, we have a qualitatively quite precise description (see [BDZ], [D] and [BDG]). $d = 2$ is actually the most delicate case. We know here that there are constants $0 < c_1 < c_2$ with

$$\lim_{N\to\infty} P_N(c_1 \log N \leq X(0) \leq c_2 \log N \mid \Omega^+) = 1.$$

For $d \geq 3$, we know precisely where the field settles (see [BDZ] and [D]). To narrow down the constants for $d = 2$ however seems to be very challenging as the interplay between the macroscopic behavior of the random membrane and its local fluctuations is rather delicate.

2. The attractive interaction between the surface and the wall. This can formally been achieved by subracting an external potential proportional to $\sum \delta_0(X_i)$ from the Hamiltonian. We can also just change the measure P_N of the harmonic crystal to

$$\widehat{P}_N(dx) = \frac{1}{\widehat{Z}_N} \exp\Big(-\frac{1}{2} \sum_{\substack{k,l \\ |k-l|=0}} (x_k - x_l)^2\Big) \prod_{k\in A_N} (e^{-J} dx_k + \delta_0(dx_k)),$$

where $J \in \mathbb{R}$ is an additional parameter, regulating the relative strength of the wall to surface interaction.

The effect of this surface to wall interaction alone, on the random surface had been investigated by Dunlop, Magnen, Rivasseau & Roche [DMRR], (with a less singular interaction, but the outcome should be the same here). What they found for $d = 2$ is that the interaction localizes the random field in the sense that the variance of X_0 stays of order 1.

3. The macroscopic volume constraint. We assume $d = 2$. The macroscopic volume V_N is defined by

$$V_N \triangleq \frac{1}{N^3} \sum_{j\in A_N} X_j.$$

We want to investigate the behavior of the random surface under the condition $V_N \geq \gamma$, where $\gamma > 0$ is a fixed number. The effect of this conditioning *alone* on the random surface is easily described due to the Gaussian nature of the field. Under P_N, V_N is normally distributed with a variance of order $1/N^2$. Therefore $P_N(V_N \geq \gamma)$ is logarithmically equivalent to $\exp(-c\gamma^2 N^2)$. It is not difficult to describe the conditional law $P_N(\ \mid V_N \geq \gamma)$ macroscopically: If ξ_N is the rescaled field where the space is scaled down by a factor $1/N$:

$$\xi_N(x) \triangleq \frac{1}{N} \sum_k X(k) 1_{\{\|k-Nx\|<1/2\}}$$

$x \in [-1, 1]^2$, $\|\|$ is the maximum norm in $\mathbb{R}^2$ here, then for some sequence $\varepsilon_N \to 0$

$$\lim_{N\to\infty} P_N(\|\xi_N - u\|_1 \geq \varepsilon_N \mid V_N \geq \gamma) = 0. \tag{3.2}$$

where $u \triangleq \gamma u_0 / \int u_0 \, dx$, and u_0 is the solution of the Poisson equation $\Delta u_0 = -1$ on $[-1,1]^2$ with zero boundary condition. u satisfies the following variational problem

$$\frac{1}{2}\int |\nabla u|^2 \, dx = \min, \qquad \int u \, dx \geq \gamma. \tag{3.3}$$

This is quite similar to the shift in the coin tossing example in section 1, and is not difficult to prove by standard Gaussian large deviation and isoperimetric estimates.

The energy on the left-hand side of (3.3) can be interpreted as an integral over a surface tension $\int (1 + |\nabla u|^2)\tau(\vec{n}(x)) \, dx$ where $\vec{n}(x)$ is the normal vector to the surface at $(x, u(x))$, and $\tau(\vec{n})$ is the "surface tension", defined by

$$\tau(\vec{n}) = \lim_{N \to \infty} \frac{1}{N^2} \log \frac{Z_{N,\vec{n}}}{Z_N}.$$

Here, $Z_{N,\vec{n}}$ is the partition function coming from the harmonic crystal with boundary conditions given by a plane perpendicular to $\vec{n}$ (in contrast to Z_N which is coming from zero boundary conditions). An easy calculation shows that this leads to the energy expression in (3.3). The existence of a surface tension has also been proved for some anharmonic crystals (see [MMR]). However, to date, this has not led to large deviation estimates which would allow to prove the analogue of (3.2).

So far we have discussed these three perturbations of the surface separately, but we are interested in their *joint* effect on the surface, i.e., we want to determine the macroscopic behavior of ξ_N under the conditioned law $\widehat{P}_N(\ \ |\Omega_+, V_N \geq \gamma)$. A discussion of the joint effect of the volume and positivity constraint, i.e. of $P_N(\ \ |\Omega_+, V_N \geq \gamma)$ has recently been given by Deuschel and Ben Arous [BD]. They show that the positivity condition has no macroscopical influence on the surface, which already satifies the volume constraint. A rough explanation for that is that the probability of the positivity event has much slower decay than that of the volume one. In our case, under $\widehat{P}_N$, i.e. with the attractive surface-wall interaction present, the volume constraint and the positivity are events which have probabilities with decay of the same order of magnitude. It is therefore to be expected that there is a macroscopically visible interplay between them. The outcome is that there is a modified variational problem which takes into account the local surface to wall interactions. This leads to a constant $\delta > 0$ which can be interpreted as the difference of the wall-vapor and the wall-liquid surface tensions. The new variational problem is then

$$\delta |\mathrm{supp}(u)| + \frac{1}{2}\int |\nabla u|^2 \, dx = \min, \qquad \int u \, dx \geq \gamma. \tag{3.4}$$

This variational problem is easily solved explicitly: For fixed $\mathrm{supp}(u)$, the optimal u satisfies the Poisson equation in this domain. If $|\mathrm{supp}(u)|$ is fixed, the optimal shape is a circle. Then one can write down the solution which is unique modulo shift

$$\bar{u}_{\gamma,\delta}(x) \triangleq \frac{2\gamma}{\alpha}\left(1 - \frac{\pi |x|^2}{\alpha}\right) \vee 0,$$

where $\alpha \triangleq (\pi\gamma^2/\delta)^{1/3}$. We assume that γ is small enough in order that $\mathrm{supp}(\bar{u}_\gamma) \subset [-1,1]^2$.

Theorem 3.1 (Bolthausen & Ioffe, [BI]). *Let*

$$\delta \triangleq J + \lim_{N\to\infty} \frac{1}{N^2} \log \frac{\widehat{Z}_N \widehat{P}_N(\Omega^+)}{Z_N}.$$

There exists a sequence $\varepsilon_N \to 0$, *with*

$$\lim_{N\to\infty} \widehat{P}_N(\min_x \|\bar{u}_{\gamma,\delta}(\cdot - x) - \xi_N\|_1 \geq \varepsilon_N \mid V_N \geq \gamma, \Omega_+) = 0.$$

The theorem states that macroscopically, the volume is concentrated in an unique droplet described by a shift of $\bar{u}_{\gamma,\delta}$. The picture resembles much the situation we had for the self-attracting random walk of the form $d\widehat{P}_T = \exp(-N_T)dP/z_T$ which was shortly discussed at the end of section 2, and where a similar droplet for the (rescaled) l_T appeared. There are, however, important differences. The statement of the Theorem 3.1 hides some important aspects and the situation is more complicated than what it suggests. Contrary to what one may expect, the size of the macroscopically wetted region $\{x \in [-1,1]^2 : \xi_N(x) > 0\}$ is *not* close to the support of a shift of $\bar{u}_{\gamma,\delta}$. Outside the main droplet described by a shift of $\bar{u}_{\gamma,\delta}$, the wall is in fact wetted partially by many microscopical droplets which don't contribute macroscopically to the volume, but which do contribute macroscopically to the size of the wetted region. We don't have a precise description of this "noisy" shallow region, which however is of crucial influence on the not explicitely given constant δ.

The most delicate part of the proof of the theorem is to discuss the separation of these two regions. The difficulty is mainly coming from the fact that the total number of possibly wetted regions is of order $\exp(cN^2)$, so a brute force counting will not do. The volume constraint induces the creation of a single macroscopic dorplet, but on the other hand, the high entropy of the counting is important for the shallow region. In order to reduce the computational complexity, one has to introduce a mesoscopic scale with a coarse graining procedure, but there are many other technical steps to perform, including expansion and resumming techniques.

References

[ACC] Alexander, K., Chayes, J.T. and Chayes, L., *The Wulff construction and asymptotics of the finite cluster distribution for two-dimensional Bernoulli percolation,* Comm. Math. Phys. **131**, 1–50 (1990).

[BD] Ben Arous, G. and Deuschel, J.-D., *The construction of the d+1 Gaussian droplet,* To appear in Comm. Math. Phys. (1995).

[B1] Bolthausen, E., *On the volume of the Wiener sausage,* Ann. Prob. **18**, 1576–1582 (1990).

[B2] Bolthausen, E., *Localization of a two-dimensional random walk with an attractive path interaction,* Ann. Prob. **22**, 875–918 (1994).

[BDG] Bolthausen, E., Deuschel, J.-D. and Giacomin, G., *Entropic repulsion for the lattice free field, III the 2-dimensional case,* Preprint (1995).

[BDT] Bolthausen, E., Deuschel, J.-D. and Tamura, Y., *Laplace approximations for large deviations for nonreversible Markov processes.* Ann. Prob. **23**, 236–267 (1995).

[BDZ] Bolthausen, E., Deuschel, J.-D. and Zeitouni, O., *Entropy repulsion for the free lattice field,* Comm. Math. Physics, 417–444, **170** (1995).

[BI] Bolthausen, E. and Ioffe, D., *The harmonic crystal on the wall: A microscopic approach,* Preprint.

[BS1] Bolthausen, E. and Schmock, U., *On the maximum entropy principle for uniformly ergodic Markov chains.* Stoch. Proc. Appl. **33**, 1–27 (1989).

[BS2] Bolthausen, E. and Schmock, U., *Self-attracting d-dimensional random walks,* to appear in Ann. Prob.

[BEF] Bricmont, J., El Mellouki, A. and Fröhlich, J., *Random surfaces in statistical mechanics: roughening, rounding, wetting.* J. Stat. Phys. **42**, 743–796 (1986).

[BS] Brydges, D.C. and Slade, G., *The diffusive phase of a model of self-interacting walks,* Prob. Theory Rel.Fields. **103**, 285–316 (1995).

[C] Csisczar, I., *Sanov property, generlized I-projection and a conditional limit theorem,* Ann. Prob. **12**, 768–793 (1984).

[CDR] de Coninck, J., Dunlop, F. and Rivasseau, V., *On the microscopic validity of the Wulff construction and of the generalized Young equation,* Comm. Math. Phys. **121**, 401–415 (1989).

[DZ] Dembo, A. and Zeitouni, O., *Refinement of the Gibbs conditioning principle,* Prob. Theory and Rel. Fields, **104**, 1–14 (1996).

[D] Deuschel, J.D., *Entropic repulsion of the lattice free field, II. The 0-boundary case* Comm. Math. Phys., to appear.

[DS] Deuschel, J.D. and Stroock, D.W., *Large Deviations,* Academic Press, 1989.

[DKS] Dobrushin, R.L., Kotecky, R., and Shlosman, S., *Wulff Construction: a Global Shape from Local Interaction,* AMS translation series, Vol. 104 (1992).

[DMRR] Dunlop, F., Magnen, J., Rivasseau, V. and Roche. Ph., *Pinning of an interface by a weak potential*, J. Stat. Phys. **66**, 71–98 (1992).

[DV] Donser, M.D. and Varadhan, S.R.S., *Asymptotic evaluation of certain Markov process expectations for large time III*, Comm. Pure Appl. Math. **29**, 389–461 (1976).

[G] Georgii, H.O., *Large deviations and maximum entropy principle for interacting random fields on* $\mathbb{Z}^d$, Ann. Prob. **21**, 1845–1875 (1993).

[H] Hall, R.R., *A quantitative isoperimetric inequality in n-dimensional space,* J. reine angew. Math. **428**, 161–176 (1992).

[I1] Ioffe, D., *Large deviations for the 2D Ising model: a lower bound without cluster expansions,* J. Stat. Phys. **74**, 411–432 (1994).

[I2] Ioffe, D., *Exact large deviation bound up to* T_c *for the Ising model in two dimensions,* Prob. Rel. Fields **102**, 313–330 (1995).

[KMSS] Khanin, K.M., Mazel, A.E., Shlosman, S.M and Sinai, Ya., *Loop condensation effects in the behavior of random walks* The Dynkin Festschrift, Mark I. Freidlin ed., Birkhäuser, 167–184 (1994).

[KT] Kusuoka, S. and Tamura, *Precise estimates for large deviations of Donsker–Varadhan type,* J. Fac. Sc. Uni. of Tokyo **38**, 533 (1991).

[L] Lanford, O.E., *Entropy and equilibrium states in classical mechanics.* in Statistical Mechanics and Mathematical Problems, A. Lenard, ed., Lecture Notes in Physics **20**, Springer 1973.

[LM] Lebowitz, J.L. and Maes, C., *The effect of an external field on an interface, entropy repulsion,* J. Stat. Phys. **46**, 39–49 (1987).

[LRS] Lebowitz, J.L., Rose, H.R. and Speer, E.R., *Statistical mechanics of the nonlinear Schrödinger equation.* J. Stat. Phys. **50**, 657–687 (1988).

[LPS] Lewis, J.T, Pfister,C.-E. and Sullivan, W.G., *Entropy, concentration of probability and conditional limit theorems.* Preprint.

[MMR] Messager, A., Miracle-Solè, S. and Ruiz, J., *Convexity properties of the surface tension and equilibrium crystals,* J. Stat. Phys. **67**, 449–470 (1992).

[P] Pfister, C.-E., *Large deviations and phase separation in the two dimensional Ising model,* Helv. Phys. Acta **64**, 953–1054 (1991).

[S] Schmock. U., *Convergence of one-dimensional Wiener sausage path measure to a mixture of Brownian taboo processes,* Stochastics **29**, 203–220 (1989).

[S1] Sznitman, A.-S., *Long time asymptotics for the shrinking Wiener sausage,* Comm. Pure Appl. Math. **43**, 809–820 (1990).

[S2] Sznitman, A.-S., *On the confinement property of two dimensional Brownian motion among Poissonian obstacles,* Comm. Pure Appl. Math. **44**, 1137 (1991).

[Wi] Winterbottom, W.L., *Equilibrium shape of a small particle in a contact with a foreign substrate,* Acta Metal. **15**, 303–310 (1967).

[Wu] Wulff, G., *Zur Frage der Geschwindigkeit des Wachstums und der Auflösung der Krystallflächen,* Z. Kryst. Mineral. **34**, 449 (1901).

Progress in Mathematics, Vol. 168, © 1998 Birkhäuser Verlag Basel/Switzerland

Renormalization Group for Fronts and Patterns

J. BRICMONT*AND A. KUPIAINEN†

J. Bricmont, UCL, Physique Théorique
B-1348, Louvain-la-Neuve, Belgium
e-mail: bricmont@fyma.ucl.ac.be

A. Kupiainen, Helsinki University, Department of Mathematics
Helsinki 00014, Finland
e-mail: ajkupiai@cc.helsinki.fi

ABSTRACT. We review recent results on the stability of stationary solutions and of moving fronts in several nonlinear parabolic partial differential equations: the Ginzburg-Landau, Swift-Hohenberg and Cahn-Hillard equations, and chemical reaction-diffusion equations. Some of these results are obtained using the Renormalization Group method.

1. Introduction

One of the most interesting aspects of nonlinear partial differential equations is the dynamical emergence of spatial and temporal structures. Namely, some special solutions, like patterns or fronts, exist and govern the asymptotic behaviour of the solution for a large class of initial data. A well-known example is given by the Kolmogorov-Petrovsky-Piskunov (KPP) [33] or Fisher equation [24] that was introduced in a biological context:

$$u_t = u_{xx} + u(1-u) \tag{1.1}$$

where $u : \mathbb{R} \times \mathbb{R} \to \mathbb{R}$. For physical or biological reasons, one is mostly interested in nonnegative solutions. This equation has two constant stationary solutions, $u(x) = 0$ and $u(x) = 1$. It is easy to see that $u(x) = 1$ is stable under small perturbations and that $u(x) = 0$ is (linearly) unstable. One may thus wonder whether there exist solutions interpolating between these stationary solutions, i.e. with $u(-\infty) = 1$, $u(\infty) = 0$. One would expect such solutions to be front-like, with the stationary solution "invading" the unstable one.

Putting as ansatz $u(x,t) = g(x-ct)$ in (1.1), we get:

$$g'' + cg' + g - g^2 = 0 \tag{1.2}$$

which, if we reinterpret the variable $z = x - ct$ as "time", and g as "position", can be seen as Newton's equation of motion of a particle of mass one subjected

*Supported by EC grant CHRX-CT93-0411.
†Supported by NSF grant DMS-9205296 and EC grant CHRX-CT93-0411.

to a friction term cg' and to a force deriving from the potential $V(g) = \frac{g^2}{2} - \frac{g^3}{3}$. It is intuitively clear and easily proved that, for c not too small, solutions exist that satisfy the required conditions, i.e. such that g tends, as "time" goes to $+\infty$, to zero, the stable critical point of the potential, and to one as "time" goes to $-\infty$. More precisely, for each $c \geq 2$, there exists a solution g_c of (1.2) that remains always positive. For $c < 2$, the solution "overshoots" the minimum at zero, i.e. g_c is no longer positive. For large "time" $z = x - ct$, $g_c(z)$ will decay exponentially, as is seen from the linearization of (1.2) around $g = 0$. One gets

$$g_c(z) \leq (C_1 + C_2 z)e^{-\gamma_c z} \tag{1.3}$$

where γ_c is given by $\gamma_c^2 - c\gamma_c + 1 = 0$ i.e.

$$\gamma_c = \frac{1}{2}(c - \sqrt{c^2 - 4}), \tag{1.4}$$

which is real for $c \geq 2$, in which case $\gamma_c \leq 1$ (actually, one can take $C_2 = 0$ in (1.3), if $\gamma < 1$).

The problem of the stability of these fronts has been extensively studied [1, 6]. Each of the solutions g_c with $c \geq 2$ is stable under real perturbations: if we start with initial data $u(x, 0)$, with $u = g_c + s$ with $0 \leq u \leq 1$, s decaying faster than $e^{-\gamma_c x}$ for $x \to +\infty$, $u(x, t)$ will converge, as $t \to +\infty$, to $g_c(x - ct)$, see [1, 6, 17].

However the solution with $c = 2, \gamma_c = 1$ is more stable than the others in the sense that *any* initial data $u(x, 0)$ with $0 \leq u \leq 1$ which decays faster than e^{-x} as $x \to +\infty$ (in particular, if $u(x, 0)$ is of compact support) will converge, as $t \to +\infty$, to $g_2(x - 2t)$ [1, 6]. These results are obtained using the maximum principle for parabolic equations [35], and their extension to equations where this principle does not hold requires new ideas.

As we see in this example, one can ask three types of questions:

a) Existence of "special" solutions (stationary, front-like etc ...). This usually amounts to proving the existence of a solution for an ordinary differential equation.

b) Local stability: if we take as initial data a small perturbation of the special solution, will the resulting solution converge to the special solution as $t \to \infty$? Of course, the answer will depend crucially on the (Banach) space of allowed perturbations: in the example above, s had to decay faster than $e^{-\gamma_c x}$ for $x \to +\infty$.

c) Global stability: can one find a large set of initial data such that the resulting solution converges to the special solution as $t \to \infty$?

Obviously, these are questions of increasing difficulty. In most cases, one can give an answer only to the first two questions. Only in special situations, like the KPP equation above, can one deal with the third question. We shall briefly review here recent results on those questions, but we shall only give these results qualitatively, and refer to the original papers for precise statements.

In a series of papers ([8, 9, 10, 11, 12, 13, 14], see also [5, 26, 39, 41]) we have developed the so-called Renormalization Group approach introduced in [29], where the special solution is viewed as a fixed point of an operator and its stability can be analyzed in term of the stability of that fixed point. For a review of that method, see [7]; the present paper can be viewed as a continuation of [7]. Other approaches use "energy functionals" to study the stability of special solutions [23], and we shall also review some new results based on that method.

2. The Ginzburg-Landau equation

The Ginzburg-Landau equation is:

$$u_t = u_{xx} + u - |u|^2 u \tag{2.1}$$

where $u : \mathbb{R} \times \mathbb{R} \to \mathbf{C}$ is complex. This equation has a two-parameter family of periodic stationary solutions

$$u_{q\theta}(x) = \sqrt{1-q^2}e^{i(qx+\theta)}. \tag{2.2}$$

These solutions (2.2) are stable, under small perturbations, for $q^2 < \frac{1}{3}$ (the Eckhaus stable domain) [19]. We shall call these solutions "spirals". It is a natural question to inquire about the time development of initial data $u(x)$ which approach two such solutions at $\pm\infty$:

$$\lim_{x\to\pm\infty} |u(x) - u_{q_\pm\theta_\pm}(x)| = 0. \tag{2.3}$$

The special case where with u is *real* and $u(-\infty) = 1$, $u(\infty) = 0$ (i.e. $q_- = 0$, $q_+ = 0$) has been extensively studied [1, 6]. It is analogous to the KPP equation (1.1): $u(x) = 1$ is a stable stationary solution, while $u(x) = 0$ is an unstable one, and there exist fronts $g_c(z)$, $z = x - ct$, for $c \geq 2$, interpolating between the two.

We shall consider two questions. The first one concerns the stability of the real front solutions of (2.1) for real or complex u. We shall consider only the critical front, i.e. the one with $c = 2$, which is the most difficult one. The second question was suggested in [20], namely we take $q_\pm$ in (2.3) belonging to the Eckhaus stable domain, but not necessarily equal. What is then the long-time asymptotics of the solution? For reviews on these questions, we refer the reader to [17, 21].

2.1. Stability of Fronts Using the Renormalization Group method, Gallay [26] was able to obtain very precise asymptotics on how a small (real) perturbation of *real fronts* diffuses to zero (this improves previous results of [31, 36]). So Gallay considers (2.1), for u real and takes as initial data $u(x,0) = g_2(x) + s(x)$, with s small. He writes

$$s(z,t) = g_2'(z)w(z,t) \tag{2.4}$$

with $z = x - 2t$, and studies the behaviour of $w(z,t)$ as $t \to \infty$. Since $g_2(z)$ goes exponentially to 1 or 0 as z goes to $-\infty$ or $+\infty$, $g_2'(z)$ will be localized around $z = 0$. The main result is that w has the following universal asymptotics [26]: let $z = \xi\sqrt{t}$, then

$$w(\xi\sqrt{t},t) \simeq At^{-3/2}f^*(\xi) \tag{2.5}$$

where

$$f^*(\xi) = \begin{cases} 1 & \text{if } \xi \leq 0 \\ e^{-\xi^2/4} & \text{if } \xi > 0 \end{cases}$$

and A depends on the initial conditions. The limit (2.5) holds in a weighted $L^1 \cap L^\infty$ norm.

The second type of result deals with complex perturbations. In [10], we write $u(x,t) = (g_2(x) + s(x,t))e^{i\phi(x,t)}$, and consider initial data in the Banach space of C^1-functions ϕ's with the norm

$$\|(\phi, s)\| = \sup_x (1+ \mid x \mid)^{3+\delta}(\mid \phi(x) \mid + \mid \phi'(x) \mid + (1+e^x)(\mid s(x) \mid + \mid s'(x) \mid)), \tag{2.6}$$

for some $\delta > 0$. We show that, for initial data with $\|(\phi, s)\|$ small, the solution $\phi(z,t), s(z,t)$ satisfies

$$\begin{aligned} |\phi(z,t)| &\leq t^{-\frac{1}{2}+\delta} \\ (1+e^z)|s(z,t)| &\leq t^{-1+\delta} \end{aligned}$$

with $z = x - 2t$.

Remark. Eckmann and Wayne [23], using a completely different (and simpler) method, namely coercive functionals, have proven similar results: they can consider a larger set of perturbations (ϕ, s) than the one defined by $\|(\phi, s)\|$ small, but they do not obtain explicit upper bounds on the decay in time.

2.2. Spirals For the second question, we considered in [9] a class of initial data satisfying (2.3) with $q_\pm$, $\theta_\pm$ small, and we showed that, for any interval I,

$$\sup_{x \in I} |u(x,t) - e^{i\sqrt{t}\phi^*} u_{q^*\theta^*}(x)| \leq \frac{C_I}{\sqrt{t}} \tag{2.7}$$

where the constants q^*, ϕ^* and θ^* depend only on the boundary conditions (2.3). For a more detailed bound, see Section 3 of [9]. Graphics of the solution can be found in [21]. This result has recently been extended by Gallay and Mielke [28] to *all* values of $q_\pm$ in the Eckhaus stable domain: $q_+ \neq q_-$, $|q_\pm^2| < \frac{1}{3}$. Note that for $q_+ \neq q_-$, one can always assume $\theta_+ = \theta_-$ by an appropriate choice of the origin of coordinates.

3. Damped hyperbolic equations

In order to incorporate the effect of damping, one modifies (1.1) to:

$$\epsilon u_{tt} + u_t = \partial_x^2 u + u(1-u) \tag{3.1}$$

with $\epsilon > 0$ (other nonlinearities could also be considered). Putting

$$u(x,t) = g(\sqrt{1+\epsilon c^2}x - ct)$$

one obtains that

$$g''(x) + cg'(x) + g(x)(1-g(x)) = 0$$

i.e. equation (1.2).

So (3.1) has front-like solutions moving with speed $\frac{c}{\sqrt{1+\epsilon c^2}}$. Using the method of energy functionals, Gallay and Raugel [27] show that these fronts are locally stable under perturbations in appropriate weighted Sobolev spaces. While this local stability result extends to equations where the space dimension $n \leq 4$, in one dimension they also obtain a global stability result for $\epsilon < 1/4$.

4. Coupled reaction-diffusion equations

4.1. Anomalous diffusion In [3], Berlyand and Xin consider the following model of a chemical reaction $A \to B$ where v is the mass fraction of reactant A and u the one of reactant B:

$$u_t = u_{xx} + vu^{p-1} \tag{4.1}$$

$$v_t = \Lambda^{-1} v_{xx} - vu^{p-1} \tag{4.2}$$

where $u = u(x,t)$, $v = v(x,t)$, $x, t \in \mathbb{R}$, $t \geq 0$, and where the constant $\Lambda > 0$ is the Lewis number. First, let us analyze this system heuristically for initial data $u(x,0)$, $v(x,0)$ that decay at infinity. If u and v are positive (which is physically necessary and is preserved by (4.1, 4.2) due to the maximum principle) then, again by the maximum principle, u is larger than the solution of the heat equation with the same initial data:

$$u(x,t) \geq \frac{A}{t^{1/2}} e^{-\frac{x^2}{4t}}. \tag{4.3}$$

where A depends on $u(x,0)$.

Then, v is less than the solution of (4.2) with u replaced by its lower bound (4.3). Calling $\bar{v}$ this upper bound on v, $\bar{v}$ solves a linear equation

$$\bar{v}_t = \Lambda^{-1} \bar{v}_{xx} - A^{p-1} t^{-\frac{p-1}{2}} e^{-\frac{x^2(p-1)}{4t}} \bar{v}. \tag{4.4}$$

We can distinguish three cases: if $p > 3$, the function $t^{-\frac{p-1}{2}}$ is integrable in time and, using a Renormalization Group analysis of [8], one can show that both u and v tend diffusively to zero as $t \to \infty$. On the other hand, for $p < 3$, using the Feynman-Kac formula, we get

$$v(x,t) \leq \bar{v}(x,t) \leq \exp(-\mathcal{O}(t^{1-\frac{p-1}{2}})) \tag{4.5}$$

for $|x| \leq \mathcal{O}(\sqrt{t})$. For $|x| \geq \mathcal{O}(\sqrt{t})$, one gets a diffusive behaviour, depending on the rate of decay, as $x \to \pm\infty$, of the initial data. Then, inserting the fast decay (4.5) of v in (4.1), one shows that the effect of the nonlinear term in (4.1) is small and that u diffuses to zero. Clearly the borderline case $p = 3$ is the most delicate and the most interesting one.

It is not hard to see, by direct substitution, that, for $p = 3$, (4.4) admits a self-similar solution of the form

$$\bar{v}(x,t) = t^{-\frac{\alpha}{2}} f_\alpha^*(\frac{x}{\sqrt{t}}). \tag{4.6}$$

Indeed, writing $f_\alpha^*(\xi) = e^{-\frac{\Lambda\xi^2}{8}} \Psi(\xi)$, Ψ solves

$$-\Lambda^{-1}\Psi'' + (\frac{\Lambda}{16}\xi^2 + \frac{A^2}{4\pi} e^{-\xi^2/2} + \frac{1}{4})\Psi = \frac{\alpha}{2}\Psi \tag{4.7}$$

We see that requiring f^*_α to be positive means that Ψ is the ground state of the operator in the LHS of (4.7) i.e. of an harmonic oscillator perturbed by a positive, rapidly decaying, potential. The exponent $\frac{\alpha}{2}$ is the corresponding ground state energy. For the harmonic oscillator, $\frac{\alpha}{2} = \frac{1}{2}$ and perturbation theory tells us that, for A small,

$$\frac{\alpha}{2} = \frac{1}{2} + \frac{A^2}{4\pi\sqrt{2\Lambda^{-1}+1}} + h.o.t. \text{ in } A > \frac{1}{2}. \tag{4.8}$$

Thus, α depends on A, i.e. on $u(x,0)$. Since $\alpha > 1$, $\bar{v}$, and hence v, decay strictly faster than the solution of the heat equation. Inserting this in (4.1) gives an upper solution for u, which solves an equation where the effect of the nonlinear term is small.

These considerations lead us to expect u to have the heat equation decay:

$$u(x,t) \simeq \frac{A}{\sqrt{t}} e^{-x^2/4t} \tag{4.9}$$

and v to have the anomalous decay:

$$v(x,t) \simeq \frac{B}{t^{\alpha/2}} f^*_\alpha\left(\frac{x}{\sqrt{t}}\right) \tag{4.10}$$

where A, B and α depend on the initial data, and this is what is proven in [13].

4.2. Front solutions Observe first that, for $\Lambda = 1$ and $p = 2$ in (4.1, 4.2), the function $f = u + v$ satisfies the heat equation. Hence, if we take initial data where $u(x) + v(x) = 1$, this will be preserved in time. Now let $w = \frac{1}{2}(1 + v - u)$; then w solves the KPP equation:

$$w_t = w_{xx} + w(1-w).$$

So, in the special case $\Lambda = 1$, $p = 2$, there are stable front solutions.

Focant and Gallay [25] consider a more general reaction-diffusion system:

$$\begin{aligned} u_t &= u_{xx} + vu + kvu^2, \\ v_t &= \Lambda^{-1} v_{xx} - vu - kvu^2, \end{aligned} \tag{4.11}$$

where u, v are nonnegative functions of $x, t \in \mathbb{R}$, $t \geq 0$, and $\Lambda > 0$, $k \geq 0$ are constant parameters. They investigate the existence and stability of front solutions for this system . These are uniformly translating solutions connecting the stable stationary solution $(u,v) = (1,0)$ at $x = -\infty$ to the unstable one $(u,v) = (0,1)$ at $x = +\infty$. Let us look for solutions of the form $u(x,t) = \alpha(x-ct)$, $v(x,t) = \beta(x-ct)$, where $c > 0$ is the velocity of the front; the nonnegative functions α, β satisfy the ODE system

$$\begin{aligned} \alpha''(x) + c\alpha'(x) + \beta\alpha + k\beta\alpha^2 &= 0, \\ \Lambda^{-1}\beta''(x) + c\beta'(x) - \beta\alpha - k\beta\alpha^2 &= 0 \end{aligned} \tag{4.12}$$

together with the boundary conditions

$$(\alpha(-\infty), \beta(-\infty)) = (1,0) \ , \quad (\alpha(+\infty), \beta(+\infty)) = (0,1) \ . \tag{4.13}$$

The existence of solutions of (4.12, 4.13), has been studied by Billingham and Needham [4] and in [25] for all $\Lambda > 0$ and $k \geq 0$. Among other results, they show that, for $0 \leq k \leq 1$ and $\Lambda > 0$, there is a nonnegative solution to (4.12, 4.13) if and only if $c \geq 2$. To study the stability of these propagating fronts, Focant and Gallay consider the situation where Λ is close to 1 and k is close to 0. Setting

$$u(x,t) = \alpha(x-ct) + f(x-ct,t)\ , \quad v(x,t) = \beta(x-ct) + g(x-ct,t)\ , \tag{4.14}$$

and inserting this into (4.11), one obtains the following system for the evolution of the perturbation (f,g) in the frame of the front:

$$\begin{aligned} f_t &= f_{xx} + cf_x + \beta(1+2k\alpha)f + \alpha(1+k\alpha)g + N(f,g)\ , \\ g_t &= \Lambda^{-1}g_{xx} + cg_x - \alpha(1+k\alpha)g - \beta(1+2k\alpha)f - N(f,g) \end{aligned} \tag{4.15}$$

where $N(f,g) = fg + k(2\alpha fg + \beta f^2 + f^2 g)$.

The stability of propagating fronts with velocity $c \geq 2$ for Λ close to 1 and k close to 0 is shown by using energy functionals: if the initial value of the perturbation (f,g) is small enough (in an appropriate Hilbert space), they obtain:

$$\lim_{t\to\infty} \sup_{x\in\mathbb{R}} (|f(x,t)| + |g(x,t)|)(1+e^{\gamma x}) = 0 \tag{4.16}$$

with $\gamma = \frac{1}{2}(c - \sqrt{c^2-4})$.

5. The Swift-Hohenberg equation

The Swift-Hohenberg equation [40] is used to describe pattern formation in Bénard's problem:

$$U_t = -(1+\partial_x^2)^2 U + \epsilon^2 U - U^3 \tag{5.1}$$

where $U : \mathbb{R}\times\mathbb{R} \to \mathbb{R}$ and $t \geq 0$. Substituting $U(x,t) = \frac{\epsilon}{\sqrt{3}} Re\left(u(\frac{\epsilon x}{2}, \epsilon^2 t)e^{ix} \right)$ and keeping (formally) the leading order in ϵ, one gets:

$$u_t = u_{xx} + u - |u|^2 u$$

i.e. the Ginzburg-Landau equation (2.1) which therefore appears as an amplitude equation for the Swift-Hohenberg equation. The solution of the Ginzburg-Landau equation approximates well the one of the Swift-Hohenberg equation over large but finite times (for t of order ϵ^{-2}), see [18, 32, 37, 38, 42]). For $\epsilon^2 < 0$, $U = 0$ is a stable stationary solution of (5.1), which becomes linearly unstable for $\epsilon^2 > 0$ and wave numbers close to 1. In fact, for $\epsilon^2 > 0$, (5.1) has stationary periodic solutions, which are analogous to the spiral solutions (2.2) of the Ginzburg-Landau equation: for any $\alpha \in \mathbb{R}$, there exists $c(\alpha)$ such that, if $0 < \epsilon^2 - (1-\alpha^2)^2 \leq c(\alpha)$, then there exists a periodic solution of (5.1):

$$u_0(x) = \pm\frac{2}{\sqrt{3}}\sqrt{\epsilon^2 - (1-\alpha^2)^2}\cos\alpha x + 0(\epsilon^2). \tag{5.2}$$

See [17] for a proof. (Putting $\alpha = 1 + \epsilon q/2$ and $\theta = 0$, this solution coincides, to leading order in ϵ, with (2.2)). Although it has been known for some time when these solutions are (marginally) linearly stable [16, 17], results on their (nonlinear) stability have been obtained only recently by Schneider [39]. Combining an analysis of (5.1) linearized around the periodic solution (5.2) and the Renormalization Group method, Schneider shows that, for ϵ small, initial data of the form $u_0(x) + v(x)$, with $u_0(x)$ linearly stable and $v(x)$ small in a suitable norm, lead to a solution $u_0(x) + v(x,t)$ where $v(x,t)$ diffuses to zero as $t \to \infty$.

6. The Cahn-Hillard equation

There is also a one-parameter family of "kink" stationary solutions of the Ginzburg-Landau equation: $u_0(x) = \tanh(\frac{x}{\sqrt{2}})$, and its translates $\tanh(\frac{x+x_0}{\sqrt{2}})$. If we write, for u real, $u(x,t) = u_0(x) + v(x,t)$, and then linearize (2.1) around u_0, we get:

$$v_t = Hv \tag{6.1}$$

where

$$H = -\partial^2 + 1 - 3\cosh^{-2}\left(\frac{x}{\sqrt{2}}\right) \tag{6.2}$$

is an operator whose ground state (proportional to $u_0'(x) = -\frac{1}{\sqrt{2}}\cosh^{-2}\left(\frac{x}{\sqrt{2}}\right)$) is isolated from the rest of the spectrum. Using this fact, it is not hard to show that, for initial data of the form $u_0(x) + v(x)$, with $v(x)$ small, the resulting solution $u(x,t) \to u_0(x + x_0)$ as $t \to \infty$, for some x_0 determined by v [30].

The Cahn-Hillard equation [15], which is often used to study the phase separation in alloys and fluids:

$$\partial_t u = -\partial_x^2(\partial_x^2 u + u - u^3),$$

admits, of course, also $u_0(x) = \tanh\left(\frac{x}{\sqrt{2}}\right)$ as a stationary solution. However, the local stability of $u_0(x)$ for this equation is much more delicate. Let us write the initial data as $u_0(x) + \partial_x v(x)$, with $v(x)$ smooth, decaying to zero at infinity and small (if we choose appropriately the origin of coordinates, then one may always assume that the integral of the perturbation vanishes, hence one may write it as $\partial_x v(x)$ with $v(x)$ as above). Then it is proven in [14] that the solution equals $u_0(x) + \partial_x v(x,t)$ with $v(x,t)$ diffusing to zero as $t \to \infty$.

Acknowledgments. We would like to thank P. Collet, T. Gallay, S. Focant, and J. Taskinen for interesting discussions. This work was supported by NSF grant DMS-9205296, and by EC grant CHRX-CT93-0411.

References

[1] D.G. Aronson, H.F. Weinberger, *Multidimensional non-linear diffusion arising in population genetics*, Adv. Math. **30** (1978), 33–76.

[2] G.I. Barenblatt, *Similarity, Self-similarity and Intermediate Asymptotics*, Cambridge Univ. Press, Cambridge 1996.

[3] L. Berlyand, J. Xin, *Large time asymptotics of solutions to a model combustion system with critical nonlinearity*, Nonlinearity, **8** (1995), 161–178.

[4] J. Billingham, D.J. Needham, *The development of travelling waves in quadratic and cubic autocatalysis with unequal diffusion rates. I. Permanent form travelling waves*, Phil. Trans. R. Soc. Lond. A**334** (1991), 1–24.

[5] J. Bona, K. Promislow, G. Wayne, *On the asymptotic behaviour of solutions to nonlinear, dispersive, dissipative wave equations*, J. Math. and Computers in Simulation, **37** (1994), 264–277; and, *Higher-order asymptotics of decaying solutions of some nonlinear, dispersive, dissipative wave equations*, Nonlinearity, **8** (1995), 1179–1206.

[6] M. Bramson, *Convergence of solutions of the Kolmogorov equation to traveling waves*, Memoirs of the Amer. Math. Soc., **44** (1983), nr. 285, 1–190.

[7] J. Bricmont, A. Kupiainen, *Renormalizing partial differential equations, in Constructive Physics*, p. 83–115, V. Rivasseau (ed), Proceedings, Palaiseau, France 1994, Lecture Notes in Physics, Springer-Verlag, Berlin 1995.

[8] J. Bricmont, A. Kupiainen, G. Lin, *Renormalization group and asymptotics of solutions of nonlinear parabolic equations*, Comm. Pure. Appl. Math., **47** (1994), 893–922.

[9] J. Bricmont, A. Kupiainen, *Renormalization Group and the Ginzburg-Landau equation*, Commun. Math. Phys., **150** (1992), 193–208.

[10] J. Bricmont, A. Kupiainen, *Stability of moving fronts in the Ginzburg-Landau equation*, Commun. Math. Phys., **159** (1994), 287–318.

[11] J. Bricmont, A. Kupiainen, *Universality in blow-up for nonlinear heat equations*, Nonlinearity, **7** (1994), 1–37.

[12] J. Bricmont, A. Kupiainen, *Stable non-Gaussian diffusive profiles*, Nonlinear Analysis, Theory, Methods and Applications, **26** (1996), 583–593.

[13] J. Bricmont, A. Kupiainen, J. Xin, *Global large time self-similarity of a thermal-diffusive combustion system with critical nonlinearity*, J. Diff. Eqns. **130** (1996), 9–35.

[14] J. Bricmont, A. Kupiainen, J. Taskinen, *Stability of Cahn-Hilliard fronts*, preprint.

[15] J.W. Cahn, J.I. Hilliard, *Free energy of a nonuniform system. 1. Interfacial free energy*, J. Chem. Phys. **28** (1958), 258–267.

[16] P. Collet, *Nonlinear dynamics of extended systems; and Leçons sur les systèmes étendus.* Lecture Notes.

[17] P. Collet, J.-P. Eckmann, *Instabilities and fronts in extended systems*, Princeton Univ. Press 1990.

[18] P. Collet, J.-P. Eckman, *The time-dependent amplitude equation for the Swift-Hohenberg problem*, Comm. Math. Phys., **132** (1990), 139–153.

[19] P. Collet, J.-P., Eckmann, H. Epstein, *Diffusive repair for the Ginzburg-Landau equation*, Helv. Phys. Acta, **65** (1992), 56–92.

[20] P. Collet, J.-P. Eckmann, *Solutions without phase-slip for the Ginzburg-Landau equation*, Commun. Math. Phys. **145** (1992), 345–356.

[21] P. Collet, J.-P. Eckmann, *Space-time behaviour in problems of hydrodynamic type: a case study*, Nonlinearity, **5** (1992), 1265–1302.

[22] M. Cross, P. Hohenberg, *Pattern formation outside of equilibrium*, Rev. Mod. Phys. **65** (1993), 851–1112.

[23] J.-P. Eckmann, C.E. Wayne, *Non-linear stability of front solutions for parabolic partial differential equations*, Commun. Math. Phys., **161** (1994), 323–334.

[24] R.A. Fisher, *The advance of advantageous genes*, Ann. of Eugenics, **7** (1937), 355–369.

[25] S. Focant, T. Gallay, *Existence and stability of propagating fronts in an autocatalytic reaction-diffusion system*, preprint.

[26] T. Gallay, *Local stability of critical fronts in nonlinear parabolic partial differential equations, Nonlinearity*, **7** (1994), 741–764; *Existence et stabilité des fronts dans l'équation de Ginzburg-Landau à une dimension*, Thèse, Univ. de Genève, Genève 1994.

[27] T. Gallay, G. Raugel, *Stability of travelling waves for a damped hyperbolic equation*, to appear in ZAMP (1997).

[28] T. Gallay, A. Mielke, in preparation.

[29] N. Goldenfeld, O. Martin, Y. Oono, *Intermediate asymptotics and renormalization group theory*, J. Sci. Comp. **4** (1989), 355–372;
N. Goldenfeld, O. Martin, Y. Oono, F. Liu, *Anomalous dimensions and the renormalization group in a nonlinear diffusion process*, Phys. Rev. Lett., **64** (1990), 1361–1364;
N. Goldenfeld, O. Martin, Y. Oono, *Asymptotics of partial differential equations and the renormalization group*, in: Proc. of the NATO ARW on Asymptotics beyond all orders, H. Segur, S.Tanveer, H. Levine, eds, Plenum (1991);
N. Goldenfeld, *Lectures on phase transitions and the renormalization group*, Addison-Wesley, Reading (1992);
L-Y. Chen, N. Goldenfeld, Y. Oono, *Renormalization group theory for global asymptotic analysis*, Phys. Rev. Lett. **73** (1994), 1311–1315.

[30] D. Henry, *Geometric Theory of Semilinear Parabolic Equations*, Lecture Notes in Mathematics **840**, Springer-Verlag 1981.

[31] K. Kirchgässner, *On the nonlinear dynamics of travelling fronts*, J. Diff. Eqns, **96** (1992), 256–278.

[32] P. Kirrmann, G. Schneider, A. Mielke *The validity of modulation equations for extended systems with cubic nonlinearities*, Proceedings of the Royal Society of Edinburgh **122A** (1992), 85–91.

[33] A.N. Kolmogorov, I.G. Petrovskii, N.S. Piskunov, *Etude de la diffusion avec croissance de la quantité de matière et son application à un problème biologique.* Moscow Univ. Math. Bull., **1** (1937), 1–25.

[34] R. Pego, M.I. Weinstein, *Asymptotic stability of solitary waves*, Commun. Math. Phys., **164** (1994), 305–349.

[35] M.H. Protter, H.F. Weinberger, *Maximum principles in partial differential equations*, Appl. Math. Sci. **44**, Springer, New York 1983.

[36] D.H. Sattinger, *Weighted norms for the stability of travelling waves*, J. Diff. Eq. **25** (1977), 130–144.

[37] G. Schneider, *A new estimate for the Ginzburg-Landau approximation on the real axis*, J. Nonlinear Science **4** (1994), 23–34.

[38] G. Schneider, *Error estimates for the Ginzburg-Landau approximation.* J. Appl. Math. Physics (ZAMP) **45** (1994), 433–457.

[39] G. Schneider, *Diffusive Stability of spatial periodic solutions of the Swift-Hohenberg equation*, Commun. Math. Phys., **178** (1996), 679–702.

[40] J. Swift, P.C. Hohenberg, *Hydrodynamic fluctuations at the convective instability*, Phys. Rev. A**15** (1977), 319–328.

[41] J. Taskinen, *Diffusion equation with general polynomial perturbation*, to appear in Differential and integral equations.

[42] A. van Harten, *On the validity of Ginzburg-Landau's equation.* J. Nonlinear Science **1** (1991), 397–422.

Progress in Mathematics, Vol. 168, © 1998 Birkhäuser Verlag Basel/Switzerland

Geometry of Tori: Riemannian versus Finsler?

Dmitri Burago*

The Pennsylvania State University, Department of Mathematics
McAllister Building, University Park, PA 16801, USA
and also
St. Peterburg, Russia

This is the text of the lecture which I would have given if I had been able to attend the Congress. I would like to thank sincerely the organizers for their kind invitation.

This lecture is based on the results obtained jointly with S. Ivanov. We also owe inspiration and ideas to many other mathematicians, whose incomplete list includes F. Nazarov, V. Bangert, M. Gromov, B. Kleiner and C. Croke. The purpose of this lecture is to present a comprehensible account of several results around the Hopf conjecture and to stress open problems. For the sake of simplicity, we will sacrifice generality and sweep non-essential technical difficulties under the carpet.

I believe that the difference between Riemannian and Finsler geometries (while our questions are formulated in purely synthetic language) is far from being understood. Although the Hopf conjecture represents a statement which has essentially different answers for these two geometries, this may just be due to an awkward way of formulating it. We will begin with an outline of the proofs of Hopf's conjecture [7](stating that a Riemannian torus without conjugate points is flat), and asymptotic volume estimate [8] (that the ratio of volumes of big balls of the same radius in the universal cover of a Riemannian torus and in standard R^n converges to a number not less than 1). We want to emphasize that the whole argument, except at one particular point, works just as well for Finsler metrics also, and we will stress the only point where the assumption that we are dealing with a *Riemannian* metric plays its crucial role.

We will be dealing with a Riemannian torus $\mathbf{T}^n$ which can be regarded as $\mathbf{R}^n/\mathbf{Z}^n$, where R^n is also supplied with some other $\mathbf{Z}^n$-invariant metric d. (Certainly, the vector structure of $\mathbf{R}^n$ is non-invariant and the choice of Euclidean structure is arbitrary.) An important *stable* (that is, changing in an obvious way when passing to a finite cover) invariant of $\mathbf{T}^n, d$ is its *stable norm*, a (unique) Banach norm $||\cdot||$ on $\mathbf{R}^n$ satisfying $|||x-y||-d(x,y)|\leq c$ for some constant c and every $a, b\in \mathbf{R}^n$. It is quite easy to show its existence for a metric d without conjugate points, as well as for a relaxed estimate $|||x-y||-d(x,y)|\leq o(d(x,y))$, while the general case is somewhat tricky [5]. We will fix the notation B and F for

*The author is partially supported by a Sloan Foundation Fellowship and NSF Grant DMS-95-05175.

the unit ball and sphere of $||\cdot||$ respectively. Presumably, one can recover important information about d from $||\cdot||$, in particular, its smoothness is closely related to the structure of certain sets of minimizing geodesics. This group of problems remains wide open and might be a topic for another lecture [4],[9]. It is also not clear whether there is any difference between Finsler and Riemannian metrics in this respect.

A natural candidate for a coordinate function in the Banach space $(\mathbf{R}^n, ||\cdot||)$ is a linear function L with (dual) norm 1, that is a function supporting B. For such L, (similar to the standard construction of Busemann functions), one can construct a Lipschitz-1 (w. r. t. d) function $\tilde{L}$ (generalized coordinate) such that the difference $L-\tilde{L}$ is Z^n-invariant and bounded (by absolute value) by $2c$. Now, a naive approach to prove, say, the asymptotic volume estimate, is to choose n functions $\tilde{L}_i$, such that the parallelepiped formed by $\tilde{L}_i^{-1}(1), \tilde{L}_i^{-1}(-1)$ has minimum volume, and work in these coordinates. Unfortunately, such a choice of coordinates gives too rough estimate [2].

Among all quadratic forms given by a finite sum $Q = \sum a_i L_i^2$ (which form a finite-dimensional space), where L_i support B and $\sum a_i = n, a_i > 0$, consider the one with the smallest volume of its unit ball. Simple differentiation shows that the unit ball of this quadratic form Q is contained in B. Actually, it is the maximum-volume ellipsoid inscribed in B. We reserve the notations a_i and $\tilde{L}_i$, $i = 1, 2 \ldots N$ for those coefficients and corresponding generalized coordinates.

With this notation, the volume growth estimate becomes quite transparent. This is a variation of Derrik's proof of the Besikovitch inequality [10]. A ball of radius $R + 2c$ in $(\mathbf{R}^n, d)$ contains $R \cdot B$, which in its turn contains a ball of radius R in $\mathbf{R}^n, Q$. A map $I : (\mathbf{R}^n, Q) \to \mathbf{R}^N$ given by $(\sqrt{a_1}L_1, \sqrt{a_2}L_2, \ldots \sqrt{a_N}L_N)$ is a linear isometry. A map

$$\tilde{I} = (\sqrt{a_1}\tilde{L}_1, \sqrt{a_2}\tilde{L}_2, \ldots \sqrt{a_N}\tilde{L}_N)$$

is volume non-increasing, due to $|grad\tilde{L}_i|_d \leq 1$, since $\tilde{L}_i$ are Lipschitz-1 functions (**just here we use that d is Riemannian!**), and the choice of Q. Thus the composition of $\tilde{I}$ and the orthogonal projection on the image $im(I)$ is also volume non-increasing.

However, the image $I(Ball_R)$ is a disc of the same radius in an n-dimensional subspace of $\mathbf{R}^N$, and the maps I and $\tilde{I}$ remain within distance $2c\sqrt{n}$. Thus the projection of $\tilde{I}(B)$ contains an n-ball of the radius at least $R - 2c\sqrt{n}$, and this completes the proof.

To prove the Hopf conjecture, we need to introduce more notation. For a unit (w. r. t. d) tangent vector $(x, v) \in U\mathbf{T}^n$, consider a lift $\gamma(t)$ of the geodesic $exp_x(tv)$ onto $\mathbf{R}^n$, and denote its rotation vector $D(x, v) = \lim \frac{1}{t}(\gamma(t) - \gamma(0))$. Due to the Birkhoff Ergodic theorem, D is correctly defined almost everywhere. By the definition of $||\cdot||$, $||D(x, v)|| \leq 1$, and it exactly equals 1 if d has no conjugate points (and hence every geodesic minimizes the distance.) Hence D maps $D : UT^n \to F$. Denote by $m = D(mes)$ the image of the normalized Liouville measure mes.

The key observation belongs to the elementary geometry of convex bodies. Let us assume that for a probability measure μ on F and every linear function L supporting F, $\int_F L^2 d\mu \leq \frac{1}{n}$. Then, adding these inequalities for L_i with coefficients

a_i, one concludes that these inequalities are actually equalities and the support of μ belongs to the unit sphere of Q.

We want to study how a generalized coordinate $\tilde{L}$ grows along γ. Let $v = grad(\tilde{L})$. Since $L - \tilde{L}$ is Z^n-invariant, the vector field v is also correctly defined on the torus. We have

$$L(D(x,v)) = \lim \frac{1}{t} L(\gamma(t)) - L(\gamma(0)) = \lim \frac{1}{t} (\tilde{L}(\gamma(t) - \tilde{L}(\gamma(0)) = \lim \frac{1}{t} \int_0^t \langle v, \frac{d\gamma}{dt} \rangle dt$$

We want to apply the Birkhoff Ergodic Theorem, however, integrating the above formula would give 0 due to the symmetry of D. To avoid this, we first apply the Cauchy inequality:

$$L^2(D(x,v)) \leq \lim \frac{1}{t} \int_0^t \langle v, \frac{d\gamma}{dt} \rangle^2 dt.$$

Now the Ergodic Theorem yields

$$\int_F L^2(p) dm(p) \leq \int_{U\mathbf{T}^n} \langle v(x), w \rangle^2 dmes(x,w).$$

Note that the normalized Liouville measure mes is the standard probability measure s on each fiber $S^{n-1} = U_x T^n$, and $|v_x| \leq 1$. Since the integral $\int_{S^n} \langle v, w \rangle^2 ds(w) = \frac{v^2}{n}$, (just add those integrals for orthogonal vectors v, v_1, ..., v_{n-1} of equal length, **this is just where we use the fact that d is Riemannian!**), fiber-wise integrations gives $\int_F L^2(p) dm(p) \leq 1/n$. Adding these formulae for L_i with coefficients a_i, we have $\int_F Q(p) dm(p) \leq 1$. Recalling that the unit ball of Q lies inside F, one concludes that the integrand is at least 1 and all the above inequalities are actually equalities. If we had the Cauchy inequality on a finite interval, it would immediately tell us that vector fields v form constant angles with all geodesics, which implies flatness. In our case, we have the asymptotic Cauchy inequality, but, taking into account Z^n-invariance of v, one shows that this angle is constant almost everywhere, which again implies flatness.

Loosely speaking, the gist of the proof is that we noticed some (integral-type) extremal property of (the uniform measure on) ellipsoids among other surfaces of central-symmetric convex bodies. Then we swept this measure together with this integral invariant from fibers of the unit tangent bundle onto the sphere of the stable norm by geodesic flow. It shows that this sphere is also an ellipsoid (or, more precisely, the support of m belongs to an ellipsoid), and whole the structure is very rigid. However, the absence of conjugate points is a purely dynamical condition on a Lagrangian system on a torus. It implies, in particular, that the geodesic flow admits a so-called Heber's foliation [21] and it is C^∞-conjugate to the geodesic flow of a flat torus on invariant tori filled by trajectories with rational rotation numbers (just those for which all difficulties occur in KAM theory). One may conjecture that the geodesic flow on a Finsler torus without conjugate points is smoothly conjugate to that of a flat one, and all known examples confirm this guess. An affirmative solution of this conjecture, in particular, would provide another proof of the Hopf conjecture by a result of C. Croke and B. Kleiner.

The situation with asymptotic volume growth is even more intriguing. All known examples of Finsler metrics without conjugate points also exhibit standard volume growth. On the other hand, it is difficult to imagine that there are Finsler metrics with slower volume growth, but none of them are without conjugate points. Indeed, one can decrease the metric in the direction of a geodesic with conjugate points without influencing the stable norm and therefore only decreasing volume growth. (However, a rigorous proof may require quite tedious considerations). By means of a (more or less standard, since the works of M. Gromov [19]) trick with embedding into (finite-dimensional approximation of) l^∞ using generalized coordinates, one can show that the volume growth inequality for Finsler tori is equivalent to the following problem: are (regions of) linear subspaces in finite-dimensional Banach spaces minimal surfaces? Note that we mean this minimality (as well as volume growth) w. r. t. any *sensitive* (i. e. strictly monotonous w. r. t. metric) measure, for instance, Hausdorff measure, or the projection of the Liouville measure (however, Gromov's *mass* is not sensitive.) Formulated in the most classical way, this problem is still open for both mentioned measures, in spite of this Busemann had already mentioned [12] that the volume form (for the Liouville measure) fails to be convex on the Grassmannian cone in the exterior algebra, (regardless of the fact that it is locally convex). One may even wonder what is the minimum Hausdorff measure of a ball for intrinsic metrics that majorize the standard Euclidean distance function on the boundary sphere, and then further about semi-continuity of the Hausdorff measure w. r. t. Gromov-Hausdorff convergence with fixed topology (of a ball).

Added in proof. The measure on a Finsler manifold obtained as the projection of Liouville (symplectic) measure from the set of co-tangent vectors whose norm is less than one will be called Liouville volume (Liouville area in dimension two). It has recently been shown that both the volume growth theorem and the minimality of affine 2-dimensional subspaces in Banach spaces are true for Liouville area (for 2-dimensional tori and surfaces). To emphasize that minimality of affine subspaces should not be perceived as an "intuitively obvious statement" which only requires a formal proof, we mention the following example. There exists a circle γ in a linear 2-plane of a 4-dimensional Banach space and a 2-cycle α whose boundary is $\partial\alpha = 10\gamma$ and whose Liouville area is less than 10 times the Liouville area of the affine disc enclosed by γ.

For a surface in Euclidean space, its Gaussian map induces a measure on the Grassmannian manifold, and the latter is embedded in the exterior algebra. If the boundary of the surface belongs to an affine plane then this measure satisfies an obvious linear integral identity. There is a close connection between other (non-linear) constraints on this measure and the minimality of affine planes in Banach spaces. In particular, the existence of such constraints is implied by the existence of Banach norms whose Liouville area form is not globally convex on the Grassmannian cone, and still affine planes minimize the Liouville area.

References

[1] A. Avez. Variétés riemanniennes sans points focaux. C.R.A.S. A-B 270 (1970) A 188–191.
[2] I. Babenko. Asymptotic volume of tori and geometry of convex bodies. Mat. Zametki, **44** (2), 1988.
[3] V. Bangert. Minimal Geodesics. Erg. Theory and Dyn. Syst. **10**, 1990, 263–286.
[4] V. Bangert. Geodesic rays, Busemann functions and monotone twist maps. Calc. Var. **2**, 49–63 (1994).
[5] D. Burago. Periodic Metrics. Advances in Soviet Math., New York, **9**, 205–210 (1992).
[6] D. Burago. Periodic Metrics. In "Seminar on Dynamical Systems", 90–96, in ser. "Progress in Nonlinear Differential Equations", vol. 12, H. Brezis editor, Birkhäuser, 1994.
[7] D. Burago and S. Ivanov. Riemannian tori without conjugate points are flat. GAFA **4:3**, 1994, 259–269 (1994).
[8] D. Burago and S. Ivanov. On asymptotic volume of tori. GAFA **5:5**, 800–808 (1995).
[9] D. Burago, S. Ivanov and B. Kleiner. On the structure of the stable norm of periodic metrics. Submitted to Research Letters.
[10] Yu. Burago, V. Zalgaller. Geometric Inequalities. Springer-Verlag 1988.
[11] H. Busemann. The Geometry of Geodesics. Acad. Press, New York, 1955.
[12] H. Busemann, G Eward and G Shephard. Convexity on Grassmannian Cones. Annals of Math, 151.
[13] C. Croke. Volumes of balls in manifolds without conjugate points. Int. J. Math. , **3**, no. 4, 455–467 (1992).
[14] C. Croke, A. Fathi. An inequality between energy and intersections. Bull. London Math. Soc., **22** 489–494 (1990).
[15] C. Croke and B. Kleiner. On tori without conjugate points. Bull. London Math. Soc. **22**, 489–494 (1990).
[16] L. Green. A theorem of E. Hopf. Mich. Math. J., **5** 31–34 (1958).
[17] M. Gromov. Dimension, non-linear spectra and width. Lecture Notes in Mathematics, vol. 1317, 132–184.
[18] M. Gromov. Asymptotic Invariants of Infinite Groups, Geometric Group Theory vol. 2, London Math. Society Lecture Notes 182, 1993.
[19] M. Gromov. Filling Riemannian manifolds. J. Diff. Geom. **18**, 1–147 (1983).
[20] M. Gromov, J. Lafontaine, P. Pansu. Structure metriques pour les varietés Riemanniennes, CEDIC\Fernand Math, Paris 1981.
[21] Jens Heber. On the geodesic flow of tori without conjugate points. Math. Z. **216**, 209–216 (1994).
[22] G. Hedlund, M. Morse. Manifolds without conjugate points. Trans. Amer. Math. Soc., **51**, 362–386 (1942).
[23] E. Hopf. Closed Surfaces Without Conjugate Points. Proc. Nat. Acad. of Sci. **34**, (1948).
[24] A. Knauf. Closed orbits and converse KAM Theory. Nonlinearity, **3**, 961–973 (1990).

Progress in Mathematics, Vol. 168, © 1998 Birkhäuser Verlag Basel/Switzerland

Counting Curves on Surfaces: A Guide to New Techniques and Results

LUCIA CAPORASO

Mathematics Department, Massachusetts Institute of Technology
Cambridge MA 02139, USA
e-mail: caporaso@severi.mit.edu

1. Introduction

1.1. Abstract and summary. A series of recent results solving classical enumerative problems for curves on rational surfaces is described. Motivation to the subject came from recent ideas from quantum field theory leading to the definition of quantum cohomology. As a by-product, formulas enumerating rational curves on certain varieties were derived from the properties of certain generating functions representing the free energy of certain topological field theories. A mathematically acceptable construction of quantum cohomology came soon afterwards ([RT] and [KM]-[K]). Different proofs of some of these formulas were provided later (in [CH1] and [CH2]) using different methods that could be generalized to cases (such as Hirzebruch surfaces) for which the quantum cohomology theory did not give enumerative results. For higher genera, the connection between enumerative geometry and quantum cohomology or quantum field theory is still largely conjectural. On the other hand, a recursive formula enumerating plane curves of any genus has recently been proved using purely algebro-geometric techniques ([CH3]). Moreover a generating function exists together with a differential equation implying such a recursion ([G]).

The enumerative problem is precisely stated in the introduction.

The second section contains a short description of the relation with Quantum Cohomology and Kontsevich's formula for plane rational curves. A discussion of enumerative problems for Hirzebruch surfaces concludes this part, which is entirely dedicated to rational curves.

The general case of plane curves of any genus is described in the third chapter, focussing on the results of [CH3]. The main recursive formula of that paper is explained, together with an outline of the proof and a description of the generating function found in [G]. At the end, there is a discussion of generalizations of the procedure of [CH3] to other varieties.

Sections 2 and 3 are completely independent from each other.

In the fourth and last section various enumerative techniques are applied to the concrete example of counting rational plane cubics through 8 general points.

1.2. Terminology. We work with complex, projective algebraic varieties. Let S be a smooth, minimal rational surface (that is, S is either the projective plane $\mathbb{P}^2$ or a Hirzebruch surface $\mathbb{F}_n$) and let D be a curve on S. The linear system $|D|$ of all curves linearly equivalent to D is a projective space whose dimension is given by the Riemann-Roch theorem (if S is the plane then $|D|$ is the $\frac{d(d+3)}{2}$-dimensional projective space of all curves of degree $d = \deg D$). Moreover the genus $p_a(D)$ of a smooth curve in $|D|$ is constant, computable by the adjunction formula (equal to $\binom{d-1}{2}$ for a smooth plane curve of degree d).

We study the geometry of certain closed subvarieties of $|D|$, the so-called "Severi varieties". The following notation is used throughout the paper. Severi varieties are denoted by the symbol "V", suitably decorated; for example, by $V_g(D)$ we denote the Severi variety defined to be the closure of the locus of all irreducible curves in $|D|$ having fixed geometric genus g (see below). All of these Severi varieties have pure dimension (that is, all of their irreducible components have the same dimension); we use a decorated "r" for the dimension of V (equally decorated); for example, $r_g(D) := \dim V_g(D)$. Analogously, we use a decorated "N" for the degree of V as a subvariety of $|D|$, for example, $N_g(D) := \deg V_g(D)$.

1.3. The problem. For a given integer g (with $0 \leq g \leq p_a(D)$ to avoid trivial cases), consider $V_g(D) \subset |D|$ defined above. Also, we denote by $V_g(d)$ the variety of irreducible plane curves of genus g and degree d. These varieties were first introduced by Severi for $\mathbb{P}^2$ (in [S]) and have been object of much study. For example, it is known that their general point represents a curve with $\delta = p_a(D) - g$ nodes (and no other singularities, a fortiori), and that their dimension satisfies $r_g(D) = -(K_S \cdot D) + g - 1 = \dim |D| - \delta$, where K_S is the canonical class of S.

This is to say that for any $r_g(D)$ general points on the surface S there is a finite number of curves of $|D|$ having geometric genus g and passing through such points. Such a number is the degree of $V_g(D)$ as a subvariety of $|D|$ and it is precisely what we want to compute. More generally, we will consider a larger class of Severi varieties and we will describe various ways to arrive at formulas for their degrees.

The simplest example is that of the plane, where we are asking:

Question. *How many plane curves of degree d and genus g pass through $3d+g-1$ general points?*

Until recently there was an explicit formula answering this classical problem, only in some special cases, namely for plane curves having few nodes (up to 6 nodes, as far as the author knows; due to [KP] and [Va]; see [DI] for a list of formulas). For example, it is not hard to deal with curves with only one node, and to show that for each d there are $3(d-1)^2$ of them through the appropriate number of points (see 4.1).

An interesting recursive procedure to compute $N_g(d)$ in general was suggested in [$R1$], (but notice that the formula there is not correct, see also [R2] and [Ch]). Ran constructs a family where the plane degenerates to a reducible surface (called a "fan"). Correspondingly he gets a family of Severi varieties of which he studies the flat limit. This procedure can be viewed as a recursion because such a degeneration of $\mathbb{P}^2$ to a reducible surface induces a family of irreducible curves specializing to a reducible one.

In fact, the common part to most ways of approaching these problems is the use of a technique where the irreducible curves that one is trying to enumerate degenerate to reducible curves, so that one obtains an inductive formula.

2. Rational curves

2.1. Relation with Quantum Cohomology. Interest and enthusiasm for these problems was revived by recent ideas from quantum field theory which led to various enumerative predictions for rational curves on varieties.

It was proposed by Gromov to study a series of new invariants of a given variety V. This was also done by Witten (cf. [W]) on the basis of physical intuition, using intersection theory on $\overline{M_{g,n}}$, the moduli space of Deligne-Mumford stable curves of genus g, with n marked points, a space well known to string theorists.

These invariants, which we call "Gromov-Witten invariants" (they are also called "topological σ-models" or "mixed invariants" depending on context) depend on the geometry of curves lying on V. For certain varieties (such as projective spaces) a subclass of them corresponds to enumerative invariants; for example, the degrees $N_0(d)$ for plane rational curves are Gromov-Witten invariants.

We restrict this brief description to the case of curves of genus zero, to avoid parts of the theory that are still at a conjectural state, and because this is where there is a clear link with algebraic geometry: the above mentioned enumerative predictions all have to do with rational curves. (See [G] for some very recent developements for genus one.)

It was conjectured (and it is now proved) that the Gromov-Witten invariants satisfy a series of properties; the most important of them is the so-called "splitting principle" or "composition law", which gives a way of computing these invariants recursively. On their existence one can base the construction of a family of quantum ring structures on the cohomology ring of V, deforming the standard cup product; the associativity of this quantum product is a consequence of the splitting principle.

2.2. A formula for plane rational curves. In 1993 Kontsevich derived a beautiful formula for rational curves in the plane, assuming the associativity of the quantum product for $\mathbb{P}^2$ (not yet proved at the time).

Kontsevich's formula. *For $d \geq 2$*

$$N_0(d) = \sum_{d_1+d_2=d} N_0(d_1)N_0(d_2)d_1d_2\left[\binom{3d-4}{3d_1-2}d_1d_2 - \binom{3d-4}{3d_1-3}d_2^2\right].$$

This, together with the basic fact $N_0(1) = 1$, (i.e. there is a unique line through two distinct points) allows one to compute degrees of all Severi varieties of rational curves in the plane.

Here is, briefly, how such a formula was discovered. Using the Gromov-Witten invariants, one defines a generating function (the "potential") on the cohomology ring of V, which completely encodes the quantum ring structure. This is part of the basic set-up of Quantum Cohomology, I will not say much about it and refer to [KM] for the details. For the special case of $\mathbb{P}^2$ such a generating function is, for $\Delta \in H^*(\mathbb{P}^2)$:

$$\Phi(\Delta) = \Phi^{cl} + \Phi^q = \frac{1}{2}(x_0^2 x_2 + x_0 x_1^2) + \sum_{d=1}^{\infty} N_0(d) \frac{x_2^{3d-1}}{(3d-1)!} e^{dx_1}$$

where the variables x_0, x_1, x_2 are the coefficients of $\Delta = x_0\Delta_0 + x_1\Delta_1 + x_2\Delta_2$, with Δ_0 the identity, and Δ_1 and Δ_2 the duals of the class of the line and the point respectively.

For each Δ one gets a quantum ring structure on $H^*(\mathbb{P}^2)$ which is defined using the rank-3 tensor of all derivatives $(\partial_j\partial_i\partial_k\Phi)_{|\Delta}$. The summand $\Phi^{cl} = \frac{1}{2}(x_0^2x_2 + x_0x_1^2)$ gives the "generating function" for the classical cup product, which of course does not depend on Δ

It is worth mentioning that a "full" potential should be defined so as to include a term for every genus (so that Φ^q above corresponds to the genus zero part), but it is still an open question how to do that in a geometrically meaningful way.

The crucial observation is that the quantum product is associative if and only if the potential Φ satisfies the WDVV differential equation, that is to say, if and only if the following identity holds:

$$\Phi^q_{222} = (\Phi^q_{112})^2 - \Phi^q_{111}\Phi^q_{122}.$$

Finally, as the reader can check by a straightforward computation, the above identity implies Kontsevich's formula.

Complete proofs of Kontsevich's formula were given independently by Ruan and Tian (in [RT]) and by Kontsevich and Manin (in [KM] and [K]), using rather sophisticated techniques. In both cases the goal was to give a mathematically rigorous definition of the Gromov-Witten invariants, so that they satisfy the required properties (especially the composition law!). In [RT] this is done using symplectic topology and the Gromov theory of pseudo-holomorphic curves. In [KM] the authors follow an algebro-geometric approach and use the existence of a good compactification of the moduli space of maps from $\mathbb{P}^1$ to V (constructed in the later paper [K]). We notice that these techniques work for a larger class of varieties; in the specific case of surfaces they give enumerative results for the plane, for $\mathbb{F}_0 = \mathbb{P}^1 \times \mathbb{P}^1$, for $\mathbb{F}_1$ and for blow-ups of $\mathbb{P}^2$ at general points.

2.3. Hirzebruch surfaces. In the First Reconstruction Theorem (in [KM]) there is a sketch for a heuristic argument to obtain the formula above. More recently we showed (in [CH1]) that such a heuristic argument could be made into a completely rigorous proof involving only classical tools, with the advantage that this old fashioned approach (which we call the "cross ratio" method) leads to formulas for rational curves on any rational surface S, that could not be found otherwise.

The methods of [KM] and [RT] do not answer our enumerative questions for the Hirzebruch surfaces $\mathbb{F}_n$. In fact $\mathbb{F}_n$ is not "convex" which implies that there does not exist a well-behaved compact moduli space of maps as it exists for $\mathbb{P}^2$ (convexity here is defined follows: a variety V is convex if for every stable map $f : \mathbb{P}^1 \to V$ we have $H^1(f^*T_V) = 0$ where T_V is the tangent bundle of V). On the other hand, the techniques of [RT] only depend on the symplectic type of the surface; the degrees $N_0(D)$ coincide with certain Gromov-Witten numbers as long as S is $\mathbb{P}^2$, $\mathbb{P}^1 \times \mathbb{P}^1$ or $\mathbb{F}_1$. But while the Gromov-Witten numbers are symplectic-invariants, the Severi degrees $N_0(D)$ are not, as we can see in the following example. There are only two symplectic types of Hirzebruch surfaces $\mathbb{F}_n$, depending on the parity of n. So that we can compare $\mathbb{F}_0$ with $\mathbb{F}_2$, which must have the same Gromov-Witten invariants. On both surfaces, consider D equal to the anticanonical class; then one can show (using any of the techniques described in 4.1 or 4.2) that $N_0(-K_{\mathbb{F}_0}) = 12$ while $N_0(-K_{\mathbb{F}_2}) = 10$.

We describe the cross ratio method in an example later (cf. 4.2). Here we just give a summary of the results that it gives. First, for $\mathbb{F}_n$, with $n \leq 2$, and blow-ups of the plane at general points one obtains an inductive formula (completely analogous to Kontsevich's formula) expressing $N_0(D)$ as a function of simply $N_0(D')$, with $D' < D$ (that is $D - D'$ effective and non-zero). This corresponds to the fact that for such surfaces one can construct degenerations of rational curves in $|D|$ whose degenerate fibers will be reducible rational curves all of whose components are general points of Severi varieties $V_0(D')$.

For $n \geq 3$ a new phenomenon complicates things: this is the occurrence in codimension one of degenerate loci that are no longer of type $V_0(D')$; these will instead be loci of curves satisfying certain tangency conditions. More precisely, for $\mathbb{F}_n$ let us define the *tangential Severi varieties* $V_0^i(D) \subset V_0(D)$ as the closure of the set of curves in $V_0(D)$ that have a point of contact of order i with the exceptional curve E of $\mathbb{F}_n$ (that is, the unique curve E having self intersection $-n$). Then the cross ratio method gives a formula for $N_0(D)$ in terms of the degree $N_0^i(D')$ of $V_0^i(D')$, for suitable i (see [CH1]). Therefore to have a complete picture one should also compute the degrees of these tangential Severi varieties.

There is another ad-hoc technique to deal with rational curves, the so-called "rational fibration method" (illustrated in example 4.2). This is described in [CH2] where it is applied to obtain a complete set of formulas for $\mathbb{F}_3$. The picture appears to be essentially the same as for the cross ratio approach: we get inductive formulas expressing $N_0(D)$ in terms of degrees of tangential Severi varieties. Such a method again focuses on the study of one-parameter families of rational curves, and it is based on the very basic fact that two line bundles on $\mathbb{P}^1$ are isomorphic if they have the same degree. See 4.2 for an example illustrating this method.

3. Higher genera

3.1. The degeneration method of [CH3]. We now consider curves of any genus, for which none of the methods described in the previous chapter seem to work so far. We shall give an answer to the Question in 1.2; from now on, we take $S = \mathbb{P}^2$. Fix a line $L \subset \mathbb{P}^2$ once and for all, the inductive technique now uses degenerations whose special fiber is forced to contain an increasing number of points of L, until it becomes reducible, having to contain L itself as a component. What remains of the special fiber is a curve of degree $d - 1$ (which may very well be reducible) so that one can then use induction.

This procedure is different from the ones used for rational curves for a crucial reason. For curves of genus 0 one uses generic degenerations (such as: the family of all rational plane curves of degree d through $3d-1$ general points). Here we use a special type of degeneration, by using the device of placing some of the points on a fixed line.

Therefore we start by considering a larger class of Severi varieties, including certain tangential loci. This is inspired by the case of Hirzebruch surfaces described above, the role of the exceptional curve E is played by the line L. For rational curves on $\mathbb{F}_n$ we were forced to consider curves satisfying tangency conditions with respect to E, getting more complicated sets of recursions. For curves of any genus in the plane we actually choose to take tangential loci from the beginning, and in this way we get a rather simple formula.

3.2. A formula for plane curves of any genus. Let us define generalized Severi varieties. Let $\alpha = (\alpha_1, \ldots, \alpha_h)$ and $\beta = (\beta_1, \ldots, \beta_k)$ be strings of nonnegative integers. Fix $\sum \alpha_j$ general points on L denoted by $\{p_j^{(i)}\}_{1 \le j \le \alpha_i}$. Assume that $\sum i\alpha_i + \sum i\beta_i = d$. The generalized Severi variety $V^{d,\delta}[\alpha, \beta]$ is defined as the closure of the locus of reduced plane curves of degree d with δ nodes (hence the geometric genus is $g = \binom{d-1}{2} - \delta$) which

(i) do not contain L,

(ii) have contact of order i with L at $p_j^{(i)}$ for $1 \le j \le \alpha_i$ (briefly: have α_i assigned points of contact of order i with L)

(iii) have β_i points of contact of order i with L at some "unassigned" points.

Notice that we do not assume the curves to be irreducible, this is why we change notation and label by the number of nodes δ instead of the geometric genus.

Some examples: $V^{3,1}[(0),(3)]$ is the variety of rational cubics, denoted by $V_0(3)$ with the notation of the previous sections (we omit the 0 entries in α and β whenever this does not create confusion); and $V^{d,\delta}[(0),(d)] = V_g(d)$; also $V^{4,3}[(0),(4)]$ the 11-dimensional variety of quartics with 3 nodes; this has two irreducible components V_1 and V_2, where $V_1 = V_0(4)$ and V_2 parametrizes all reducible quartics made of the union of a line and a cubic.

One can show that

(1) each irreducible component of $V^{d,\delta}[\alpha,\beta]$ has the expected dimension

$$r^{d,\delta}[\alpha,\beta] = \dim V^{d,\delta}[\alpha,\beta] = \frac{d(d+3)}{2} - \delta - \sum i\alpha_i - \sum (i-1)\beta_i = 2d + g - 1 + |\beta|;$$

(2) the general point of $V^{d,\delta}[\alpha,\beta]$ parametrizes a curve having only nodes as singularities and smooth along L.

Set now $N^{d,\delta}[\alpha,\beta] := \deg V^{d,\delta}[\alpha,\beta]$. Let us introduce the notation $\alpha! = \alpha_1!\alpha_2!\ldots$, $\binom{\alpha}{\alpha'} := \binom{\alpha_1}{\alpha_1'} \cdot \binom{\alpha_2}{\alpha_2'} \cdot \ldots$ and if $S = \{s_1, s_2, \ldots\}$ is any ordered set (for example, $S = \mathbb{N}$ the set of positive integers) $S^\alpha := s_1^{\alpha_1} s_2^{\alpha_2} s_3^{\alpha_3} \cdot \ldots$; let also $\epsilon^{(j)}$ be defined as the string of integers having 1 at the jth place and 0 elsewhere. We can then state the main enumerative result (See Theorem 1.1 in [CH3]):

Theorem.
$$N^{d,\delta}[\alpha,\beta] = \sum_{j:\beta_j>0} jN^{d,\delta}[\alpha+\epsilon^{(j)},\beta-\epsilon^{(j)}]$$
$$+\sum \mathbb{N}^{\beta'-\beta}\binom{\alpha}{\alpha'}\binom{\beta'}{\beta}N^{d-1,\delta'}[\alpha',\beta']$$

where the second sum is taken over all α',β' *and* $\delta'\geq 0$ *satisfying* $\alpha'\leq\alpha$, $\beta'\geq\beta$, $\delta'\leq\delta$ *and* $\delta-\delta'+|\beta'-\beta|=d-1$.

The proof of this theorem uses techniques of deformation theory together with semistable reduction. The degeneration technique is as follows. Let $V=V^{d,\delta}[\alpha,\beta]$; if $p\in\mathbb{P}^2$ is a point we denote by H_p the hyperplane of $|D|$ parametrising curves through p. Let now $p_1,\ldots,p_t$ be points on L. We consider the scheme-theoretic intersection
$$V_t := V\cap(\cap_{i=1}^t H_{p_i})$$
which has the same degree as V. The formula above can be read as a statement describing the hyperplane section $V\cap H_{p_1}$ as a scheme. Notice in fact that $V^{d,\delta}[\alpha+\epsilon^{(j)},\beta-\epsilon^{(j)}]$ has codimension 1 in V, while the last condition $\delta-\delta'+|\beta'-\beta|=d-1$ is saying precisely that the codimension of $V^{d-1,\delta'}[\alpha',\beta']$ in V must be 1. The coefficients of the formula have different meanings. We will illustrate the procedure on an example (cf. 4.3). All details can be found in [CH3].

We conclude with a nice and hopefully inspiring new way of writing the above formula; this was found by Getzler and appears in [G]. Let z be a variable and let $u=(u_1,u_2,\ldots)$ and $v=(v_1,v_2,\ldots)$ be sets of variables. Then we define a generating function using the degrees $N^{d,\delta}[\alpha,\beta]$
$$G=\sum\frac{u^\alpha}{\alpha!}v^\beta N^{d,\delta}[\alpha,\beta]\frac{z^{r^{d,\delta}[\alpha,\beta]}}{r^{d,\delta}[\alpha,\beta]!}$$
then an easy computation shows that the recursion in the above theorem is equivalent to the following identity
$$\frac{\partial G}{\partial z}=\sum_{k=0}^{\infty}kv_k\frac{\partial G}{\partial u_k}+\mathrm{Res}_{t=0}\exp(\sum_{k=0}^{\infty}\frac{u_k}{t^k}+\sum_{k=0}^{\infty}kt^k\frac{\partial}{\partial v_k})G$$
where clearly the first summand $\sum_{k=0}^{\infty}kv_k\frac{\partial G}{\partial u_k}$ corresponds to the first summand in the above formula and the remaining part corresponds to the second summand. We just notice that taking the residue for $t=0$ (that is, the coefficient of t^{-1}) is the analog of the codimension-one condition $\delta-\delta'+|\beta'-\beta|=d-1$.

3.3. Generalizations. The recursive procedure that we have just described can easily be applied to Hirzebruch surfaces, by replacing the line L with the exceptional curve E. Then one obtains a similar formula for curves of any genus (see [V]). A more subtle problem is generalizing the method to higher dimensional projective spaces. More precisely, one could ask for the number of (smooth) curves of given degree and genus in $\mathbb{P}^n$ that pass through the appropriate number of points, or more generally which satisfy a given set of intersection conditions with respect to linear subspaces of varying dimension. For example, it is not hard to show that, in $\mathbb{P}^3$, there are finitely many rational curves of degree d satisfying $4d$ linear conditions (ie, passing through $4d$ points, or meeting $4d$ lines, and so on).

The natural way of generalizing our technique is to place the "linear conditions" on a fixed hyperplane one at the time so as to get a recursion. This has been shown to work by Vakil, for curves of genus 0 in $\mathbb{P}^n$. The higher genus case is still under investigation.

4. Examples

Consider the following well-known

Proposition. *There are 12 rational plane cubics passing through 8 general points.*

In symbols: $N_0(3) = N^{3,1}(0,3) = 12$. We give here a few different ways to prove such a result, to illustrate how the techniques we talked about work. We first prove it by classical arguments (two of them) in 4.1, then we use the cross ratio method and the rational fibration method in 4.2 (these are ad-hoc techniques for rational curves, inspired by the First Reconstruction Theorem of [KM]). Finally, we use the recursive procedure of [CH3] which led to the Theorem stated in the previous section.

We let $V = V_0(3) \subset \mathbb{P}^9$ and $N = N_0(3)$; we have $\dim V = 8$.

Remark. The following fact about the geometry of V will be needed (cf. [CH3] and loc. cit.). V is irreducible and smooth at its general point (corresponding to an irreducible cubic with one node). V contains a codimension 1 irreducible subvariety W whose general points parametrize reducible cubics given by the union of a conic and a line. This is the unique degenerate locus of V having codimension 1 and parametrizing reducible curves. V is singular along W, looking like two smooth sheets crossing transversally. The two sheets correspond to deformations of the reducible curves that are locally trivial at either one of the nodes.

4.1. Classical proofs. These proofs are well-known. We actually prove the more general fact that the degree of the Severi variety of curves of degree d with one node is $3(d-1)^2$. So, let $V = V^{d,1}[0,d]$ and N be its degree. V has codimension 1 in the space $\mathbb{P}^r$ of all curves of degree d, so that $N = \ell \cap V$ where ℓ is a general line in $\mathbb{P}^r$.

We can identify ℓ with a general pencil of curves of degree d; that is, a family given by a polynomial equation $F(t_0, t_1; x_0, x_1, x_2) = 0$ where F is homogeneous of degree 1 in t_0, t_1 and of degree d in x_0, x_1, x_2. We obtain in this way a family

$$\mathcal{Y} \longrightarrow \mathbb{P}^1 \cong \ell$$

where $\mathcal{Y} \subset \mathbb{P}^1_{t_0,t_1} \times \mathbb{P}^2_{x_0,x_1,x_2}$. Then N corresponds exactly to the total number of nodes of the fibers of such a family.

There are now two ways of computing such a number, an algebraic way and a topological way. Algebraically, the nodes of the fibers are given by the (non-zero) solutions of the system

$$F_{x_0} = F_{x_1} = F_{x_2} = 0,$$

that is, by the number of points in which the three surfaces given by $F_{x_i} = 0$ in $\mathbb{P}^1_{t_0,t_1} \times \mathbb{P}^2_{x_0,x_1,x_2}$ intersect. Now, these surfaces are of class $h_1 + (d-1)h_2$, where h_i is the pull-back of the generator of $\mathrm{Pic}(\mathbb{P}^i)$, $i = 1, 2$. Hence the basic relations $h_1^3 - h_1^2 h_2 = h_2^3 = 0$ and $h_1 h_2^2 = 1$ imply that $(h_1 + (d-1)h_2)^3 = 3(d-1)^2$, which is what we wanted to prove. We obtain the proposition as a special case.

The same result can be obtained by computing the Euler characteristic of $\mathcal{Y}$ in two different ways. First, $\mathcal{Y}$ is the blow up of $\mathbb{P}^2$ at the d^2 base points of the pencil, hence

$$\chi(\mathcal{Y}) = 3 + d^2.$$

On the other hand, the family $\mathcal{Y} \longrightarrow \mathbb{P}^1$ has general fiber F a Riemann surface of topological characteristic $\chi(F) = 2 - 2g = 2 - 2\binom{d-1}{2}$. If N is the total number of nodes, we have (cf. [GH] Chapter 4)

$$\chi(\mathcal{Y}) = \chi(F)\chi(\mathbb{P}^1) + N$$

which implies $N = 3(d-1)^2$.

4.2. The cross ratio and the rational fibration methods. Here we prove the proposition by using the fact that the curves in question are rational (see also [DI]). Fix 7 general points in the plane and let $\Gamma \subset V$ be the irreducible curve parametrizing all nodal cubics through such points. Let $\mathcal{X} \longrightarrow \Gamma$ be the corresponding family, so that $\mathcal{X} \subset \Gamma \times \mathbb{P}^2$. Observe that the family has exactly $\binom{7}{2}$ reducible fibers, corresponding to all reducible cubics of type $C_1 \cup C_2$, where C_1 is a line through two of the base points and C_2 is the conic through the remaining five points. If $t \in \Gamma$ is a point such that the fiber X_t is one of these reducible curves, then t will be a node of Γ by the introductory remark. We then let B be the normalization of Γ and we let $\mathcal{Y}$ be the normalization of the base change family $\mathcal{X} \times_\Gamma B$. It turns out that $\mathcal{Y}$ is a smooth surface and that all of the fibers of $\mathcal{Y} \longrightarrow B$ are at most nodal. We do not really need this; if it were not the case that all fibers were nodal (there may be a priori some cuspidal curves) we could make a base change and perform semistable reduction; if $\mathcal{Y}$ were not smooth, we could take its minimal desingularization. None of these two operations would affect the rest of the procedure.

Finally $\mathcal{Y} \longrightarrow B$ is a family of generically smooth rational curves, having $2\binom{7}{2}$ reducible nodal fibers made of two (smooth, rational) components, and no other singular fiber. We let $\pi : \mathcal{Y} \longrightarrow \mathbb{P}^2$ be the natural map.

This part of the set-up is common to both the cross ratio method and the rational fibration method. Now we concentrate on the first. From $\mathcal{Y} \to B$ we want to obtain a family of rational curves having 4 sections. We do this as follows: the first two sections will be two of the 7 base points, call them p_1 and p_2; the other two sections will be given by intersecting the curves of the family with 2 fixed lines L_3 and L_4 (which are chosen to be general with respect to the base points of the family). This will clearly be possible after a base change of order 9; in fact, L_3 intersects each curve of the family in 3 points, hence $\pi^* L_3$ is a curve in $\mathcal{Y}$ which is a 3 to 1 cover of B. Therefore if we perform the base change of degree 3 given by $\pi^* L_3 \to B$ we can define a (single-valued) section of $\mathcal{Y} \times_B \pi^* L_3$ representing the intersection of L_3 with the curves of the family. Then we repeat the same process to obtain a (single-valued) section out of the intersection with L_4. In total, we made a degree 9 base change $A \to B$ and we get a new family $\mathcal{W} \to A$ where $\mathcal{W} := \mathcal{Y} \times_B A$. We call again π the natural map from $\mathcal{W}$ to $\mathbb{P}^2$. By construction, this is a family of generically smooth rational curves, all of whose fibers are at most nodal, and such that there are 4 sections $p_i : A \to \mathcal{W}$, for $i = 1, 2, 3, 4$. These sections are such that the first two correspond to p_1 and p_2 respectively, while $\pi(a) \in L_i$ for $i = 3, 4$ and for all $a \in A$. The reader familiar with the theory of moduli of curves will immediately see that this family has a canonical morphism $\phi : A \longrightarrow \overline{M_{0,4}}$ where $\overline{M_{0,4}}$ denotes the Deligne-Mumford moduli space of stable curves of genus 0 with 4 marked points. There is an equivalent way of describing this map. Define $\varphi : A \longrightarrow \mathbb{P}^1$ via the cross ratio:

$$\varphi(a) = \frac{\big(p_1(a) - p_3(a)\big)\big(p_2(a) - p_4(a)\big)}{\big(p_1(a) - p_2(a)\big)\big(p_3(a) - p_4(a)\big)}.$$

Notice that $\overline{M_{0,4}} \cong \mathbb{P}^1$ so that if we want ϕ to coincide with φ we just have to identify the 3 boundary points of $\overline{M_{0,4}}$ with the degenerate values 0, 1, and ∞ of the cross ratio as follows. The value 0 is identified with the point corresponding to the isomorphism class of stable, nodal, rational curves having p_1 and p_3 on one component and p_2, p_4 on the other component (we call these curves of type $(13, 24)$). The value ∞ with the stable nodal rational curve having p_1 and p_2 on one component and p_3, p_4 on the other component.

Now we prove the Proposition using the basic fact that $\deg\varphi^*(0) = \deg\varphi^*(\infty)$.

To compute the degree of the zero divisor of φ, we know that the value 0 is achieved on curves of type $(13, 24)$ (notice that, by the genericity assumption, the sections p_1 and p_3 are disjoint, and so are the sections p_2 and p_4). To count them, we observe that there are 10 reducible cubics through the seven base points, such that p_1 and p_2 lie on different components: 5 have p_1 on the line (because we need one more point, out of the remaining 5 base points, to pin down the line) and 5 have p_1 on the conic (as before the line is determined by p_2 and any one of the remaining 5 base points). The corresponding 10 points of Γ will be nodes, by the remark at the beginning of the section, correspondingly we get 20 points on B. Then we have to take into account the base change of order 9, and, more important, the fact that L_i meets the conic in two points. Finally we get

$$\deg \varphi^*(0) = 9 \cdot 2 \cdot 20.$$

The poles of φ are assumed on points of A corresponding to reducible fibers of type $(12, 34)$ and also on points of A where the section p_3 crosses the section p_4 (the sections $p_1$and p_2 do not intersect). Notice that p_3 and p_4 intersect in all points $a \in A$ such that the fiber W_a is mapped by π to a nodal cubics passing through the 7 base points and through the point of intersection of L_3 and L_4. There are N such plane curves, and hence $9N$ corresponding points in A, where φ has a pole. To count the fibers of type $(12, 34)$ we proceed as before: there is a unique plane reducible cubic of our family having p_1 and p_2 on a line, then we have to account for the node of Γ, for the base change of order 9, and for the fact that L_3 and L_4 meet the conic in two points. We get a contribution of $2 \cdot 2 \cdot 2 \cdot 9$ to $\deg \varphi^*(\infty)$. Finally, we find $\binom{5}{3}$ curves having p_1 and p_2 on a conic, and a total contribution of $\binom{5}{3} \cdot 2 \cdot 9$ to $\deg \varphi^*(\infty)$. We get

$$\deg \varphi^*(\infty) = 9N + 2 \cdot 2 \cdot 2 \cdot 9 + \binom{5}{3} \cdot 2 \cdot 9 = \deg \varphi^*(0) = 9 \cdot 2 \cdot 20$$

which yields $N = 12$.

The rational fibration method is somewhat simpler, because it does not involve any further construction, once we arrive at the above family $\mathcal{Y} \longrightarrow B$. We let Y be the class of the fiber, so that $Y^2 = 0$. Then we let A be the class of a section corresponding to one of the seven base points, call it q. Then $(A \cdot Y) = 1$. Let $B' \subset B$ be the set of points such that the corresponding fiber is reducible (so that B' contains exactly $2 \cdot \binom{7}{2}$ points), for $b \in B'$ let Z_b and W_b be the two components. Choose the names Z and W so that $(A \cdot Z_b) = 1$ and $(A \cdot W_b) = 0$ (in other words, A always intersects the Z_b component, for every $b \in B'$). Then the classes Y, A, $\{W_b\}_{b \in B'}$ generate the Néron-Severi group of Y, which is to say that we can compute intersection numbers by expressing every other class as a combination of these generators. Notice that we have $W_b^2 = -1$ and $(A \cdot W_b) = (Y \cdot W_b) = 0$.

Let L be the hyperplane class in $\mathbb{P}^2$. Then we prove the proposition using

$$(\pi^* L \cdot \pi^* L) = (\deg \pi)(L \cdot L) = N.$$

Let us write

$$\pi^* L = c_Y Y + c_A A + \sum_{b \in B'} c_b W_b$$

and now notice that

(i) $(\pi^* L \cdot Y) = 3 \implies c_A = 3$

(ii) $(\pi^* L \cdot A) = 0 \implies c_Y = -3A^2$

(iii) $(\pi^* L \cdot W_b) = (L \cdot \pi_* W_b) = \deg \pi_* W_b \implies c_b = -\deg \pi_* W_b$.

Using all of these relations we obtain

$$(\pi^* L \cdot \pi^* L) = -9A^2 - \sum_{b \in B'} (\deg \pi_* W_b)^2.$$

To compute A^2, pick another of the base points, call it q', and let A' be the corresponding section of the family. Since $A^2 = (A')^2$ and $(A \cdot A') = 0$ we get $2A^2 = (A - A')^2$. Now $A - A'$ is supported exactly on those reducible fibers such that q and q' lie on different components (in other words, $q' \in Z_b$). The number of them is $2 \cdot 10$, in fact there will be 5 curves of the family such that q lies on a line and q' on a conic, and 5 such that q lies on a conic and q' on a line; the factor of 2 comes from the fact that Γ has a node at such curves. So we conclude that $A^2 = 10$.

The last thing to compute is $\sum(\deg \pi_* W_b)^2$. This amounts to counting how many reducible fibers have q on the line, and how many have q on the conic. Clearly q is on a line for $2\cdot 6$ fibers and on a conic for $2\cdot\binom{6}{2}$ fibers. Hence $\sum(\deg \pi_* W_b)^2 = 2\cdot 6\cdot 4+2\cdot\binom{6}{2}=78$ and we conclude $N=9\cdot 10-78=12$.

4.3. The proof using the Theorem of [CH3]. As we said, for the inductive technique we fix a line L in $\mathbb{P}^2$ and we pick points $p_1,p_2,p_3,p_4,\dots$ on L; then we study successive scheme-thoretic intersections

$$V_t = V\cap(\cap_{i=1}^t H_{p_i})$$

for which, of course, $\deg V=\deg V_t$.

By dimension count we have $V\cap H_{p_1}=V^{3,1}[(1),(2)]$ and $V_2=V\cap H_{p_1}\cap H_{p_2}=V^{3,1}[(2),(1)]$ and these intersections are transverse, hence there are no extra factors in the degree computation. The next step gives two components of dimension 5, namely

$$V_3 = V^{3,1}[(3),(0)]\cup W$$

where the general point of W is a reducible cubic $X=L\cup C$ where C is a conic. Hence $\deg W=1$, but there will be a coefficient of 2 in the degree computation. This is because V_2 is singular along W, by the remark at the beginning of this chapter. This coefficient 2 corresponds to $\binom{\beta'}{\beta}$ in the formula.

Now we have to compute the degree of $V' := V^{3,1}[(3),(0)]$. By intersecting with a fourth hyperplane we now get five different irreducible components of dimension 4, three of which are of the same type. Let us write

$$V'\cap H_{p_4}=U_1\cup U_2\cup U_3\cup W'\cup W''.$$

The general point of U_i is a curve $X=L\cup C$ where C is a conic through p_i. To understand this degeneration in more detail, consider a generic one-parameter family of curves of V' degenerating to X. This is a family whose general fiber has one node while the special fiber has two, one of which is p_i.

We claim that the limit on X of the node of the general fiber must be p_i. To see this, we normalize the total space of the family so as to obtain a family of generically smooth curves. Let Y be the curve lying over X, so that Y is the partial normalization at the limit node. Now the total space of this normalized family has an obvious map to $\mathbb{P}^2$, call it π. The preimage $\pi^{-1}L$ of the fixed line L cannot have isolated points, because L is irreducible of codimension 1. Hence $\pi^{-1}L=L'\cup S_1\cup S_2\cup S_3$ where L' is a subcurve of Y and S_i is a curve such that $\pi S_i=p_i$. Let $q\in Y$ be the point lying over the node of X such that $q\notin L'$. Clearly $q\in\pi^{-1}L$, therefore there must be a curve of $\pi^{-1}L$ passing through q. Such a curve can only be one of the S_i, and this implies our claim.

We conclude that, in the degree computation, we shall find no further coefficient, as the intersection is transverse and V' is smooth along U_i. We have of course $\deg U_i=N^{2,0}[(1),(1)]=1$.

The general curve parametrized by W' is the union of L and a conic C tangent to L at some unassigned point. The degree of the variety of conics tangent to L is

2, and H_{p_4} is tangent to V' along W', hence we get a factor of 2 which correspond to $\mathbb{N}^{\beta'-\beta}$ in the theorem.

Finally W'' parametrizes curves of type $X = L \cup L_1 \cup L_2$ where L_i is a line; the same argument used before shows that the limit node of such a curve can only be $L_1 \cap L_2$, so that V' is smooth along W'' and the intersection is transverse. The degree of the variety of nodal conics is 3.

We can summarize the above description of the (scheme intersection) $V^{3,1}[(3)(0)] \cap H_{p_4}$ by looking at the following three possibilities for the limit of the node of the general fiber. If such a limit is an assigned point, say p_i, then the limit curve lies in U_i; if the limit node is an unassigned point of L, then the limit curve is in W'; finally, if the limit node is not on L, we get a curve in W''.

We conclude that $N^{3,1}[(3)(0)] = 3 + 2 \cdot 2 + 3 = 10$ and hence $N_0(3) = 12$. The reader might find the following explicit formula useful

$$N_0(3) = 2 \cdot N^{2,0}[(0),(2)] + 3 \cdot N^{2,0}[(1),(1)] + 2 \cdot N^{2,0}[(0),(0,1)] + N^{2,1}[(0),(0)]$$

where the coefficient 3 in front of the second summand corresponds to $\binom{\alpha}{\alpha'}$.

References

[CH1] L. Caporaso and J. Harris, *Parameter spaces for curves on surfaces and enumeration of rational curves.* (1995) Preprint, alg-geom 9608023. To appear in Compositio Mathematica.

[CH2] L. Caporaso and J. Harris, *Enumeration of rational curves: the rational fibration method.* (1995) Preprint, alg-geom 9608024. To appear in Compositio Mathematica.

[CH3] L. Caporaso and J. Harris, *Counting plane curves of any genus* (1996) Preprint, alg-geom 9608025. To appear in Inventiones Mathematicae.

[Ch] Y. Choi, *Severi degrees in cogenus 4* (1996) Preprint, alg-geom 9601013.

[CM] B. Crauder and R. Miranda, *Quantum cohomology of rational surfaces.* The Moduli Space of Curves, Progress in mathematics 129, Birkhäuser (1995) pp. 33–80.

[DI] P. Di Francesco and C. Itzykson, *Quantum intersection rings.* The Moduli Space of Curves, Progress in mathematics 129, Birkhäuser (1995) pp. 81–148.

[FP] W. Fulton and R. Pandharipande, *Notes on stable maps and quantum cohomology.* (1995) Preprint.

[G] E. Getzler, *Intersection theory on* $\overline{M}_{1,4}$ *and Gromov-Witten invariants in genus 1.* MPI Preprint (1996).

[H] J. Harris, *On the Severi problem.* Invent. Math. 84 (1986), pp. 445–461.

[KP] S. Kleiman and R. Piene, Private communication.

[Ko] J. Kollár, *Rational Curves on Algebraic Varieties*, Springer 1996.

[K] M. Kontsevich, *Enumeration of rational curves via torus action.* The Moduli Space of Curves, Progress in mathematics 129, Birkhäuser (1995) pp. 335–368.

[KM] M. Kontsevich and Y. Manin, *Gromov-Witten classes, quantum cohomology and enumerative geometry.* Commun.Math.Phys. 164 (1994), pp. 525–562 and hep-th/9402147.

[P] R. Pandharipande, *Intersections of* $\mathbb{Q}$*-divisors on Kontsevich's moduli space* $\overline{M}_{0,n}(\mathbb{P}^r, d)$ *and enumerative geometry.* 1995, Preprint.

[R1] Z. Ran, *Enumerative geometry of singular plane curves.* Invent. Math. 97 (1989), pp. 447–465.

[R2] Z. Ran, *On the quantum cohomology of the plane old and new.* (1995) Preprint, alg-geom 9508011.

[RT] Y. Ruan and G. Tian, *A mathematical theory of Quantum Cohomology* J. Diff. Geom. 42 No. 2 (1995) pp. 259–367.

[S] F. Severi, *Vorlesungen über algebraische Geometrie* Anhang F. Leipzig: Teubner 1921.

[V] R. Vakil, *Curves on rational ruled surfaces* (1996) Preprint.

[Va] I. Vainsencher, *Counting divisors with prescribed singularities.* Trans AMS 267 (1981), 399–422.

[W] E. Witten, *Two dimensional gravity and intersection theory on moduli space.* Surveys in Diff. Geom. 1 (1991) pp. 243–310.

Progress in Mathematics, Vol. 168, © 1998 Birkhäuser Verlag Basel/Switzerland

Minimal Surfaces in Singular Spaces

ULRICH DIERKES

Gerhard-Mercator-Universität
Fachbereich Mathematik
Lotharstraße 65, D 47057 Duisburg
Germany

In 1690 the contest to determine the *catenary* started with Jakob Bernoulli's proposal in the Acta Eruditorum: "*To find the curve assumed by a loose string hung freely from two fixed points.*" (In fact the history of the catenary started with Galilei, but after some faulty attempts the matter fell asleep for at least fifty years.) Jakob Bernoulli's proposal was also the starting point of a long quarrel between the Bernoulli brothers, since the younger, Johann B., succeeded in solving this problem, while there is no evidence that Jakob himself knew the solution in 1690 (Although from 1697 to 1698 he succeeded in obtaining the general equation for a flexible line). Besides Johann Bernoulli the problem of the catenary was solved by Leibniz and Huygens (using some important previous work by Pardies); for further information on the interesting history we refer to the excellent exposition in Truesdell ([54] in particular p. 69–88), or Euler [25].

Nowadays, the catenary problem is treated in the standard literature on the calculus of variations, and may be formulated as an isoperimetric problem (see e.g. Bolza [6]) : Given two points A and B in $\mathbb{R}^2$ and a number $L >\mid A-B \mid$. Find a curve $(x(t), y(t)), 0 \leq t \leq 1$ of given length L and connecting A and B such that the y-coordinate of the centre of gravity

$$y_c := \frac{1}{L} \cdot \int_0^1 y\,(t)\,\sqrt{(\dot{x})^2 + (\dot{y})^2}\;dt$$

assumes a stationary or minimal value. It is well known, that solution curves are of the form $y(x) = \alpha \cosh(\frac{x-\beta}{\alpha})$, $\alpha, \beta \in \mathbb{R}$, i.e. the catenaries.

Here we are interested in the higher-dimensional mathematical analogue of the catenary problem, i.e. to determine the shape and describe the properties of a surface M of prescribed area and boundary which is of constant mass density, such that the $x_{n+1}-$ coordinate of the centre of gravity assumes a stationary value. Suppose that the surface M is given as a graph $u : \Omega \to \mathbb{R}$, $\Omega \subset \mathbb{R}^n$, $n \geq 2$, and that the gravitation force acts in the $-x_{n+1}$ direction, this problem amounts to finding stationary values of the integral

$$E(u) = \frac{1}{A_0} \int_\Omega u\,\sqrt{1+ \mid Du \mid^2}\;dx$$

in a class of functions, which fulfill prescribed boundary conditions as well as the subsidiary condition

$$A(u) = \int_\Omega \sqrt{1+ | Du |^2} \, dx \; = \; A$$

for some given value of A.

We observe here, that, in contrast to the one-dimensional situation, the corresponding minimization problem for $n \geq 2$ in general has no solution, even if A is close to the area of the corresponding minimal surface. In fact, as was pointed out by Nitsche [42], for given $\epsilon > 0$, $\Omega = B_1(0) \subset \mathbb{R}^2$ and arbitrary constant boundary data, there are functions f_n which are constant along $\partial\Omega$, such that $A(f_n) = \pi(1+\epsilon)$, $E(f_n) \to -\infty$, as well as $f_n(0,0) \to -\infty$.

Introducing a Lagrange multiplier λ, the problem is reduced to the free varational problem

$$\int_\Omega (u+\lambda) \sqrt{1+ | Du |^2} \, dx \; \to \; \text{stationary}$$

in a class of functions which is defined by boundary conditions. The corresponding Euler equation is given by

$$\operatorname{div} \left\{ \frac{(u+\lambda) Du}{\sqrt{1+ | Du |^2}} \right\} = \sqrt{1+ | Du |^2} \tag{1}$$

or equivalently if $(u+\lambda) > 0$,

$$\operatorname{div} \left\{ \frac{Du}{\sqrt{1+ | Du |^2}} \right\} = \frac{1}{(u+\lambda)\sqrt{1+ | Du |^2}} \, . \tag{1'}$$

Thus the higher-dimensional analogue of the (one-dimensional) catenary problem may be formulated as follows:

Let $\Omega \subset \mathbb{R}^n$, $\varphi \in C^0(\overline{\Omega})$ and $A \in \mathbb{R}$ be given. Find a function $u \in C^{2,\alpha}(\Omega) \cap C^0(\overline{\Omega})$ and some $\lambda \in \mathbb{R}$ such that

$$(P) \quad \begin{cases} \sqrt{1+ | Du |^2} \, \operatorname{div} \dfrac{Du}{\sqrt{1+ | Du |^2}} = \dfrac{1}{u+\lambda} \quad \textit{in } \Omega \, , \\[2ex] u = \varphi \textit{ on } \partial\Omega \, , \quad \textit{and} \\[2ex] A(u) = \displaystyle\int_\Omega \sqrt{1+ | Du |^2} \quad dx = A \end{cases}$$

Clearly, there is an obvious necessary condition on the number A, namely that $A \geq A_0$, A_0 denoting the infimum of area of all graphs bounded by φ. But, surprisingly, and in contrast to the one dimensional situation, there is a further necessary condition, namely that $A \leq a_1(\varphi)$, $a_1(\varphi)$ denoting some specific number depending on the boundary values φ. Indeed, it was shown by Nitsche [42] that problem (P) in the corresponding rotationally symmetric case has no solution, provided $A > a_1(\varphi)$.

In the light of the above remarks the following existence result is natural (and, in a sense, optimal).

Theorem. ([21]) *Let $\Omega \subset \mathbb{R}^n$ be a bounded, mean-convex domain of class $C^{2,\alpha}$ and suppose $\varphi \in C^{2,\alpha}$. Then there exists some number $A_1 > A_0$ depending only on $n, \Omega, |\varphi|_{2,\alpha}$ such that for all numbers $A \in (A_0, A_1]$ there is some $\lambda \in \mathbb{R}$ and a function $u = u_\lambda \in C^{2,\alpha}(\overline{\Omega})$ which solves the problem (P).*

Let us also remark that the Dirichlet problem corresponding to $(1')$ is always solvable under natural assumptions on Ω and φ, cf. [20].

A parametric approach in the two-dimensional case is given in Böhme-Hildebrandt-Tausch [5].

It is worth mentioning that in this form, equation (1) or $(1')$ has been derived by Lagrange [37], pp. 158–162, Cisa de Gresy [13], pp. 274–276, and also Jellett [34], pp. 349–354, as the equilibrium condition for a *heavy, inextensible* and *flexible* surface of constant mass density, which is exposed to a vertical gravitational field. Later on a different approach to describing the equilibrium of a flexible surface in a force field was given by Poisson [46], pp. 173–187, who uses direct arguments from mechanics. For surfaces in $\mathbb{R}^3$ he introduces two independent "*tensions*" T and T', which describe the forces inside the surface. Let the external force field be given by X, Y, Z then he deduces the following equilibrium conditions

$$(A) \quad Xk + \frac{\partial}{\partial x}\left[\frac{T(1+q^2)}{k}\right] - \frac{\partial}{\partial y}\left[\frac{T'pq}{k}\right] = 0,$$

$$(B) \quad Yk - \frac{\partial}{\partial x}\left[\frac{Tpq}{k}\right] + \frac{\partial}{\partial y}\left[\frac{T'(1+p^2)}{k}\right] = 0,$$

$$(C) \quad Zk + \frac{\partial}{\partial x}\left[\frac{Tp}{k}\right] + \frac{\partial}{\partial y}\left[\frac{T'q}{k}\right] = 0,$$

where we have put $p := \frac{\partial u}{\partial x}$, $q = \frac{\partial u}{\partial y}$ and $k = \sqrt{1+p^2+q^2}$.

Poisson is especially interested in the case where $T = T'$ i.e. when the tensions coincide. In fact, in this case one easily deduces from (A), (B) and (C) the equation

$$(D) \quad Z - pX - qY + \frac{T}{k^2}\left[(1+q^2)\frac{\partial^2 u}{\partial x^2} - 2pq\frac{\partial^2 u}{\partial x \partial y} + (1+p^2)\frac{\partial^2 u}{\partial y^2}\right] = 0 \text{ as well}$$

as the condition

(E) $X dx + Y dy + Z dz + dT = 0$, i.e. the external force must have a potential $\mathcal{U}$ and $T = \mathcal{U} + c$.

From (D) and (E) Poisson deduces:

(I) The *minimal surface* equation, by taking $X = Y = Z = 0$ and $T = \text{const.}$

(II) The equation for *capillary surfaces* by taking

$$X = -\frac{pN}{k}\,,\; Y = -\frac{qN}{k}\,,\; Z = \frac{N}{k}\,,$$

where $N = a + bz$, and hence $T = \text{const.}$

Thus (D) reduces to

$$a + bz + \frac{T}{k^3}\left[(1+q^2)\,\frac{\partial^2 u}{\partial x^2} - 2pq\,\frac{\partial^2 u}{\partial x \partial y} + (1+p^2)\,\frac{\partial^2 u}{\partial y^2}\right] = 0$$

which is the familiar equation for capillary surfaces. According to Poisson, this is also the equilibrium condition of a flexible surface which is covered by a heavy fluid (provided we assume $T = T'$).

(III) The equation of a *heavy surface in a gravitational field*, by taking $X = Y = 0\,,\, Z = g\epsilon\,,\, g =$ gravitational constant, and ϵ denoting the density of the surface. The tension is then given by $T = c - g\,\epsilon z$ and hence (D) yields the relation

$$g\epsilon + \frac{c - g\,\epsilon z}{k^2}\left[(1+q^2)\,\frac{\partial^2 u}{\partial x^2} - 2pq\,\frac{\partial^2 u}{\partial x \partial y} + (1+p^2)\,\frac{\partial^2 u}{\partial y^2}\right] = 0$$

which is equivalent to (1) or (1′) if one takes $g\,\epsilon = 1$.

Furthermore it is interesting to observe that equation (1) or (1′) describes the shape of a *"hanging roof"*, which in turn is of importance for the construction of perfect domes. In fact, turning a "hanging solution" upside down gives the optimal form of a cupola – a construction principle which was already known to R. Hooke in 1675: Ut pendet continuum flexile, sic stabit contiguum rigidum inversum, i.e. as hangs the flexible line, so but inverted will stand the rigid arch. For a modern derivation of (1) using principles from mechanics we refer to the monograph of the architect F. Otto [44], p. 290. Architects have coined the name "*Gleichgespannte Membranen unter Eigenlast*" (membranes of equal tensions under their own weight) or "*heavy surfaces of minimal surface type*" for solutions of (1).

There is still another interpretation of equation (1). To this end let us consider somewhat more generally the integral

$$E_\alpha = \int\limits_\Omega u^\alpha \sqrt{1 + |\,Du\,|^2}\, dx \quad , \quad \alpha \in \mathbb{R}\,,$$

and the corresponding Euler equation

$$\operatorname{div} \frac{Du}{\sqrt{1 + |\,Du\,|^2}} = \frac{\alpha}{u\sqrt{1 + |\,Du\,|^2}}\,. \tag{2}$$

Furthermore suppose that $\mathbb{R}^{n+1}$ (or $\mathbb{R}^n \times \mathbb{R}^+$, $\mathbb{R}^+ = \{x_{n+1} \geq 0\}$) is endowed with the singular metric $g = (g_{ij}) = | x_{n+1} |^{\frac{2\alpha}{n}} \delta_{ij}$, for $i, j = 1, \ldots, n+1$. Let $\Omega \subset \mathbb{R}^n$ be a bounded domain and $u : \Omega \to \mathbb{R}^+$ be a C^1-function in Ω. Then the area of $graph(u) = \{(x, u(x));\ x \in \Omega\}$ with respect to the metric g over any compact subset K of Ω is given by

$$E_{\alpha,K}(u) = \int_K u^\alpha \sqrt{1 + | Du |^2}\, dx\,.$$

Thus it follows that $graph(u)$ is a minimal hypersurface in $(\mathbb{R}^n \times \mathbb{R}^+ , g)$ provided u satisfies (2). Note that there is an important fact about E_α, namely that

$$E_{\alpha,K}(u) = \lambda^{n+\alpha} E_{\alpha,K_\lambda}(u_\lambda)\,,$$

where $u_\lambda = \lambda u(x/\lambda)$ and $K_\lambda = \{\lambda x; x \in K\}$. Thus u_λ satisfies (2) whenever u does. Observe also that $\alpha = -n$ corresponds to the hyperbolic metric on the upper half-space, while $\alpha = 0$ gives the euclidean case.

Although the integral $E_\alpha(u)$ may be defined for functions $u \in \mathcal{BV}(\Omega)$ (cf. Anzellotti [2]), it is not possible to find a minimum of E_α in $\mathcal{BV}(\Omega)$, since the integral is merely singular elliptic. It turns out, that a natural class to work with is the function space

$$\mathcal{BV}^+_{1+\alpha}(\Omega) = \{u \in L_{1+\alpha}(\Omega) : u \geq 0\,, u^{1+\alpha} \in \mathcal{BV}(\Omega)\}\,.$$

For $u \in \mathcal{BV}^+_{1+\alpha}(\Omega)$ we define the area E_α as a Radon measure as follows:

$$\int_\Omega u^\alpha \sqrt{1 + | Du |^2} := \sup \Big\{ \int_\Omega \Big(u^\alpha g_{n+1} + \frac{1}{1+\alpha} u^{1+\alpha} \sum_{i=1}^n D_i g_i\Big)\, dx;$$

$$g \in C^1_c(\Omega, \mathbb{R}^{n+1})\,, \quad | g(x) | \leq 1\Big\}\,.$$

Motivated by the trace formula for functions of bounded variation one is then led to the following Dirichlet problem: Let $\Omega \subset \mathbb{R}^n$ be a Lipschitz domain and suppose $\varphi \in L_{1+\alpha}(\partial\Omega)$, $\varphi \geq 0$.

Minimize

$$\int_\Omega u^\alpha \sqrt{1 + | Du |^2} + \frac{1}{1+\alpha} \int_{\partial\Omega} | u^{1+\alpha} - \varphi^{1+\alpha} |\, dH_{n-1}$$

in the class $\mathcal{BV}^+_{1+\alpha}(\Omega)$.

It turns out that this problem is solvable (see [3]). Generally speaking solutions are analytic on the set $\{u > 0\}$ (see [3], [19]) while singularities may occur on the boundary $\partial\{u = 0\}$. In addition the coincidence set $\{u = 0\}$ of a minimizer is nonempty if we assume $\sup_{\partial\Omega} \varphi < \frac{|\Omega|}{|\partial\Omega|}$ (in the case $\alpha = 1$ e.g.). In fact, it has been shown in [15] that for all $\alpha > 0$ and $n \geq 2$ there are minima which are merely of Hölder class $C^{0,1/2}$. Also it would be desirable to prove that any minimizer is hölder continuous with exponent $1/2$. So far, only partial answers are known, cf. [3].

Now we turn to the global behaviour of the smooth solutions of the Euler equation (2). Here it is necessary and convenient to also consider the parametric counterpart of E_α, namely

$$\mathcal{E}_\alpha(M) = \int_M |\, x_{n+1} \,|^\alpha \, dH_n \,,$$

where $M \subset \mathbb{R}^{n+1}$ denotes some hypersurface of $\mathbb{R}^{n+1}$ and H_n stands for the n-dimensional Hausdorff-measure.

We say that a Lipschitz-boundary $M = \partial U\,,\, U \subset \mathbb{R}^{n+1}$ has least area (or mass) $\mathcal{E}_\alpha$ in a ball $B_r(\zeta) \subset \mathbb{R}^{n+1}\,,\, \zeta \in M\,$, if

$$\mathcal{E}_\alpha(M \cap B_r(\zeta)) = \int_{M \cap B_r(\zeta)} |\, x_{n+1} \,|^\alpha \, dH_n \leq \int_{\overline{M} \cap B_r(\zeta)} |\, x_{n+1} \,|^\alpha \, dH_n,$$

whenever $\overline{M} = \partial \mathcal{U}\,,\, \mathcal{U} \subset \mathbb{R}^{n+1}$, is a Lipschitz manifold and $(U - \mathcal{U}) \cup (\mathcal{U} - U)$ has compact closure in $B_r(\zeta)$. M minimizes $\mathcal{E}_\alpha$ globally, provided this holds for any r. Obviously, a C^2-boundary M of least mass $\mathcal{E}_\alpha$ in $B_r(\zeta)$ satisfies the *stability inequality*

$$\int_M |\, x_{n+1} \,|^\alpha \, (H^2 + \alpha \,|\, A \,|^2)\eta^2 \, dH_n \leq \alpha \int_M |\, x_{n+1} \,|^\alpha \,|\, \nabla \eta \,|^2 \, dH_n \tag{3}$$

for all $\eta \in C_c^1(M \cap B_r(\zeta))$, cp [16]

Note that (3) expresses the fact that the second variation of the area $\mathcal{E}_\alpha$ is nonnegative and that we have assumed M to be *stationary*, i.e. the mean curvature H fulfills

$$H = \alpha \, x_{n+1}^{-1} \, \nu_{n+1} \,, \tag{4}$$

$\nu = (\nu_1\,, \ldots\,, \nu_{n+1})$ denoting the unit normal, while $|\, A \,|$ stands for the length of the second fundamental form. The mean curvature equation (4) is just the parametric version of equation (2).

It is now a very natural and important question, whether equation (4) admits a *Bernstein theorem*. By this we mean a theorem of the type:

global existence ⇒ triviality of the solution in a specified sense (e.g. linearity)

Recall that the classical minimal surface equation

$$\operatorname{div} \frac{Du}{\sqrt{1 + |\, Du \,|^2}} = 0\,, \ \text{ in } \mathbb{R}^n\,, \tag{5}$$

does have this property, provided $n \leq 7$.

In fact, this was shown to hold by Bernstein ($n = 2$, [4], see also Hopf [32]), De Giorgi ($n = 3$), [14], using important previous work by Fleming [29], Almgren ($n = 4$), [1] and Simons [53] ($n = 5, 6, 7$). Bombieri-De Giorgi-Giusti [7] in a celebrated paper then proved that the Bernstein property fails in dimension $n \geq 8$. In fact it is the existence of 7-dimensional singular minimizing cones in $\mathbb{R}^8$ which admit the construction of non-linear entire solutions of the equation (5) in $\mathbb{R}^8$.

Imposing suitable boundedness or growth conditions on $| Du |$ a Bernstein theorem holds without any dimensional restriction (see Moser [39], Bombieri-DeGiorgi-Miranda [8], Nitsche [43], Simon [52], Caffarelli-Nirenberg-Spruck [10], Ecker-Huisken [23]). In addition, there are several important variants and extensions of Bernstein's result to surfaces of arbitrary dimension and codimension in euclidean $\mathbb{R}^N$ as well as in riemannian manifolds $\mathcal{M}^{\mathrm{N}}$ (cf. Heinz [30], Schoen-Simon-Yau [49], Hildebrandt-Jost-Widman [31], Fischer-Colbrie [27], Chern-Osserman [11]). Bernstein-type results for solutions of the non-parametric Euler equation of a $C^{2,\alpha}$ parametric elliptic variational integral with integrand not depending explicitly on the spatial variables were given by Jenkins [35] ($n = 2$) and L. Simon [51] for $n \geq 3$. It turns out that a Bernstein result always holds for such equations if $n = 2$ or 3 and, additionally, up to $n \leq 7$ provided the associated parametric integrand is sufficiently close to the area integrand (in the C^3-topology).

Further extensions of the Bernstein theorem to maximal space-like hypersurfaces in the Lorentz-Minkowski spaces were given by Calabi [9] ($n \leq 4$) and Cheng-Yau [12] (general n).

Now we return to the equation (2) or its parametric counterpart (4). It turns out that there are entire (i.e. globally defined), non-trivial solutions of (2). These solutions may be obtained by a careful analysis of the ordinary differential system corresponding to rotationally symmetric solutions of (2) see Keiper [36]. However the situation drastically changes, if we additionally require stability of the solutions (i.e. the inequality (3) where $M = graph(u)$).

Theorem 1. ([17]). *Suppose* $\alpha > 0$, $\alpha + n < 4 + 2\sqrt{\frac{2}{n+\alpha}}$ (i.e. $\alpha + n < 5.23$). *Then there is no entire, stable, solution* $u \in C^2(\mathbb{R}^n)$ *of the Euler equation* (2).

In particular observe that there are no *minimizers* for the area functional E_α under the condition required on n and α.

Thus for $n = 5$ and $\alpha = 0.2$ for example there are no entire stable solutions of equation (2), while, it turns out, that in the case $n = 5$ and $\alpha = 2$ e.g. there are rotationally symmetric, classical and globally stable solutions of (2) defined on all of $\mathbb{R}^5$ (cf. the discussion in Keiper [36]). Furthermore the rotationally symmetric solution in the case $\alpha = 1, n = 5$ turn out not to be globally stable, thus we conjecture that the optimal bound on $\alpha + n$ in the theorem is somewhere between 6 and 7.

Note that there is no Bernstein theorem for negative values of α, as follows for example from [38] in the hyperbolic case $\alpha = -n$. In fact there are non totally geodesic minimizing graphs with asymptotic boundary at infinity.

Now we turn to the parametric version of Theorem 1, and, as an additional assumption, we have to require a minimizing property of the surface.

Theorem 2. ([17]) *Suppose* $\alpha > 0$, $\alpha + n < 4 + 2\sqrt{\frac{2}{n+\alpha}}$ *and let* M *be a smooth (at least* C^2*) boundary of least area* $\mathcal{E}_\alpha$. *Then* M *has to be a hyperplane which is either perpendicular to the coordinate plane* $\{x_{n+1} = 0\}$ *or is identical to* $\{x_{n+1} = 0\}$.

Note that the smoothness assumption "$M \in C^2$" is necessary in Theorem 2, since there are Lipschitz boundaries which globally minimize $\mathcal{E}_\alpha$ and which are not hyperplanes, e.g. any Lipschitz hypersurface which is the union of two half-planes E_1, E_2, where $E_1 \subset \mathbb{R}^n \times \{x_{n+1} \geq 0\}$ is perpendicular to $\{x_{n+1} = 0\}$ and $E_2 \subset \{x_{n+1} = 0\}$ is one of the half-planes which are determined by $E_1 \cap \{x_{n+1} = 0\}$.

A quantitative version of the Bernstein theorem for the minimal surface equation in two dimensions was first provided by E. Heinz [30]. He proved the existence of an absolute constant c such that the principal curcatures κ_1, κ_2 of the graph of any solution u fulfill the inequality

$$(\kappa_1^2 + \kappa_2^2)\,(x_0) \leq c\,r^{-2}\,, \tag{6}$$

provided u satisfies (5) on a ball $B_r(x_0) \subset \mathbb{R}^2$.

Clearly, the Bernstein theorem in case $n = 2$, follows from this estimate by letting $r \to \infty$. Many authors have refined the curvature estimate (6) and its proof, in particular Hopf [33], Nitsche [40, 41], Osserman [45], Finn-Osserman [26] and others. R. Schoen [47] proved an analogue of Heinz's inequality for stable minimal surfaces in three-dimensional manifolds, extending earlier results of doCarmo-Peng [22] and Fischer-Colbrie and Schoen [28], namely that a complete stable minimal surface in $\mathbb{R}^3$ has to be a plane. So far, the results mentioned are all two-dimensional.

The step from two dimensions to the higher-dimensional case requires completely different techniques and was first performed by Schoen-Simon-Yau [49], $n \leq 5$ and Simon [50], $n \leq 7$. In fact the estimate of Schoen-Simon-Yau also applies to stable minimal hypersurfaces in riemannian manifolds, provided suitable curvature assumptions are satisfied by the ambient manifold (for a very general estimate see also Schoen and Simon [48]).

Concerning the equation (2) (or its parametric counterpart (4)) we have

Theorem 3. ([18]) *Let* $u \in C^2(B_r(x_0))$, $B_r(x_0) \subset \mathbb{R}^n$, *be a solution of (3) such that* $M = graph(u)$ *is stable in* $B_r(\xi) \subset \mathbb{R}^{n+1}$, $\xi = (x_0, u(x_0))$. *Suppose* $\alpha + n < 4 + 2\sqrt{\dfrac{2}{n+\alpha}}$ *and* $\alpha > 0$. *Then there is an absolute constant* $\varepsilon_0 = \varepsilon_0\,(n, \alpha) \in (0, 1)$ *such that for* $r \leq \varepsilon_0\,\xi_{n+1}$ $(= \varepsilon_0\,u\,(x_0))$ *we have the estimate.*

$$(H^2 + \alpha\,|\,A\,|^2)\,(\xi) \leq c_1\left(n, \alpha, \frac{\xi_{n+1}}{r}\right) r^{-2}\,, \tag{7}$$

while for $r \geq \varepsilon_0\, \xi_{n+1}$ the inequality

$$(H^2 + \alpha \mid A \mid^2)\,(\xi) \leq c_2(n,\alpha,q) r^{-2} \left(\frac{r}{\xi_{n+1}}\right)^{\frac{n+\alpha}{2+q}} \left[1 + \left(\frac{\xi_{n+1}}{r}\right)^{\alpha}\right]^{\frac{1}{2+q}} \tag{8}$$

holds true for all $q \in [0, \sqrt{2/(n+\alpha)})$ with some constant c_2 depending only on n, α and q.

Remark. Note that under the assumption of Theorem 1 we find $n + \alpha < 2(2+q)$ for some $q \in [0, \sqrt{2/(n+\alpha)})$. Thus Theorem 3 implies Theorem 1.

Again, this theorem extends to the parametric case, assuming in addition a minimizing property of the surface.

Theorem 4. ([18]) *Let $M = \partial\mathcal{U}\,, \mathcal{U} \subset \mathbb{R}^{n+1}$, be a C^2-boundary of least area $\mathcal{E}_\alpha$ in a ball $B_r(\xi) \subset \mathbb{R}^{n+1}$ centered at $\xi \in M$ with $\xi_{n+1} \neq 0$. Suppose $\alpha > 0\,,\ \alpha + n < 4 + 2\sqrt{2/(n+\alpha)}$. Then we have the estimates (7) & (8) as in Theorem 3.*

These results are proved by first obtaining integral curvature estimates for stable hypersurfaces $M \subset \mathbb{R}^{n+1}$ of the kind that were established for stable *minimal* hypersurfaces by Schoen-Simon-Yau [49]. To this end one first has to derive a generalized "Simons" inequality for the Laplacian of the length of the second fundamental form $\mid A \mid$ and the Laplacian of the mean curvature H. Then one uses this inequality together with the stability estimate (3) to derive a curvature estimate of the type

$$\int_M \mid x_{n+1} \mid^\alpha \,[\alpha \mid A \mid^2 + H^2]^{2+q}\, \xi^{4+2q}\, dH_n \leq$$
$$C(n,\alpha) \int_M \mid x_{n+1} \mid^\alpha \mid \nabla\xi \mid^{4+2q}\, dH_n$$

for all $\xi \in C^1_c\,(M, \mathbb{R})$ and all positive $q < \sqrt{\dfrac{2}{n+\alpha}}$.

Theorem 1 & 2 then immediately follow by proving suitable growth estimates for the area

$$\int_{M \cap B_R(\xi)} \mid x_{n+1} \mid^\alpha \, dH_n \qquad \text{for large values of } R.$$

Finally theorems 3 and 4 are proved by using a Moser type iteration argument on the surface M in combination with the generalized Simons inequality and a Sobolev type estimate holding for hypersurface M which satisfy (4).

For details we refer to [17] and [18].

References

[1] Almgren, F.J., Some interior regularity theorems for minimal surfaces and an extension of Bernstein's theorem. *Ann. Math.* **84** (1966), 277–292.

[2] Anzellotti, G., Dirichlet problem and removable simgularities for functionals with linear growth. *Boll. U. M. I.* **18** (1981), 141–159.

[3] Bemelmans, J., Dierkes, U., On a singular variational integral with linear growth. *Arch. Ration. Mech. Anal.* **100** (1987), 83–103.

[4] Bernstein, S., Sur un théoreme de géometrie et ses applications aux équations aux dérivées partielles du type elliptique. *Commun. Soc. Math. Kharkov* **15** (2) (1915–1917), 38–45.

[5] Böhme, R., Hildebrandt, S., Tausch, E., The two-dimensional analogue of the catenary. *Pac. J. Math.* **88** (1980), 247–278.

[6] Bolza, O., Vorlesungen über Variationsrechnung. Teubner 1909.

[7] Bombieri, E., De Giorgi, E., Giusti, E., Minimal cones and the Bernstein problem. *Invent. Math.* **88** (1969), 243–268.

[8] Bombieri, E., De Giorgi, E., Miranda, M., Una maggiorazione a priori relativa alle ipersuperfici minimali non parametriche. *Arch. Ration. Mech. Anal.* **32** (1969), 255–267.

[9] Calabi, E., Examples of Bernstein problems for some nonlinear equations. *Proc. Sym. Global Analysis* Univ. of Calif., Berkeley (1968).

[10] Caffarelli, L., Nirenberg, L., Spruck, J., On a form of Bernstein's Theorem. (Preprint)

[11] Chern, S.S., Osserman, R., Complete minimal surfaces in euclidean n-space. *J. Anal. Math.* **19** (1967), 15–34.

[12] Cheng, S.Y., Yau, S.T., Maximal space-like hypersurfaces in the Lorentz - Minkowski spaces. *Ann. Math.* **104** (1976), 407–419.

[13] Cisa de Gresy, Considération sur l'équilibre des surfaces flexibles et inextensibles. *Mem. Reale. Accad. Sci. Torino* **23** (1) (1818), 259–294.

[14] De Giorgi, E., Una estensione del teorema di Bernstein. *Ann. Sc. Norm. Super.* Pisa, (III) **19** (1965), 79–85.

[15] Dierkes, U., Minimal hypercones and $C^{0,1/2}$ minimizers for a singular variational problem. *Indiana Univ. Math. J.* **37** (4) (1988), 841–863.

[16] Dierkes, U., On the non-existence of energy stable minimal cones. *Ann. Inst. Henri Poincaré, Anal. Non Linéaire* **7** (6) (1990), 589–601.

[17] Dierkes, U., A Bernstein result for energy minimizing hypersurfaces. *Calc. Var.* **1** (1993), 37–54.

[18] Dierkes, U., Curvature estimates for minimal hypersurfaces in singular spaces. *Invent. Math.* **122** (1995), 453–473.

[19] Dierkes, U., On the regularity of solutions for a singular variational problem. Math. Z. **225** (1997), 657–670.

[20] Dierkes, U., Huisken, G., The N-dimensional analogue of the catenary: existence and non-existence. *Pac. J. Math.* (I) **141** (1990), 47–54.

[21] Dierkes, U., Huisken, G., The N-dimensional analogue of the catenary: Prescribed area. In: Calculus of Variations and Geometric Analysis. *Ed. J. Jost.* International Press (1996), 1–13.

[22] doCarmo, M., Peng, C.K., Stable complete minimal surfaces in $\mathbb{R}^3$ are planes. *Bull. AMS* (1) (1979), 903–905.

[23] Ecker, K., Huisken, G., A Bernstein result for minimal graphs of controlled growth. *J. Differ. Geom.* **31** (1990), 397–400.

[24] Ecker, K., Huisken, G., Interior curvature estimates for hypersurfaces of prescribed mean curvature. *Ann. Inst. Henri Poincaré. Anal. Non Linéaire* (4) **6** (1989), 251–260.

[25] Euler, L., Opera omnia, vol X et XI, seriei secundae.

[26] Finn, R., Osserman, R., On the Gauss curvature of non-parametric minimal surfaces. *J. Analyse Math.* **12** (1964), 351–364.

[27] Fischer-Colbrie, D., Some rigidity theorems for minimal submanifolds of the sphere. *Acta Math.* **145** (1980), 29–46.

[28] Fischer-Colbrie, D., Schoen, R., The structure of complete stable minimal surfaces in 3-manifolds of nonnegative scalar curvature. *Comm. Pure Appl. Math.* **33** (1980), 199–211.

[29] Fleming, W.H., On the oriented Plateau problem. *Rend. Circ. Mat. Palermo* **2** (1962), 1–22.

[30] Heinz, E., Über die Lösungen der Minimalflächengleichung. *Nachr. Akad. Wiss. Gött. Math. Phys. Kl.* (II) (1952), 51–56.

[31] Hildebrandt, S., Jost, J., Widman, K.O., Harmonic mappings and minimal submanifolds. *Invent. Math* **62** (1980), 269–298.

[32] Hopf, E., On S. Bernstein's theorem on surfaces $z(x, y)$ of nonpositive curvature. *Proc. Amer. Math. Soc.* **1** (1950), 80–85.

[33] Hopf, E., On an inequality for minimal surfaces $z = z(x, y)$. *J. Rat. Mech. Anal.* **2** (1953), 519–522.

[34] Jellett, F.H., Die Grundlehren der Variationsrechnung. Braunschweig, Verlag der Hofbuchhandlung v. Leibrock, 1860.

[35] Jenkins, H., On two-dimensional variational problems in parametric form. *Arch. Ration. Mech. Anal.* **8** (1961), 181–206.

[36] Keiper, J.B., The axially symmetric n-tectum. (Preprint)

[37] Lagrange, J.L., Mécanique analytique quatrième édition. Œuvres tome onzième.

[38] Lin, F.H., On the Dirichlet problem for minimal graphs in hyperbolic space. *Invent. Math.* **96** (1989), 593–612.

[39] Moser, J., On Harnack's theorem for elliptic differential equations. *Commun. Pure Appl. Math.* **14** (1961), 557–591.

[40] Nitsche, J.C.C., Über eine mit der Minimalflächengleichung zusammenhängende analytische Funktion und den Bernstein'schen Satz. *Arch. d. Math.* **7** (1956), 417–419.

[41] Nitsche, J.C.C., On an estimate for the curvature of minimal surfaces $z = z(x, y)$. *J. Math. Mech.* **7** (1958), 767–770.

[42] Nitsche, J.C.C., A nonuniqueness theorem for the two-dimensional analogue of the catenary. *Analysis* **6** (1986), 143–156.

[43] Nitsche, J.C.C., Lectures on minimal surfaces. Vol. 1. *Cambridge University Press* 1989.

[44] Otto, F., Zugbeanspruchte Konstruktionen. Bd. I, II. Berlin Frankfurt/M. Wien: Ullstein 1962, 1966.

[45] Osserman, R., On the Gauss curvature of minimal surfaces. *Trans. Amer. Math. Soc.* **96** (1960), 115–128.

[46] Poisson, Sur les surfaces élastique. *Mem. Cl. Sci. Math. Phys. Inst.* France 1812, deux. p. 167–225.

[47] Schoen, R., Estimates for stable minimal surfaces in three dimensional manifolds. In: Seminar on Minimal Submanifolds, Princeton University Press 1983, 111–126.

[48] Schoen, R., Simon, L., Regularity of stable minimal hypersurfaces. *Comm. Pure Appl. M.* **34** (1981), 741–797.

[49] Schoen, R., Simon, L., Yau, S.T., Curvature estimates for minimal hypersurfaces. *Acta Math.* **134** (1975), 275–288.

[50] Simon, L., Remarks on curvature estimates for minimal hyperfaces. *Duke Math. J.* **43** (3) (1976), 545–553.

[51] Simon, L., On some extensions of Bernstein's theorem. *Math. Z.* **154** (1977), 265–273.

[52] Simon, L., Entire solutions of the minimal surface equation. Preprint, CMA-R25-87 *Austral. Nat. Univ. Canberra* 1987.

[53] Simons, J., Minimal varieties in Riemannian manifolds. *Ann. Math.* **88** (1968), 62–105.

[54] Truesdell, C., The rational mechanics of flexible or elastic bodies, 1638–1788. Introduction to Leonhardi Euleri Opera omnia, vol. X et XI, seriei secundae.

Progress in Mathematics, Vol. 168, © 1998 Birkhäuser Verlag Basel/Switzerland

Surfaces in 3-Torus: Geometry of Plane Sections

I. DYNNIKOV*

Dept. of Mech. and Math., Moscow State University
Moscow 119899, Russia
e-mail: dynnikov@mech.math.msu.su

Introduction

Let M be a closed smooth surface in the 3-dimensional torus $\mathbb{T}^3$, and let $\widehat{M}$ denote the covering surface of M in the 3-space $\mathbb{R}^3$: $\widehat{M} = p^{-1}(M)$, where p: $\mathbb{R}^3 \to \mathbb{T}^3 = \mathbb{R}^3/\mathbb{Z}^3$ is the natural projection. The surface $\widehat{M}$ is *3-periodic*, which means that it is invariant under the action of the lattice $\mathbb{Z}^3$. Given a non-zero vector $H = (H^1, H^2, H^3)$ consider all intersections of $\widehat{M}$ with planes orthogonal to the vector H. For almost any plane orthogonal to H such an intersection is a union of regular curves which we call *trajectories*. Some of them may be non-closed, and we will be interested in the *asymptotic behaviour* of non-closed trajectories.

If the surface M is defined as a level surface of a function f, that is $M = \{x \mid f(x) = C\}$, then the trajectories are the integral curves of the equation

$$\dot{x} = v(x) = H \times \nabla f(x)$$

where $\times$ is vector product and ∇f denotes the gradient of the function f.

This purely geometric and simply formulated problem posed in 1982 by S.P. Novikov [7] arises in solid state physics and the asymptotic properties of the non-closed trajectories for certain surfaces turned out to play a very important role in explaining some effects in conductivity theory (see [6, 1]). Here we will concentrate only on the topological and geometric aspects of the problem.

Of course, for some types of surfaces M, for example if M is a sphere, all the trajectories are closed no matter which vector H. But it is not difficult to construct a surface having non-closed intersection curves with some (or even with any) plane. We will see below a number of examples.

To begin with let us consider some special cases and present some simple ideas. Suppose that the vector H is *rational*, that is a scalar multiple of some integer one: $H = \lambda\mathbf{l}$, where $\mathbf{l} \in \mathbb{Z}^3$. Then, evidently, the intersection of the surface $\widehat{M}$ with any plane P perpendicular to the vector H is invariant under shifts by vectors from the 2-dimensional sublattice

$$\{\mathbf{m} \in \mathbb{Z}^3 \mid \mathbf{m} \perp \mathbf{l}\} \cong \mathbb{Z}^2.$$

*Work is supported by RFFI grant (96-01-01404).

Figure 1: Periodic trajectory

We express this by saying that the intersection $M \cap P$ is *2-periodic.*

It easily follows from 2-periodicity that each non-closed regular connected component of $M \cap P$ is *periodic*, which means that it is invariant under the shift by a non-zero vector $\mathbf{v} \in \mathbb{Z}^3$, $\mathbf{v} \perp \mathbf{l}$. Such a curve γ can be parametrized so that the difference

$$(\gamma(t) - \mathbf{v}t)$$

is a periodic function of t, see Figure 1.

Paper [6] by I. M. Lifschitz and V. G. Peschansky contains an interesting example of a surface M on which the non-closed trajectories, though not being periodic, lie in finite-wide strips when vector H is sufficiently close to some special one. We refer to this property of trajectories as *the strip assertion.* We illustrate the idea of [6] below using a small modification of that example.

Let $\widehat{M}$ be the surface defined by the equation

$$\cos(2\pi x_1) + \cos(2\pi x_2) + \cos(2\pi x_3) = 0,$$

where x_1, x_2, x_3 are the coordinates corresponding to the standard guage in 3-space. Denote the inside domain

$$\{(x_1, x_2, x_3) \in \mathbb{R}^3 \mid \cos(2\pi x_1) + \cos(2\pi x_2) + \cos(2\pi x_3) \leq 0\}$$

of the surface $\widehat{M}$ by $\widehat{I}$. We have: $\widehat{M} = \partial\widehat{I}$ and $\widehat{M} \cap P = \partial(\widehat{I} \cap P)$ for any plane P.

Let P_k be the plane defined by the equation $x_3 = k/2$, where k is an integer. Then for an even k the intersection set $\widehat{I} \cap P_k$ is the disjoint union of a 2-periodic family of identical disks as shown in Figure 2.a. For an odd k this intersection set looks like a plane with the union of a 2-periodical family of disks cut out (Figure 2.b).

Now let us consider the plane P defined by an equation of the form $\varepsilon x_1 + \delta x_2 + x_3 + c = 0$ where $(\varepsilon, \delta) \neq (0, 0)$, that is a plane perpendicular to the vector $H = (\varepsilon, \delta, 1)$ which is close to $(0, 0, 1)$. Such a plane intersects each of P_k along a straight line and if ε and δ are sufficiently small, the intersection of a wide strip in the plane P containing the line $P \cap P_k$ with the set $\widehat{I}$ will look like that of the corresponding strip in the plane P, see Figure 2.c. It readily follows from this that there is at least (in fact, exactly) one unbounded connected component of the intersection $P \cap \widehat{M}$ in the strip between each two straight lines $P \cap P_k$ and $P \cap P_{k+1}$ and each of the non-closed curves is contained in such a strip as a whole.

In 1982 S. P. Novikov [7] proposed to study the foliation defined by the restriction of the closed 1-form

$$\Omega = H^1 dx_1 + H^2 dx_2 + H^3 dx_3$$

Figure 2:

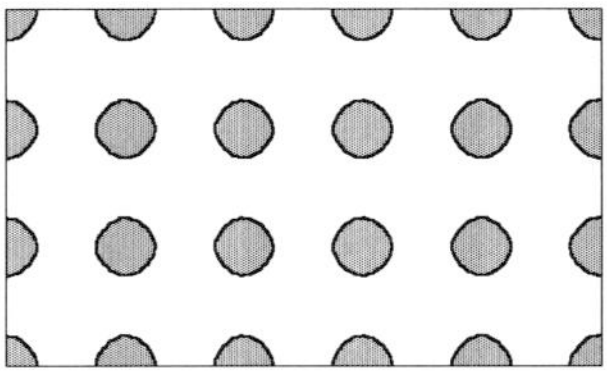

a) $\widehat{I} \cap P_{2l}$

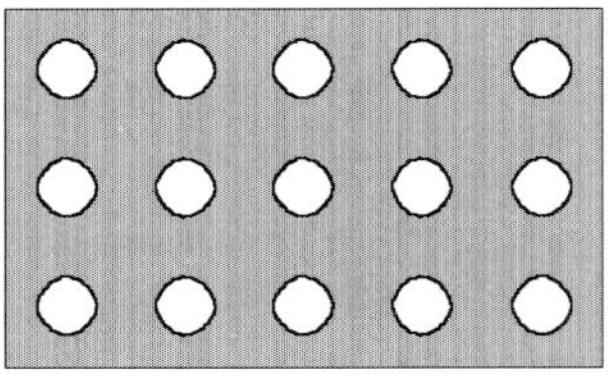

b) $\widehat{I} \cap P_{2l+1}$

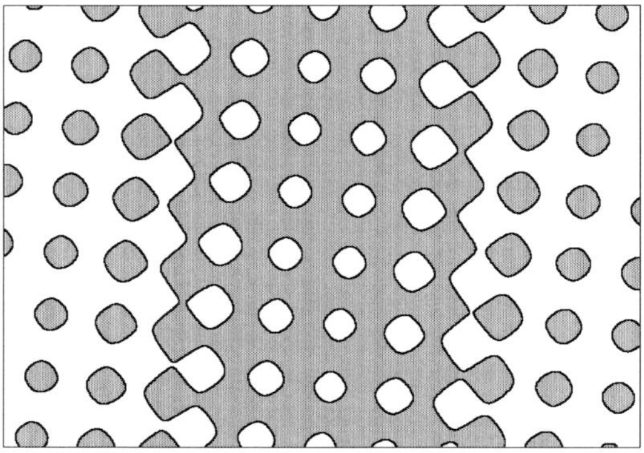

c) $\widehat{I} \cap P$

to a surface M embedded in the 3-torus, where x_1, x_2, x_3 are coordinates on the 3-torus regarded up to integers. The regular non-separatrix leaves of this foliation are just images of the trajectories in the 3-space we considered above. The first advantage taken of this approach is due to A. Zorich [10] who proved the following statement.

THEOREM 1 (A. V. ZORICH) *Let M be a fixed surface embedded in $\mathbb{T}^3$ and H_0 be a rational vector such that the restriction ω_0 of the 1-form*

$$\Omega_0 = H_0^1 dx_1 + H_0^2 dx_2 + H_0^3 dx_3$$

to the surface M is a Morse one and the foliation defined by ω has no saddle connections between different saddle critical points. Then for a vector H sufficiently close to H_0 any trajectory of the covering surface $\widehat{M}$ lies in a finite-wide strip.

Though the assertion of the theorem is similar to one stated above for a concrete example, the argument is quite different. The main observation made by Zorich was that under the hypothesis of the theorem one could find a family of closed trajectories $\gamma_1, \gamma_2, \ldots, \gamma_k$ homologous to zero in $\mathbb{T}^3$ which cut the surface M into pieces of genus 0 and 1. The pullbacks of these trajectories in the 3-space cut the covering surface $\widehat{M}$ into parts each of which lies between two parallel planes. The intersection of such a part with a plane of any other direction is contained in a strip of finite width.

Figure 3:

a) b)

The following two questions appear naturally. The first is whether each non-closed trajectory lies in a strip for any surface and *arbitrary* vector H? It follows from Zorich's theorem that for a generic surface M and a vector H from a small neighbourhood of a rational vector the strip assertion is valid. But these neighbourhoods of rational directions may not cover all the sphere S^2, and, moreover, the union of such neighbourhoods may be of arbitrarily small measure.

The second question concerns the behaviour of the non-closed trajectories in the case when each of them lies in a strip. *A priori* one expects two possible principally different cases shown on Figure 3. A non-closed trajectory can go through the strip (Figure 3.a) or come back to that "infinity" it comes from (Figure 3.b).

It was conjectured by S. Novikov [8] that the non-closed trajectories always lie in finite-wide strips and come through the strips in one direction. It turned out that the strip assertion is not valid for an arbitrary pair (M, H). The first counterexample was found by S. Tsarev [9], in which each non-closed trajectory has an asymptotic direction in a common sense but cannot be squeezed in a strip. Examples of surfaces on which the non-closed trajectories have no asymptotic direction were constructed by the author in [3].

However, the assertion of Novikov's conjecture is valid "with probability one". As shown in papers [2, 3], a family of closed trajectories homologous to zero in $\mathbb{T}^3$ cutting the surface into pieces of genus ≤ 1 exists for almost any suface M and vector H. The following result was proved.

THEOREM 2 *For almost any surface M and vector H one of the following conditions holds:*

i) *all the trajectories on $\widehat{M}$ are closed;*

ii) *each non-closed trajectory on $\widehat{M}$ is contained in a strip of finite width.*

As for the behaviour of non-closed trajectories lying in strips, the case in which the non-closed trajectories come back is impossible. If a non-closed regular trajectory lies in a strip, then it has the form of a finitely disturbed straight line in the sense of the following definition.

Definition 0.1 We say that a curve $r(t)$, $t \in (-\infty, +\infty)$ has a *strong asymptotic direction* if there exist a parametrization $\tau(t)$ (possibly, different from the original one), $d\tau/dt \geq 0$, $\tau \to \pm\infty$ $(t \to \pm\infty)$, a vector u, $|u| = 1$, and a constant C such that

$$|r(\tau) - \tau u| < C \quad \text{for all } \tau.$$

The vector u is called then *the strong asymptotic direction vector.*

As shown in [2, 3], not any vector can be the strong asymptotic direction vector of a trajectory. The strong asymptotic direction is always perpendicular to an integer vector: $u \perp \mathbf{k}$, $\mathbf{k} \in \mathbb{Z}$.

Acknowledgements. I would like to express my gratitude to S. P. Novikov for formulating the problem and A. V. Zorich for stimulating discussions and help in studying the theory of closed 1-forms on surfaces.

1. Closed 1-forms on surfaces

Our problem may be regarded as a very special case in the existing theory of closed 1-forms on surfaces or, what is not the same but closely related, the theory of foliations with invariant transversal measure on surfaces. In this section we recall the notions of level curve and irrationality degree and discuss the relation of our problem with the general theory of closed 1-forms on surfaces.

Definition 1.1 For a closed 1-form ω on a surface we define its *level curve* passing through a point X_0 as the set of points X such that there exists a curve γ with ends X_0, X such that the restriction of ω to γ vanishes, $\omega\Big|_\gamma = 0$. A level curve is called *regular* if it contains no critical points of the 1-form ω, and it is called *singular* otherwise.

Remark. Note that the notion of a regular level curve coincides with the notion of a regular trajectory. A level curve passing through a critical point (stationary trajectory) also contains all the separatrices attached to it.

Definition 1.2 *The irrationality degree* of a closed 1-form ω is the dimension over the field $\mathbb{Q}$ of the space generated by the integrals of ω over all the integer cycles. In particular, the irrationality degree is zero iff the 1-form is the differential of a function. A closed 1-form is called *rational,* if its irrationality degree equals 1.

It is evident that any oriented surface M can be represented as a connected component of a level surface of some function $f : \mathbb{T}^3 \to \mathbb{R}$. In what follows we will always assume that the surface M is a level surface of a function. We will also consider the whole family of level surfaces of a fixed smooth function f and then we will use the notation

$$M_c = \{x \in \mathbb{T}^3 \mid f(x) = c\}.$$

In the physical theory where our problem arises the surfaces of interest are level surfaces by construction.

For a connected and oriented surface M to be a level surface of a function is equivalent to the condition that the image of the fundumental class $[M]$ in $H_2(\mathbb{T}^3, \mathbb{Z})$ is zero. If the surface M is not homologous to zero in $\mathbb{T}^3$ then our problem becomes much simpler. Indeed, if $\alpha = i_*([M]) \neq 0$, then there exists a homology class $\beta \in H_1(\mathbb{T}^3, \mathbb{Z})$ such that $\alpha \cap \beta \neq 0$. Hence the sublattice $L = i_*(H_1(M, \mathbb{Z}))$ of the lattice $H_1(\mathbb{T}^3, \mathbb{Z}) \cong \mathbb{Z}^3$ does not contain β and its dimension does not exceed 2.

We denote by $d(M)$ the dimension of the sublattice $L = i_*(H_1(M, \mathbb{Z}))$. If $d(M) \leq 2$ then any connected component of the covering surface $\widehat{M}$ lies between two planes parallel to the sublattice. Hence for any vector H not perpendicular to L each trajectory is contained in a finite-wide strip. For a vector H orthogonal to L the strip assertion is also valid because of the rationality of the vector.

Asymptotic properties for dynamical systems with invariant transversal measure on surfaces have been studied in a large number of papers. A frequently considered asymptotic characteristic, *asymptotic cycle*, is defined as follows.

Let $\gamma(t)$ be an integral curve starting at a point X_0. Let us fix some simply connected neighbourhood W of the point X_0. If γ is not a separatrix then it returns to the neighbourhood W infinitely many times. Let $t_1, t_2, \ldots$ be an increasing sequence tending to ∞ such that the points $\gamma(t_i)$ lie in W. Consider the loops l_k represented by the segments $t \in [0, t_k]$ of the trajectory $\gamma(t)$. Take some scalar product in $H_1(M, \mathbb{R})$ and consider the homology classes

$$\sigma_k = \frac{\alpha_k}{|\alpha_k|}, \quad \text{where } \alpha_k = [l_k] \in H_1(M, \mathbb{R}).$$

If the limit

$$\lim_{k \to \infty} \sigma_k$$

exists, it is called *the asymptotic cycle* of the trajectory.

For an arbitrary closed 1-form ω on a surface M the following assertion is valid with probability one (see [5]): the corresponding dynamical system is uniquely ergodic and all the non-separatrix trajectories have an asymptotic cycle which is up to a multiplication constant the Poincaré dual to $[\omega] \in H^1(M, \mathbb{R})$.

Unfortunately, such theorems cannot be applied to our situation because we consider a very special class of 1-forms ω on the surface M, namely, the 1-forms which are pullbacks of constant 1-forms on the torus $\mathbb{T}^3$ under the embedding $i : M \to \mathbb{T}^3$. Such 1-forms have an irrationality degree not more than 3, whereas a generic one has an irrationality degree equal to $2g$, where g is the genus of M.

As noted in [11], if our dynamical system is uniquely ergodic then the non-closed trajectories should have the asymptotic cycle which has zero image in $H_1(\mathbb{T}^3, \mathbb{R})$ since $i_*([M]) = 0$. This means that for a trajectory $\gamma(t)$ in the 3-space we would have

$$\lim_{T \to \infty} \frac{\gamma(T) - \gamma(0)}{l(T)} = 0,$$

where $l(T)$ denotes the length of the segment $\gamma(t), t \in [0, T]$.

2. Local parameters

As we work with infinite-dimensional functional space of pairs (M, H) it should be explained how words like "almost all" and "codimension one" make sense for sets of objects in that space. The point is that in a small neighbourhood of a generic point (M, H) of the space of all pairs (M, H) one can find a *finite* number of paramters which completely define the asymptotic behaviour of the trajectories and depend smoothly (in some natural sense) on M and H. In the case of a generic closed 1-form on a surface this statement follows from a result by A. Katok [4] a modification of which is given below.

Let ω be a closed 1-form on a compact surface M, and let A be the set of critical points of ω which are supposed to be non-degenerate. Let us take a closed 1-form ω' which is close (in the C^1-metric) to ω. Then the set A' of the critical points of ω' is a small disturbance of A. Thus, there is a natural isomorphism between the relative cohomologies: $H^1(M, A) \cong H^1(M, A')$.

THEOREM 3 *In the situation described above, if the relative cohomology classes* $[\omega]$ *and* $[\omega']$ *in* $H^1(M, A) \cong H^1(M, A')$ *coincide then a diffeomorphism* $\varphi : M \to M$ *exists such that* $\omega' = \varphi^*(\omega)$. *In particular, the foliations defined by these two 1-forms are topologically equivalent.*

Applied to our situation this theorem translates into the following, almost trivial, statement. Suppose that the 1-form ω has only Morse critical points denoted by $A_1, \ldots, A_m$. Consider a small deformation of the surface M, under which the number and types of the critical points of the restriction of the 1-form Ω do not change. Let $A_1 + dA_1, \ldots, A_m + dA_m$ denote the critical points on the deformed surface.

PROPOSITION 1 *If* $(H, dA_i) = 0$ *for all* $1 \le i \le m$ *then the foliation on the deformed surface is isomorphic to the original one. Moreover, for any plane* P *perpendicular to* H *the intersection set of* $\widehat{M}$ *with* P *and that of the deformed surface differ from each other in a small deformation.*

Let us comment on the previous statement. First, consider a deformation of the surface M inside a small ball not containing critical points of the 1-form ω. It is clear that in 3-space this will result in a small deformation of the trajectories encountering the preimage of the ball.

Now consider a deformation near a critical point A_i of the 1-form ω. Let $A_i + dA_i$ be the critical point after deformation. Denote by Δ a small line segment connecting A_i and $A_i + dA_i$ and by $\widehat{\Delta}$ its preimage in $\mathbb{R}^3$. If the segment Δ is not orthogonal to H, and P is a plane perpendicular to H, then as a result of the deformation we will have a surgery of $P \cap \widehat{M}$ near each point from the set $P \cap \widehat{\Delta}$. The type of the surgery depends on the type of the critical point and deformation. Appearing and collapsing of circles corresponds to maxima and minima points, whereas near a saddle point we will have a surgery shown in Figure 4. But if Δ is perpendicular to the vector H then no surgery of the intersection $P \cap \widehat{M}$ will happen.

Thus, locally the only parameters responsible for the behaviour of the trajectories regarded up to a small deformation are: 1) the components of the vector H and 2) the "*altitudes*" of the critical points, that is the numbers $(\tilde{A}_i, H)$, where $\tilde{A}_i$ denotes a point in the preimage $p^{-1}(A_i) \subset \mathbb{R}^3$ and $(\,,\,)$ denotes the scalar product.

Figure 4:

The number of these local parameters is different in different parts of the space of pairs (M, H) since so is the number of the critical points. But there exists a nice linear combination of them which has a purely geometric interpretation.

Suppose that the surface M is defined by an equation of the form $f(x) = 0$, where $f\colon \mathbb{T}^3 \to \mathbb{R}$ is a generic smooth function. Suppose also that the vector H is not rational. Consider the *closed* trajectories lying in a plane $P \perp H$. We call a closed trajectory *positive* if the vector $\nabla(f\big|_P)$ is directed outside the disk bounded by the trajectory and *negative* otherwise. Let us count the *algebraic number* of closed trajectories contained in a square of side R in the plane P. This number has the form $\rho R^2 + o(R^2)$ $(R \to \infty)$ where ρ is a constant not depending on the plane P. We call ρ *the algebraic density* of closed trajectories.

Assign to each critical point A_i a *weight* w_i defined as the index of the vector field $v(x) = H \times \nabla f(x)$ at the point A_i, multiplied by $\operatorname{sgn}(\nabla f(A_i), H)$.

THEOREM 4 *One can choose a point $\tilde{A}_i$ in the preimage $p^{-1}(A_i)$ of each critical point so that the following equality will hold*

$$(\sum_i w_i \tilde{A}_i, H) = |H|\rho.$$

For a continuous family of pairs (M, H) the points $\tilde{A}_i$ can be chosen in a continuous way.

For the proof of this statement see [3]. One can show that the sum of the w_i over all the critical points is equal to zero. This means that the sum $\sum_i w_i \tilde{A}_i$ is a well-defined vector in the 3-space. Thus, the algebraic density of the closed trajectories multiplied by $|H|$ locally is a non-trivial *linear* function of the relative cohomology class $[\omega] \in H^1(M, A)$.

3. Stable and non-stable dynamics

In this section we determine some stable and non-stable properties of the foliation defined by the restriction of constant 1-form Ω to the surface $M \subset \mathbb{T}^3$.

A generic closed 1-form on a surface typically does not have homologically non-trivial closed level curves since the necessary condition for the existence of such a curve is the vanishing of the integral of the 1-form along the corresponding cycle. In our case, if the genus of the surface M is greater than 1 then there should

Figure 5:

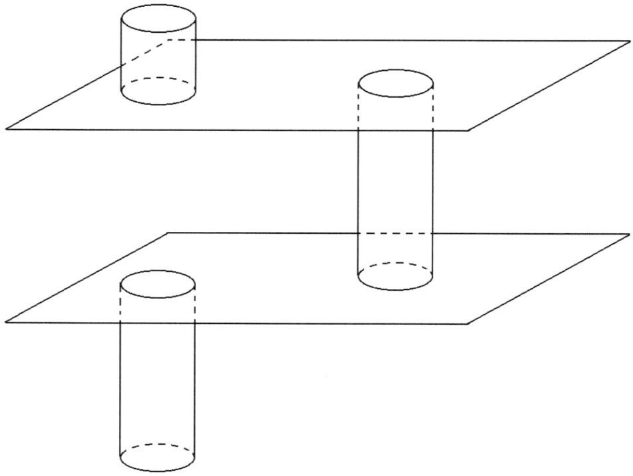

be cycles which are non-trivial in the homology of the surface but trivial in that of the 3-torus. The integral of the 1-form ω over such a cycle vanishes.

As a consiquence, the existence of closed trajectories becomes a stable property in our situation as opposed to the "generic" case. Note, that we call a compact trajectory on the surface M closed iff it defines a zero cycle in $H_1(\mathbb{T}^3, \mathbb{Z})$. The preimage of such a trajectory in the 3-space is a $\mathbb{Z}^3$-periodic family of copies of the trajectory.

1. Consider a suface M, consisting of an even number of 2-tori embedded in $\mathbb{T}^3$ so that the covering surface $\widehat{M}$ has the form of a union of a periodic family of parallel planes. Let H be a vector not perpendicular to those planes. Then the foliation defined by the 1-form $\omega = \Omega\Big|_M$ has the form of a linear winding and in the covering space $\mathbb{R}^3$ all the trajectories are just straight lines with the direction given by the vector product $H \times n$ where n is the normal vector to the surface $\widehat{M}$.

In this simplest case the 1-form ω has no critical points, so after a small deformation of the surface M we get an equivalent foliation and the same asymptotic behaviour of the trajectories. Moreover, small perturbation of the surface will cause a *small* perturbation of the trajectories.

2. Let us take once again a surface M consisting of 2-tori whose preimages in $\mathbb{R}^3$ are parallel planes. For definiteness, suppose that the planes are orthogonal to the vector $(0, 0, 1)$. Now let us cut out small disks from the tori and connect the holes by thin tubes as shown in the Figure 5. If we take a vector H sufficiently close to $(0, 0, 1)$ then each tube will be cut by some plane perpendicular to H along a closed curve. In this case it is clear that the non-closed trajectories will differ from the ones considered above in finite deformation and each non-closed trajectory will have a strong asymptotic direction. The direction is given again by the vector product $H \times n$, here $n = (0, 0, 1)$.

In this case a small deformation of the surface M does not necesserely result in a small deformation of the non-closed trajectories in the covering space $\mathbb{R}^3$. In

Figure 6:

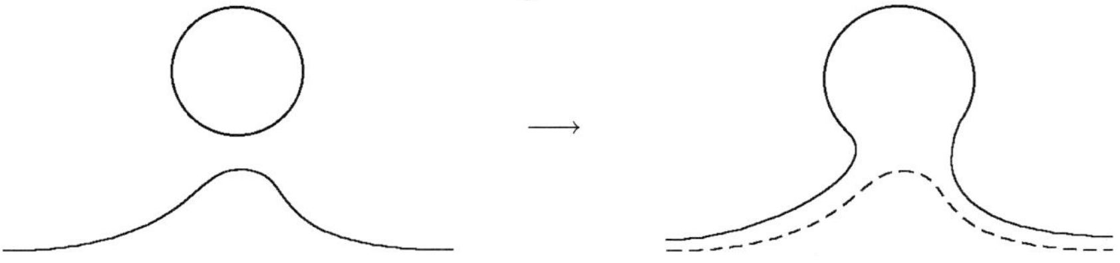

the planes perpendicular to H we may see surgeries like the one shown in Figure 6 The surgeries can occur near the saddle points of the 1-form ω.

The deformation can now lead to a foliation topologically not equivaliant to the original one. However, the non-closed trajectories on the deformed surface $\widehat{M}$ can be received from correspondnig ones on the original surface by *finite* deformation. Thus, in this case, the asymptotic behaviour of the non-closed trajectories is also stable.

3. Now, let us consider the case of a connected suface M of genus > 1 such that for some fixed vector H all the trajectories are open. Suppose also that the 1-form ω has only non-degenerate critical points and there are no saddle connections between saddles. Then any non-stationary trajectory is everywhere dense on the surface M.

Since the genus g of the surface M is greater than 1 and there are no closed trajectories, the 1-form ω should have $2g-2>0$ critical points of saddle type. Let us deform our surface near a saddle point so that the altitude of the point changes. According to Section 2. such a deformation results in a change in the density of the closed trajectories which was equal to zero before deformation. Therefore, after deformation some trajectories are closed and the closure of any non-closed trajectory will no longer be the whole surface M but only a part which has *less genus* than M.

The following statement can be proved in a similar way.

PROPOSITION 2 *Let the vector H be fixed. If there is no closed trajectory on a level surface $M_{c_0} = \{x \in \mathbb{T}^3 \mid f(x) = c_0\}$ then any other level surface M_c, where c is close to c_0, $c \neq c_0$, contains closed trajectories.*

We see that in this case a small deformation leads to a foliation with principally different structure. It is natural to expect global changes in the asymptotic behaviour of the non-closed trajectories.

Remark. One can show that in the situation just considered the non-closed trajectories have no asymptotic direction even in the sense of the usual definition. And the result of changing the altitude of one of the saddle points is quite surprizing: *all* the trajectories are closed on the deformed surface. Deforming the whole surface M, not only within a small heighbourhood of a critical point, one can get an infinite number of topologically different foliations. Non-closed trajectories for some of them have a strong asymptotic direction and we can get an infinite number of different strong asymptotical directions by an arbitrarily small deformation.

4. Reduction procedure

To determine the asymptotic behaviour of a non-closed trajectory it is important to know which part of the surface M the trajectory "fills up". Keeping in mind Zorich's result [10] it is natural to conjecture that typically a non-closed trajectory does not fill up the whole surface M but only a part of genus 1. In fact, Zorich shows that such a situation is stable under small perturbations. We have seen that filling up the whole surface M of genus > 1 by one non-closed trajectory is an unstable (codimension 1) situation.

The objective of this section is to show how, given a surface M and vector H, one can construct another surface M^r called *the reduced surface* such that:

i) the non-closed connected components of the intersection sets $P \cap \widehat{M}$ and $P \cap \widehat{M^r}$ are in one-to-one correspondence for any plane $P \perp H$;

ii) the non-closed component of $P \cap \widehat{M^r}$ corresponding to a regular component of $P \cap \widehat{M}$ is also regular and can be obtained from it by a finite deformation;

iii) the intersection $P \cap \widehat{M^r}$ has no compact connected component.

Existence of the reduced surface means that in order to observe all possible asymptotic behaviours it is sufficient to consider only pairs (M, H) such that all the trajectories are non-closed.

Before describing the reduction procedure let us consider the structure of the set of closed trajectories. Clearly, the trajectories in a small neighbourhood of a closed one are also closed. Thus, we have a number of cylinders consisting of closed trajectories, having as their bases singular level curves of the 1-form ω or parts of singular level curves.

We always suppose that the number of critical points of ω on M is finite. It easily follows that the number of closed trajectory cylinders is then also finite. Note that a small perturbation of the surface M and vector H can cause only an increase in the number of closed trajectory cylinders because of the stability of closed trajectories.

We will describe the reduction procedure in the generic case, namely when the vector H is *totally irrational*, the 1-form ω has only non-degenerate critical points, and there is no saddle connection between two different saddle points. The reduction procedure can be generalized to work in the case of an arbitrary surface M with a finite number of singular points and such that the 1-form ω has critical points which are not "too bad". More details can be found in [3].

It will be more convenient for us to deal with 3-periodic surfaces rather than compact surfaces in the 3-torus, so the reduction procedure will be described in terms of the Abelian $\mathbb{Z}^3$-covering surface. The procedure consists of three steps.

Step 1. Remove all the compact level curves of the 1-form ω, that is closed trajectories, the maxima and minima points, and "eight" figures consisting of a saddle point and separatrices attached to it. We will get a surface with boundary each connected component of which is a separatrix coming from an ending at the same saddle point of the 1-form ω (see Figure 7). Such a separatrix bounds a flat disk in a plane perpendicular to the vector H.

Step 2. Attach a plane disk to each component of the surface obtained in Step 1. We will get a piecewise smooth surface *embedded* in 3-space. Indeed, the

Figure 7: Reducing the surface M

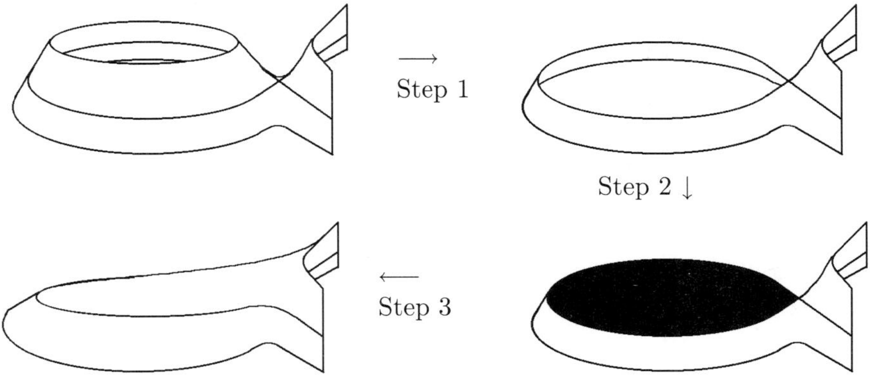

Figure 8:

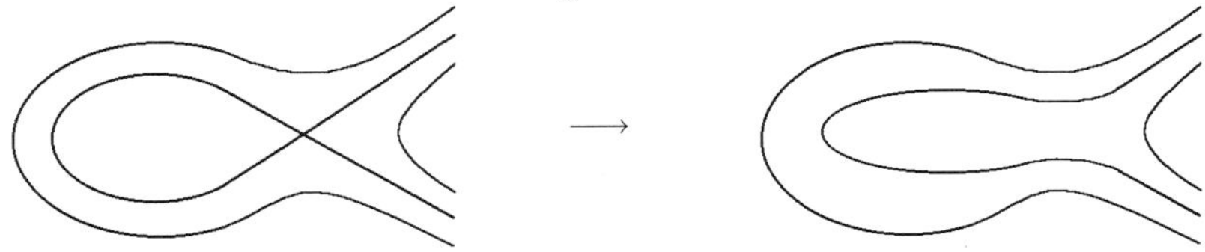

interior of each of the disks can intersect the original surface $\widehat{M}$ only along a closed trajectory or a singular compact level curve all of which were discarded in Step 1. Note that the non-closed trajectories on the surface obtained are exactly the same as on the original one and we get rid of all the closed trajectories.

Step 3. Now we deform the surface in a small neighbourhood of each disk that we attached in the Step 2, as shown in Figure 7. As a result, each intersection curve with any plane perpendicular to the vector H will be *finitely deformed* and we get regular trajectories instead of the unbounded level curves coming through saddle points (see Figure 8).

We denote by M^r the surface obtained from M by the reduction procedure. Note that M^r depends on the vector H and there is also some arbitrariness in the Step 3 of the procedure.

By construction, the dimension $d(M^r)$ of the image of $H_1(M^r, \mathbb{Z})$ in the lattice $H_1(\mathbb{T}^3, \mathbb{Z}) \cong \mathbb{Z}^3$ is greater than 1, that is $d(M^r) = 2$ or 3. Indeed, if $d(M^r) = 1$ then each connected component of the surface $\widehat{M^r}$ lies in a finite neighbourhood of a straight line having a rational direction. The intersection of such a component with a plane of totally irrational direction must be compact.

It is also clear that M^r cannot have connected components of genus 0, hence $\chi(M^r) \leq 0$. The proof of the following statement can be found in [3].

THEOREM 5 *Let H be a fixed totally irrational vector, and M be a surface on which some of the trajectories are non-closed. Then the following statements are equivalent.*

i) *At least one non-closed trajectory has a strong asymptotic direction.*
ii) *All the non-closed trajectories have a strong asymptotic direction.*
iii) *The image of $H_1(M^r, \mathbb{Z})$ in $H_1(\mathbb{T}^3, \mathbb{Z})$ is 2-dimensional.*
iv) *The surface M^r has a form of the union of 2-tori.*
v) *The following equality holds*

$$\chi(M^r) = 0.$$

If these equivalent conditions are sutisfied then all the connected components of the surface M^r define the same up to sign indivisible homology class $\alpha \in H_2(\mathbb{T}^3, \mathbb{Z})$. The strong asymptotic cycle of any non-closed trajectory has the image in $H_1(\mathbb{T}^3, \mathbb{Z})$ given by the element $\alpha \cap [\Omega]$.

Using the standard identifications $\mathbb{R}^3 \cong H_1(\mathbb{T}^3, \mathbb{R}) \cong H_2(\mathbb{T}^3, \mathbb{R}) \cong H^1(\mathbb{T}^3, \mathbb{R})$ and $\mathbb{Z}^3 \cong H_1(\mathbb{T}^3, \mathbb{Z}) \cong H_2(\mathbb{T}^3, \mathbb{Z}) \cong H^1(\mathbb{T}^3, \mathbb{Z})$ one can rephrase the last assertion of the theorem saying that the components of the surface M^r define up to sign the same integer vector $\mathbf{k} \in \mathbb{Z}^3$ with relatively prime coordinates and the strong asymptotic direction vector is a scalar multiple of $H \times \mathbf{k}$.

THEOREM 6 *Let H be a fixed totally irrational vector.*
i) *If M can be represented as a level surface of a function then so can M^r.*
ii) *Let f be a generic function on $\mathbb{T}^3$, M_c be the level surface $M_c = \{x \in \mathbb{T}^3 \mid f(x) = c\}$, ω_c be the restriction of the 1-form Ω to M_c. Suppose that for all c from an interval $[c_1, c_2]$ the 1-forms ω_c have the same number and types of critical points and the same number of closed trajectory cylinders and there are no critical values of the function f in the interval $[c_1, c_2]$. Then the surfaces M_c^r, $c \in [c_1, c_2]$ can be chosen so that they will form a family of level surfaces of another function $g : \mathbb{T}^3 \to \mathbb{R}$, that is $M_c^r = \{x \in \mathbb{T}^3 \mid g(x) = c\}$.*
iii) *For any two different level surfaces M_c and M_d of the same function the reduced surfaces M_c^r and M_d^r can be chosen non-intersecting.*

Proof. Item i) follows from the fact that the reduction procedure does not change the fundamental homology class of the surface. Items ii) and iii) are obtained directly from the construction. □

5. Main results

We have seen that absence of closed trajectories on a surface M having connected components of genus > 1 is unstable. More precisely, the set of pairs (M, H) for which all the level curves of the 1-form ω are unbounded (in the 3-space) is contained in the codimention 1 set defined by the equation $\rho = 0$ where ρ is the algebraic density of closed trajectories. Proposition 6 allows this to be generalized as follows.

Let f be a generic function and H a totally irrational vector. As before, we denote the level surface $\{x \in \mathbb{T}^3 \mid f(x) = c\}$ by M_c.

PROPOSITION 3 *For almost any c we have $\chi(M_c^r) = 0$.*

Proof. We will prove here a weaker statement, namely that the set of c such that $\chi(M_c^r) \neq$ is nowhere dense. It is sufficient for our purposes.

Suppose that we have $\chi(M_c^r) = k < 0$ for $c \in [c_1, c_2]$ and the interval $[c_1, c_2]$ does not contain critical values of the function f. Then according to Theorem 6 we can reduce the surfaces M_c, $c \in [c_1, c_2]$, so that the surfaces M_c^r obtained are level surfaces of the same function. By construction, there are no closed trajectories on M_c, $c \in [c_1, c_2]$, which contradicts Proposition 2. □

PROPOSITION 4 *If we have $\chi(M_{c_0}^r) = 0$ for at least one level surface M_{c_0} then it is true for all M_c.*

Proof. It follows from Theorem 5 that each connected component of M_{c_0} defines a non-trivial homology cycle in $H_2(\mathbb{T}^3, \mathbb{Z})$. According to Proposition 6 for any c we can chose $M_{c_0}^r$ and M_c^r so that $M_{c_0}^r \cap M_c^r = \emptyset$. It follows that $d(M_c^r) \leq 2$ and it sufficies only to apply Theorem 5 once again. □

Thus, we see that there may be two possibilities: 1) $\chi(M_c^r) = 0$ for all c and all the non-closed trajectories have a strong asymptotic direction; 2) the set of c such that the level surface M_c contains non-closed trajectories is nowhere dense. The following statement allows to be strengthened this result.

Denote by U the set of $c \in \mathbb{R}$ such that the form $\omega_c = \Omega\Big|_{\widehat{M_c}}$ has unbounded level curves.

PROPOSITION 5 *The set U is either a closed interval $[c_1, c_2]$ or it is a point $\{c_0\}$.*

Remark. The vector H is not supposed to be totally irrational in this statement.

We omit the elementary but tiresome proof of this statement. The sketch proof can be found in [3].

Summarizing, we get the following result.

Let H be again a totally irrational vector, and f be a generic fixed function.

THEOREM 7 *If the set U contains more than one point, that is $U = [c_1, c_2]$, $c_1 < c_2$, then:*

i) *for any $c \in [c_1, c_2]$ we have $\chi(M_c^r) = 0$, that is the surface M_c^r is the union of 2-tori;*
ii) *the vector $\mathbf{k}(M_c, H)$ defined in the previous section is the same for all $c \in [c_1, c_2]$, so we will use the notation $\mathbf{k}(f, H)$*
iii) *the vector $\mathbf{k}(f, H)$ is stable under small perturbations of f and H.*

6. Dependence on the vector H

We saw in the previous sections that typically the reduced surface M^r, if non-empty, is the union of 2-tori which define the same up to sign indivisible homology class from $H_2(\mathbb{T}^3, \mathbb{Z})$, that is an element $\mathbf{k}$ of the lattice $\mathbb{Z}^3$. It follows from Theorem 7, that the vector $\mathbf{k}$ is a locally constant function of M and H. Let the surface M now be fixed and the vector H be normalized by the condition $|H| = 1$, so

Figure 9: $\mathbf{k}(M,H)$, $\widehat{M} = \{x \in \mathbb{R}^3 \mid \cos(2\pi x_1) + \cos(2\pi x_2) + \cos(2\pi x_3) = 0\}$, $H = (1, H_2, H_3)$

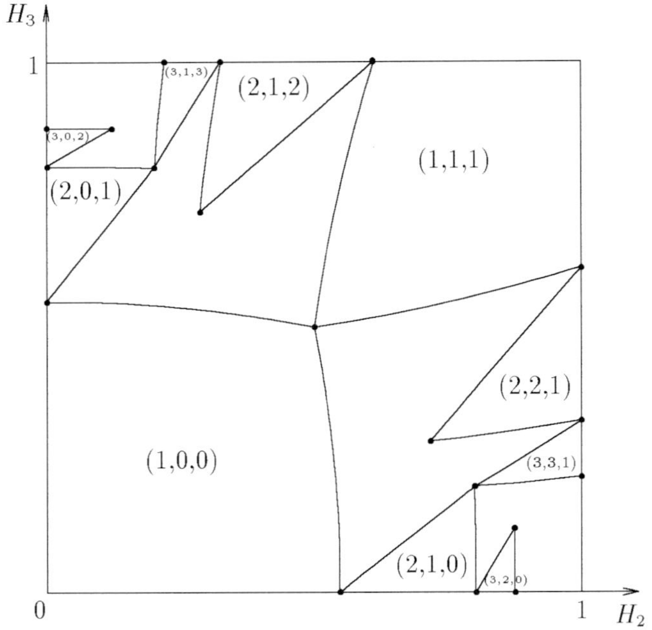

$\mathbf{k}$ is a locally constant function (not everywhere defined) on the sphere S^2. An interesting question is what kind of picture we will get if we draw the domains on the sphere S^2 on which the vector $\mathbf{k}$ is well-defined and constant. We denote the domain corresponding to an integer indivisible vector $\mathbf{k}$ by $D(\mathbf{k})$.

We present here the results of computations carried out for the surface $\cos(2\pi x_1) + \cos(2\pi x_2) + \cos(2\pi x_3) = 0$ and different vectors H. Because of the symmetry of this surface, unbounded level curves exist for any vector H. It is also clear that it is sufficient to consider only vectors H of the form $H_1 \geq H_2 \geq H_3 \geq 0$. Figure 9 shows the projections of some of the domains $D(\mathbf{k})$ to the plane $H_1 = 1$. There are an infinite number of the domains $D(\mathbf{k})$ for this surface and more than 40 of them were found in the region $H_1 \geq H_2 \geq H_3 \geq 0$ using a computer. Only a few of them are shown in the picture because the others are of too small size to be indicated precisely.

In this specific case, a complicated geometric analysis allows us to conclude the following about the whole picture.

i) Each domain $D(\mathbf{k})$ has a piecewise smooth boundary.

ii) The boundaries of two different domains $D(\mathbf{k}_1)$ and $D(\mathbf{k}_2)$ can have at most one point in common.

iii) The common point of the boundaries of domains $D(\mathbf{k}_1)$ and $D(\mathbf{k}_2)$ should be a breakpoint of at least one of the boundaries and the corresponding vector

Figure 10:

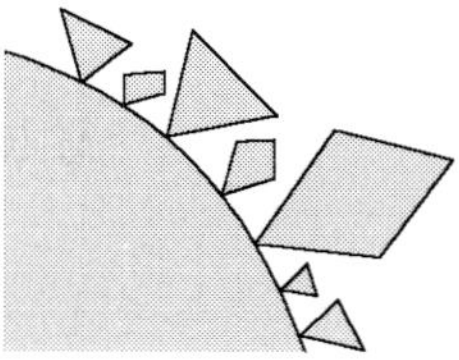

H is a linear combination of $\mathbf{k}_1$ and $\mathbf{k}_2$. If it is a breakpoint of them both then the vector H is a scalar multiple of $\mathbf{k}_1 \pm \mathbf{k}_2$.

iv) Any point H on the boundary of a domain $D(\mathbf{k}_1)$ such that the vector $H \times \mathbf{k}$ is rational also belongs to the boundary of another domain $D(\mathbf{k}_2)$. Thus, an infinite number of the other domains are attached to the boundary of $D(\mathbf{k}_1)$ (see Figure 10).

The final conclusion which can be made from this statement is that the union of all the domains $D(\mathbf{k})$ cannot cover the whole sphere S^2.

References

[1] A. A. Abrikosov, *Fundamental theorey of metals* (Russian), Moscow, Nauka, 1987.

[2] I. A. Dynnikov, *A proof of Novikov's conjecture on semi-classic motion of an electron* (Russian), Mat. Zametki, **53** (1993), 5, pp. 57–68.

[3] I. A. Dynnikov, *Semi-classical Motion of the electron. Proof of Novikov's conjecture in general position and counterexamples*, Advances in Soviet Math., 1996, to appear.

[4] A. V. Katok, *Invariant measures on flows on oriented surfaces* (Russian), Soviet Doklady, **211** (1973), 4, pp. 775–778.

[5] S. P. Kerckhoff, *Simplicial systems for interval exchange maps and measured foliations*, Ergodic Theory & Dynamical Systems (1985), 5, 257–271.

[6] I. M. Lifschitz, V. G. Peschansky, *Galvanometric characteristics of metals with open Fermi surfaces. I* (Russian), JETF, **35** (1958), 5 (11), pp. 1251–1264.

[7] S. P. Novikov, *Hamiltonian formalism and a multivalued analogue of Morse theory* (Russian), Uspekhi Mat. Nauk, **37** (1982), 5, pp. 3–49.

[8] S. P. Novikov, *Quasiperiodic Structures in Topology*, Proc. of Conference "Topological methods in mathematics", 1991, Stony Brook.

[9] S. P. Tsarev, Private communication.

[10] A. V. Zorich, *Novikov's problem on the semi-classical motion of an electron in a homogeneous magnetic field close to rational* (Russian), Uspekhi Mat. Nauk, **39** (1984), 5, 235–236.

[11] A. V. Zorich, *Asymptotic Flag of an Orientable Measured Foliation on a Surface*, Proc. of the International Symposium/Workshop on Geometric Study of Foliations, Nov. 1993, Tokyo, World Scientific Pub., 1994.

Progress in Mathematics, Vol. 168, © 1998 Birkhäuser Verlag Basel/Switzerland

One-Dimensional Quasi-Periodic Schrödinger Operators — Dynamical Systems and Spectral Theory

L.H. ELIASSON

Depart. of Math., Royal Institute of Technology
S-10044 Stockholm, Sweden

ABSTRACT. We shall describe two types of results that have been obtained for the one-dimensional quasi-periodic Schrödinger operator, both continuous and discrete. The technique is perturbation theory of KAM-type and the results connect in a fruitful way ideas from dynamical systems and from spectral theory. Both results are developments of original works of Dinaburg, Sinai, Fröhlich, Spencer, Wittver from the middle 70's and late 80's.

1. Introduction

The quasi-periodic Schrödinger operator in one dimension is determined by a *potential function*

$$E : \mathbf{T}^d \to \mathbf{R}, \qquad \mathbf{T} = \mathbf{R}/2\pi\mathbf{Z},$$

and by a *frequency vector* $\omega \in \mathbf{R}^d$. We shall assume (unless otherwise specified) that E is smooth ($\mathcal{C}^\infty$ or real analytic) and that ω is *Diophantine*[1], i.e. there exist $\kappa, \tau > 0$ such that

$$\inf_{p\in\mathbf{Z}} |< n,\omega > -2\pi p \mid \geq \frac{\kappa}{\mid n \mid^\tau} \qquad \forall n \in \mathbf{Z}^d \setminus 0. \qquad DC(\kappa,\tau)$$

($<\ ,\ >$ is the usual Euclidean product in $\mathbf{R}^d$.) The *potential* is obtained by evaluating E over a "straight line" in $\mathbf{T}^d$, continuous or discrete:

$$\theta + \mathbf{R}\omega \quad \text{or} \quad \theta + \mathbf{Z}\omega.$$

Due to the rational independence of ω both the continuous and the discrete lines are dense in $\mathbf{T}^d$.

The *Schrödinger operator* $\mathcal{L}_\theta$ is now

$$\mathcal{L}_\theta^c : \quad -\frac{d^2u}{dx^2}(x) \ + \ E(\theta + x\omega)u(x), \qquad x \in \mathbf{R} \qquad (*)^c$$

or

$$\mathcal{L}_\theta^d : \quad -(u(x+1) + u(x-1)) \ + \ E(\theta + x\omega)u(x), \qquad x \in \mathbf{Z}, \qquad (*)^d$$

[1]We give a formulation that is sufficient both in the continuous and the discrete case.

in the continuous and discrete cases respectively. Considered as a differential or difference equation

$$\mathcal{L}_\theta u = \lambda u,$$

the study of this operator is part of dynamical systems, and considered as a linear operator

$$u \longrightarrow \mathcal{L}_\theta u$$

on the Hilbert space $L^2(\mathbf{R})$ or $l^2(\mathbf{Z})$ it is part of spectral theory. Both these approaches have proven to be fruitful for the study of the operator.

2. Dynamical properties

Any second order differential or difference equation can be written as a first order system. For example, we can write $(\mathcal{L}^c_\theta - \lambda)u = 0$ as

$$\begin{cases} \begin{pmatrix} u \\ u' \end{pmatrix}' = \begin{pmatrix} 0 & 1 \\ E(\theta) - \lambda & 0 \end{pmatrix} \cdot \begin{pmatrix} u \\ u' \end{pmatrix}, & ' = \frac{d}{dx} \\ \theta' = \omega. \end{cases} \qquad (**)^c_\lambda$$

As such it is just a special case of a general skew-product on $\mathbf{T}^d$:

$$\begin{cases} \bar{u}' = A(\theta)\bar{u}, & \bar{u} \in \mathbf{R}^n \\ \theta' = \omega, \end{cases} \qquad (***)$$

and the spectral parameter λ is from this point of view just a parameter like any other.

The *skew-product flow* is

$$(\bar{u}, \theta) \mapsto (\mathcal{U}(x,\theta)\bar{u}, \theta + x\omega)$$

where $\mathcal{U}(x,\theta) \in Gl(n, \mathbf{R})$ is the *fundamental solution.*[2] Important and much studied questions for skew-product flows concern asymptotic behavior described by different well known concepts.

Reducibility. This means that there is a transformation of the form

$$\begin{pmatrix} \bar{u} \\ \theta \end{pmatrix} \mapsto \begin{pmatrix} \bar{v} = Y(\theta)\bar{u} \\ \theta \end{pmatrix}$$

that transforms $(***)$ to

$$\begin{cases} \bar{v}' = B\bar{v} \\ \theta' = \omega, \end{cases}$$

where $B \in gl(n, \mathbf{R})$ is independent of θ — the eigenvalues of B are called Floquet exponents.[3] In this case the fundamental solution has *Floquet representation*:

$$\mathcal{U}(x,\theta) = Y(\theta + x\omega)^{-1} \cdot e^{Bx} \cdot Y(\theta). \qquad (1)$$

---

[2]Or monodromy matrix, or time evolution operator, or propagator, or ...

[3]The real parts of the Floquet exponents are uniquely determined, but not the imaginary parts.

The conditions on Y may vary, but we shall assume that it is continuous and

$$Y : (2\mathbf{T})^d \to Gl(n, \mathbf{R}).$$

(By this we mean that the periods of Y are $2 \cdot 2\pi$ in each variable.)

Hyperbolicity. This means that there exist invariant and continuously varying subspaces $W^{\pm}(\theta)$ such that $W^+(\theta) + W^-(\theta) = \mathbf{R}^n$ for all $\theta \in \mathbf{T}^d$ and

$$\lim_{x \to \mp\infty} \frac{1}{|\, x \,|} \log |\, \mathcal{U}(x, \theta)\bar{u} \,| \ \exists \text{ and } \leq -\gamma < 0, \quad \forall \bar{u} \in W^{\pm}(\theta) \setminus 0 \tag{2}$$

for all $\theta \in \mathbf{T}^d$. Hyperbolicity in this sense is more special than reducibility because it can be shown [1] that when $n = 2$

$$(***) \text{ is hyperbolic} \iff$$
$$(***) \text{ is reducible and no Floquet exponents are purely imaginary.}$$

Non-uniform hyperbolicity. This means that the two invariant subspaces $W^{\pm}(\theta)$ are measurable and not defined for all θ but for a.e. $\theta \in \mathbf{T}^d$, and that (2) holds only for a.e. θ. It is clear that non-uniform hyperbolicity is inconsistent with reducibility[4].

Biasymptotic solutions. These occur in the non-uniform case if (2) holds for some θ and for some $0 \neq \bar{u} \in W^+(\theta) \cap W^-(\theta)$.

The particular skew-product $(**)^c_\lambda$ has a fundamental solution in $SL(2, \mathbf{R})$ and as such a special property [2] — a *rotation number.* In fact

$$\lim_{x \to \pm\infty} \frac{1}{x} \arg(\mathcal{U}_\lambda(x, \theta)\bar{u})$$

exists for all $\lambda, \theta, \bar{u} \neq 0$ (for some continuous choice of the argument). The number $\beta(\lambda)$ is independent of θ and $\bar{u}$ and it is continuously increasing in λ (if the argument is counted clockwise). If E vanishes identically one can compute $\beta(\lambda)$ explicitly:

$$\beta(\lambda) = \begin{cases} 0 & \lambda \leq 0 \\ \sqrt{\lambda} & \lambda \geq 0. \end{cases}$$

It is clear that if $(**)^c_\lambda$ is reducible, then the Floquet exponents equal $\pm\beta(\lambda)$ modulo $\{\frac{<n,\omega>}{2} : n \in \mathbf{Z}^d\}$. The rotation number even completely characterizes hyperbolicity [2] because

$$(**)^c_\lambda \text{ is hyperbolic} \iff \beta \text{ is constant near } \lambda.$$

There is also a discrete version of this — a *skew-product mapping*

$$\begin{cases} \bar{u} \to A_\lambda(\theta)\bar{u}, & A_\lambda(\theta) = \begin{pmatrix} 0 & 1 \\ -1 & E(\theta) - \lambda \end{pmatrix} \\ \theta \to \theta + \omega, \end{cases} \tag{$(**)^d_\lambda$}$$

[4]It is consistent with a weaker form of reducibility that only requires that the matrix $Y(\theta)$ is measurable.

whose fundamental solution is

$$\mathcal{U}_\lambda(x,\theta) = A_\lambda(\theta + (x-1)\omega)\dots A_\lambda(\theta+\omega)A_\lambda(\theta).$$

There are corresponding concepts for such mappings. For example, $(**)^d_\lambda$ is reducible if $\mathcal{U}_\lambda(x,\theta)$ can be written as (1), and it has a rotation number $\beta(\lambda)$[3,4] which for $E = 0$ is

$$\beta(\lambda) = \begin{cases} 0 & \lambda \leq -2 \\ \arctan(-\frac{\sqrt{4-\lambda^2}}{\lambda}) & -2 \leq \lambda \leq 2 \\ \pi & 2 \leq \lambda. \end{cases}$$

3. Spectral properties

In this approach $\mathcal{L}_\theta$ is a self-adjoint operator acting on the Hilbert space $H = L^2(\mathbf{R})$ or $l^2(\mathbf{Z})$. In the discrete case we have a bounded self-adjoint operator

$$\mathcal{L}^d_\theta : l^2(\mathbf{Z}) \to l^2(\mathbf{Z}),$$

and in the continuous case we have the closure of the operator

$$\mathcal{L}^c_\theta : C''_0 \to L^2(\mathbf{R}),$$

still denoted $\mathcal{L}^c_\theta$.

The *resolvent set* $\rho(\mathcal{L}_\theta)$ is the set of all $\lambda \in \mathbf{C}$ for which $(\mathcal{L}_\theta - \lambda)$ is one-to-one and onto — in this case the inverse $(\mathcal{L}_\theta - \lambda)^{-1}$ will be bounded. The complement of the resolvent set is the *spectrum*

$$\sigma(\mathcal{L}_\theta) = \mathbf{C} \setminus \rho(\mathcal{L}_\theta).$$

It is a basic fact that $\sigma(\mathcal{L}_\theta)$ is a closed non-void subset of $\mathbf{R}$ which is independent of θ.

An *eigenvalue* is a $\lambda \in \sigma(\mathcal{L}_\theta)$ for which $(\mathcal{L}_\theta - \lambda)$ is not one-to-one. If the set of eigenfunctions is complete in H, $\mathcal{L}_\theta$ is said to have *pure point spectrum.* If there are no eigenvalues, $\mathcal{L}_\theta$ is said to have a pure continuous spectrum — this case is in turn analyzed in terms of *absolutely continuous spectrum* and *singular continuous spectrum.* (In general, a self-adjoint operator may have both point, absolutely continuous and singular continuous spectrum simultaneously.)

The time-dependent Schrödinger equation. One way to think of the difference between point spectrum and continuous spectrum is in terms of the Schrödinger equation for a wave function $\psi(t,x)$:

$$\begin{cases} -i\frac{\partial\psi}{\partial t} + \mathcal{L}_\theta\psi & = 0 \\ \psi(0,x) & = \psi_0(x). \end{cases}$$

The properties of the spectrum are related to the spread of the wave function as measured, for example, by the quantity

$$r^2(t) =: \int_{\mathbf{R}} |\, x\psi(t,x)\,|^2 \, dx \quad \text{or} \quad \sum_{x\in\mathbf{Z}} |\, x\psi(t,x)\,|^2 .$$

If for example

$$r^2(t) \leq const \qquad \text{(Anderson localization)}$$

it follows from the RAGE theorem [5,6] that $\mathcal{L}_\theta$ has pure point spectrum. So continuous spectrum leads to an increase in $r(t)$ — absolutely continuous spectrum is expected to give the strongest increase. If E vanishes this is easy to analyze. In that case $\sigma(\mathcal{L}_\theta) = [0, \infty[$ and $\mathcal{L}_\theta$ has pure absolutely continuous spectrum. Using the Fourier transform one verifies easily that

$$\psi(t,x) = \frac{1}{2\pi} \int_{-\infty}^{\infty} e^{+i\xi x} e^{-i\xi^2 t} \hat{\psi}_0(\xi) d\xi.$$

Hence

$$\begin{aligned} r^2(t) = \int_{-\infty}^{\infty} \mid x\psi(t,x) \mid^2 dx &= \int_{-\infty}^{\infty} \mid i\frac{d}{d\xi}\hat{\psi}(t,\xi) \mid^2 d\xi \\ = t^2 \int_{-\infty}^{\infty} \mid -i2\xi\hat{\psi}_0(\xi) + \frac{1}{t}\hat{\psi}_0'(\xi) \mid^2 d\xi \ &\sim \ const.t^2. \end{aligned}$$

There are many relations between dynamical properties and spectral properties of the Schrödinger operator:

$$\lambda \in \rho(\mathcal{L}_\theta) \iff (**)_\lambda \text{ is hyperbolic}$$

(for a proof in the continuous case see [2]);

$$(**)_\lambda \text{ is reducible } \forall\lambda \Longrightarrow \mathcal{L}_\theta \text{ has pure absolutely continuous spectrum } \forall\theta$$

(for a proof see [7]);

$$\mathcal{L}_\theta \text{ has pure point spectrum for a.e. } \theta \Longrightarrow$$
$$(**)_\lambda \text{ is non-uniformly hyperbolic for a.e. } \lambda \in \sigma(\mathcal{L}_\theta)$$

(this is a famous result of Kotani, see [8]);

$$\text{biasymptotic solutions for } (**)_\lambda \Longrightarrow \lambda \text{ is an eigenvalue of } \mathcal{L}_\theta \text{ for some } \theta$$

(this is obvious). There are many other relations, for example an interpretation of the rotation number as the integrated density of states.

4. Periodic case

Even though these general results give relations between dynamical and spectral properties, they do not say anything about what are the actual dynamical or spectral properties of the operator. For the periodic case these properties are well known.

We shall discuss shortly the continuous periodic case ($d = 1$). It was proven by Floquet that the fundamental solution always is of the form:

$$\mathcal{U}_\lambda(x,\theta) \;=\; Y_\lambda(\theta + x\omega)\cdot e^{B_\lambda x}\cdot Y_\lambda(\theta)^{-1}$$

with

$$Y_\lambda : 2\mathbf{T} \to SL(2,\mathbf{R})$$

as smooth as is E, and with $B_\lambda \in sl(2,\mathbf{R})$. The system is hyperbolic if and only if $\det B_\lambda < 0$ which determines the gaps in the spectrum. The spectrum consists of bands and non-hyperbolic Floquet solutions exist for all λ in the bands. This implies that the spectrum is absolutely continuous.

There is a completely analogous description in the discrete case.

The quasi-periodic case is much more delicate and we shall only discuss it under Diophantine assumptions on ω and smoothness assumptions on the potential function E. We mention however two other results that are known.

If $\omega \in \mathbf{R}$ is Liouville, i.e.

$$\inf_{q,p} \mid q\omega - 2\pi p \mid\mid k \mid^{q} \;<\; 1,$$

for all non-zero k, and if $E(\theta) \;=\; K\cos(\theta),\;\; K > 2$, then $\mathcal{L}^d_\theta$ has pure singular continuous spectrum for a.e. θ [9].

If $E(\theta) = \chi_I(\theta)$, the characteristic function for an interval I in $\mathbf{T}$, then $\mathcal{L}^d_\theta$ has pure singular continuous spectrum for a.e. ω and a.e. θ [10,11].

5. Small quasi-periodic potentials

For small potentials the dynamical system approach is basic just as in the periodic case.

THEOREM 1. [12–14] *Assume that $\omega \in DC(\kappa,\tau)$ and that E is real analytic in $\mid \Im\theta \mid < r$. Then there is a constant $\varepsilon_0 = \varepsilon_0(\kappa,\tau,r)$ such that if*

$$\sup_{|\Im\theta|<r} \mid E(\theta) \mid = \varepsilon < \varepsilon_0$$

then $(**)^c_\lambda$ is reducible for a.e. λ.

For generic E (in the supremum-norm over $\mid \Im\theta \mid < r$) the set

$$\{\lambda \in \mathbf{R} : (**)^c_\lambda \textit{ is not hyperbolic}\}$$

is a Cantor set.

Also for generic E there exist λ and θ such that the fundamental solution $\mathcal{U}_\lambda(x,\theta)$ of $(**)^c_\lambda$ is unbounded and

$$\lim_{x\to\pm\infty} \frac{1}{\mid x \mid} \mid \mathcal{U}_\lambda(x,\theta) \mid = 0.$$

This gives immediately that $\mathcal{L}^c_\theta$ has purely continuous spectrum and that generically $\sigma(\mathcal{L}^c_\theta)$ is a Cantor set. The statement itself does not imply that the spectrum is purely absolutely continuous since we do not get, and cannot get, reducibility for all, but only for a.e. λ. The information obtained from the proof is however sufficient to conclude also this.

THEOREM 1′. [14] *Under the same assumptions as in Theorem 1, if*

$$\sup_{|\Im\theta|<r} \mid E(\theta) \mid = \varepsilon < \varepsilon_0$$

then the spectrum of $\mathcal{L}_\theta^c$ is purely absolutely continuous.

There is an interesting variant for non-small potentials but for large λ.

THEOREM 2. [12–14] *Under the same assumptions as in Theorem 1, there is a $\Lambda_0 = \Lambda_0(\kappa, \tau, r, \sup \mid E \mid)$ such that $(**)_\lambda^c$ is reducible for a.e. $\lambda > \Lambda_0$.*

For generic E (in the supremum-norm over $\mid \Im\theta \mid < r$) the set

$$\{\lambda > \Lambda_0 : (**)_\lambda^c \text{ is not hyperbolic}\}$$

is a Cantor set.

Also for generic E there exist $\lambda > \Lambda_0$ such that the fundamental solution $\mathcal{U}_\lambda(x, \theta)$ of $(**)_\lambda^c$ is unbounded for some θ and

$$\lim_{x\to\pm\infty} \frac{1}{\mid x \mid} \mid \mathcal{U}_\lambda(x, \theta) \mid = 0.$$

In particular, $\sigma(\mathcal{L}_\theta^c) \cap [\Lambda_0, \infty[$ is a Cantor set for generic E, and one can also prove that $\mathcal{L}_\theta^c$ has pure absolutely continuous spectrum in $[\Lambda_0, \infty[$.

Idea of proof. Finding a Floquet representation (1) consists in solving the non-linear differential equation

$$< \nabla Y(\theta), \omega > = -Y(\theta)A(\theta) + BY(\theta) \tag{3}$$

with

$$A(\theta) = A_0 + F(\theta) = \begin{pmatrix} 0 & 1 \\ -\lambda & 0 \end{pmatrix} + \begin{pmatrix} 0 & 0 \\ E(\theta) & 0 \end{pmatrix}.$$

This is a transformation problem with small divisors, which one may try to solve by a Newton iteration of KAM-type.

Since F is small it is natural to look for a transformation Y that is close to the identity and, hence, a B that is close to A_0. Linearizing equation (3) then gives

$$< \nabla \hat{Y}_1(\theta), \omega > = [A_0, \hat{Y}_1(\theta)] - F(\theta) + \hat{B}. \tag{4}$$

In order to get a solution, and a small one, of this equation we need a Diophantine condition not only on the frequency vector ω (which we have assumed) but also on the eigenvalues $\pm i\alpha_0$ of the unperturbed part A_0. For example if we want to solve (4) up to an approximation of order $\varepsilon^{\frac{3}{2}}$ we need that

$$|< n, \omega > \pm 2\alpha_0 | \geq \frac{\varepsilon^{\frac{1}{2}}}{\mid n \mid^\tau}, \qquad 0 < \mid n \mid < \frac{1}{r_0} \log(\frac{1}{\varepsilon}). \tag{5}$$

Such a condition is not fulfilled in general but since α_0 depend on the parameter λ we may force (5) to hold simply by excluding those λ for which $\alpha_0(\lambda)$

violates this condition — this is a small set when ε_0 is small. Hence, for the good λ's that remain we can solve (4), and transforming $(**)^c_\lambda$ by $Y_1(\theta) = I + \hat{Y}_1(\theta)$ gives

$$\begin{cases} \bar{u}' &= (A_1 + F_1(\theta))\bar{u}, \quad A_1 = A_0 + \hat{B} \\ \theta' &= \omega, \end{cases}$$

where $F_1(\theta)$ is of size $\varepsilon^{\frac{3}{2}}$, i.e. much smaller than F. Then we repeat this procedure with A_1 (whose eigenvalues $\pm i\alpha_1$ also depend on λ) instead of A_0. In this way we construct a sequence of transformations $\{Y_j \cdot \ldots \cdot Y_1\}_{j=1}^{\infty}$ that transforms $(**)^c_\lambda$ more and more closely to a constant skew-product. Since the Y_j's are closer and closer to the identity, the sequence converges to a transformation $Y(\theta)$ that solves (3).

This procedure (which was originally used in [12]) requires that one restricts the parameters λ, and an essential point is to control the dependence on λ in order to ensure that we do not exclude to many or even all λ's. This is a central point in all small divisor problems.

The restriction of parameters turns out to be imposed only by the use of transformations close to the identity, and the way to overcome this restriction is to enlarge the class of transformations. An exponential over the (double) torus is a matrix-valued mapping $Z : (2\mathbf{T})^d \to SL(2, \mathbf{R})$ of the form

$$Z(\theta) =: C^{-1} e^{z(\theta)} C, \quad z(\theta) = \frac{< m, \theta >}{2} \begin{pmatrix} +i & 0 \\ 0 & -i \end{pmatrix}.$$

The effect of transforming the constant skew-product

$$\begin{cases} \bar{u}' &= A_0 \bar{u} \\ \theta' &= \omega \end{cases}$$

by an exponential $\bar{u} \to Z(\theta)\bar{u}$ that commutes with A_0 is simply to replace A_0 by

$$\tilde{A}_0 = (\alpha_0 + \frac{< m, \omega >}{2}) \frac{1}{\alpha_0} A_0.$$

In particular, if $2\alpha_0 + < m\omega > \sim 0$ so that condition (5) is violated for A_0, then this condition will hold for $\tilde{A}_0$.

Hence, using transformations $Y_j(\theta) = Z_j(\theta) + \hat{Y}_j(\theta)$ that are close to exponentials, one can overcome the restriction imposed by (5). The perturbation theory of course becomes more complicated since the estimates are less good but this can be handled. The approach works up to any order for all λ without exception, but when we want to go to the limit we run into problems. The reason is that an infinite product of exponentials may not converge, and it is only for a.e. rotation number $\beta(\lambda)$ that one can prove the convergence of the sequence of transformations

$$Y_j \cdot \ldots \cdot Y_1 \to Y(\theta).$$

In order to show that the exceptional rotation numbers $\beta(\lambda)$ are avoided by a.e. λ, one must control the dependence on λ. This is a difficult problem in principle but it turns out to be easy in the $SL(2,\mathbf{R})$-case since one can use a priori information about the function $\lambda \to \beta(\lambda)$.

The exponential transformations were originally introduced in [13] and they were used in [14] in a systematic way to prove Theorem 1 and 2. □

We now discuss some related results for other skew-products. Reducibility of quasi-periodic skew-products close to constants was obtained by KAM-arguments already in the 60's. The first result was proven under the assumption of sufficiently many parameters [15]. That in general only one parameter is needed became obvious with the ideas in [16] — where in particular the case $Sp(n,\mathbf{R})$ is treated — and it was proven in general in [17]. These results give reducibility for all parameter values except a small set. The ideas developed in [14] suggest that the systems should be reducible for a.e. parameter value but in general not for all.

The best studied example besides $SL(2,\mathbf{R})$ is $SO(3,\mathbf{R})$ and we shall describe two results in this case. Assume that $\omega \in DC(\kappa,\tau)$, $A_0 \in so(3,\mathbf{R})$ and that

$$A : \mathbf{T}^d \to so(3,\mathbf{R}) \text{ is analytic in } \mid \Im\theta \mid < r.$$

THEOREM 3. [18] *There exists an* $\varepsilon_0 = \varepsilon_0(\kappa,\tau,r,A_0)$ *such that for the generic* $A(\theta)$ *in*

$$\sup_{|\Im\theta|<r} \mid A(\theta) \mid < \varepsilon_0$$

the skew-product

$$\begin{cases} \bar{u}' &= (A_0 + A(\theta))\bar{u} \\ \theta' &= \omega \end{cases}$$

is uniquely ergodic.

THEOREM 4. [19] *If* $A_0 \neq 0$, *then there exists an* $\varepsilon_1 = \varepsilon_1(\kappa,\tau,r,A_0)$ *such that if*

$$\sup_{|\Im\theta|<r} \mid A(\theta) \mid < \varepsilon_0,$$

then the skew-product

$$\begin{cases} \bar{u}' &= (\lambda A_0 + A(\theta))\bar{u} \\ \theta' &= \omega \end{cases}$$

is reducible for a.e. λ.

There are also results on more general compacts groups [20].

All the preceding results seem to translate almost immediately into discrete versions.

6. Large potentials

For large potentials the situation is completely different from the periodic one. From the spectral point of view, a large potential part is equivalent to a small differential or difference part and in the limit the operator reduces to a multiplication operator which, in the discrete case, has a pure point spectrum. Using a discrete version of [12] applied to the almost Mathieu equation

$$-\varepsilon(u(x+1)+u(x-1))+\cos(\theta+x\omega)u(x), \qquad x\in\mathbf{Z}$$

it was possible to prove that this operator had some point spectrum for small ε when ω was Diophantine [21]. This suggests that the spectral approach gives the right insight, something which has been strongly confirmed by later works.

Assume that $E:\mathbf{T}\to\mathbf{R}$ is $\mathcal{C}^\infty$ and satisfies

$$|E|_{\mathcal{C}^\nu}\le C(\nu!)^2K^\nu \qquad \forall\nu\ge 0, \tag{6}$$

and that

$$\max_{0\le\nu\le s}|\partial_x^\nu(E(\theta+x)-E(\theta))|\ge\xi>0 \qquad \forall\theta\forall x \tag{7}$$

$$\max_{0\le\nu\le s}|\partial_\theta^\nu(E(\theta+x)-E(\theta))|\ge\xi\inf_p|x-2\pi p| \qquad \forall\theta\forall x. \tag{8}$$

THEOREM 5. [24–26] *Assume $\omega\in DC(\kappa,\tau)$ and (6–8). Then there exists a constant $\varepsilon_0=\varepsilon_0(C,K,\xi,s,\kappa,\tau)$ such that if $|\varepsilon|<\varepsilon_0$, then the operator*

$$\mathcal{L}_\theta^d: \quad -(u(x+1)+u(x-1))+\frac{1}{\varepsilon}E(\theta+x\omega)u(x)$$

has a pure point spectrum for a.e. θ. Moreover, the Lebesgue measure of

$$[\inf\frac{1}{\varepsilon}E,\sup\frac{1}{\varepsilon}E]\setminus\sigma(\mathcal{L}_\theta^d)$$

is $\frac{o(\varepsilon)}{\varepsilon^2}$ as $\varepsilon\to 0$.

As a consequence we obtain that the skew-product $(**)_\lambda^d$ is non-uniformly hyperbolic for a.e. $\lambda\in\sigma(\mathcal{L}_\theta^d)$.

Idea of Proof. Multiplying the operator by ε — which does not change its spectral properties — we can think of the operator as an infinite-dimensional symmetric matrix which is a perturbation of a diagonal matrix

$$D(\theta)+\varepsilon F(\theta) = \begin{pmatrix} \ddots & & & & 0 \\ & E_{n-1} & -\varepsilon & & \\ & -\varepsilon & E_n & -\varepsilon & \\ & & -\varepsilon & E_{n+1} & \\ 0 & & & & \ddots \end{pmatrix},$$

with $E_n=E(\theta+n\omega)$ — the matrix decays exponentially off the diagonal

$$|(D+\varepsilon\hat{F})_{m,n}|\le e^{-|m-n|r}. \tag{9}$$

We want to construct an orthogonal matrix U such that

$$U^*(D+\varepsilon F)U \;=\; D_1 - \text{diagonal}. \tag{10}$$

Hence, this is a transformation problem, and an analysis of the linear equation will reveal a "small divisor" problem. Since the matrix is close to a diagonal it is reasonable to look for a U close to the identity. Putting $U = e^{\varepsilon X}$, X anti-symmetric, and linearizing (10) then gives

$$[D,X]+F \;=\; \hat{D} - \text{diagonal}. \tag{11}$$

The solution of this equation, up to an approximation of order $\varepsilon^{\frac{3}{2}}$, requires a condition on the eigenvalues of type

$$|\, E_n - E_m \,| \geq \frac{\sqrt{\varepsilon}}{|\, m-n \,|^{\tau}} \quad \text{for} \quad 0 < |\, m-n \,| < \frac{1}{r_0}\log(\frac{1}{\varepsilon}). \tag{12}$$

Under this condition we can find a solution $X_1, \hat{D}_1$ of (11), and using $U_1 = e^{\varepsilon X_1}$ we transform $D+\varepsilon F$ to

$$U_1^*(D+\varepsilon F)U_1 \;=\; D_1 + F_2,$$

where $D_1 = D + \varepsilon\hat{D}_1$,and $F_2 \sim \varepsilon^{\frac{3}{2}}$. Then we repeat the procedure with D_1 instead of D, requiring a condition of type (12) for the eigenvalues of D_1. If such a condition always holds we get an infinite sequence of transformations $\{U_1 \cdot \ldots \cdot U_j\}_{j=1}^{\infty}$ that transform $D+\varepsilon F$ more and more closely to a diagonal matrix. Since each U_j is more and more close to the identity, this sequence is easily seen to converge.

This is essentially the procedure that was used in [25,26] to construct examples of weakly quasi-periodic discrete Schrödinger operators with pure point spectrum.

In the problem we consider, however, (12) is never fulfilled. If for example $E(\theta) = \cos(\theta)$ (or any other function with exactly two critical non-degenerate points) then there are n's for which (12) is violated. The way to overcome this obstacle is to consider a larger class of transformations. What transformations? The important thing to notice is that for any n there is at most one m violating (12) — this can be shown by using the differentiability properties of the potential function E and the Diophantine property of ω. This implies that the obstruction in solving (11) occurs in blocks of dimension at most 2. Instead of assuming U close to the identity we shall therefore take U close to a block matrix with blocks of dimension ≤ 2 — such blocks may diagonalize any symmetric 2×2-matrix in a non-perturbative manner.

Of course using such transformations we get less good estimates and, more seriously, it gets more complicated to control the differentiability in θ of the spectrum of $D_1(\theta) = D(\theta) + \varepsilon\hat{D}(\theta)$. This control is shown to be possible in the work [24,25], and one can therefore continue the diagonalization procedure up to any order, for all θ. Of course an infinite product of block matrices may not converge and convergence can be obtained only for a.e. θ.

Theorem 5 is proven by enlarging the class of transformations even more — instead of block matrices of dimension ≤ 2 one must use matrices with blocks of arbitrary but finite dimension [26]. This approach, which makes it possible to diagonalize $D + \varepsilon F$ up to any approximation for all $\theta \in \mathbf{T}$, also makes it possible to analyze $D + \varepsilon F$ for the exceptional θ's. It suggests that for generic E and for generic θ the operator $\mathcal{L}_\theta^d$ has pure singular continuous spectrum. (Such results have been obtained in [27] with different methods.) □

It should be noted that Theorem 5 is one-dimensional and discrete. For $E : \mathbf{T}^d \to \mathbf{R}$, $d > 1$, the situation is somewhat different. Though a similar result is expected it is not possible to fix ω in advance by some Diophantine condition, but the choice of ω must depend on E as is analyzed in [28].

In the continuous case there cannot be a pure point spectrum because of Theorem 2 — but one expects the operator to be purely singular continuous on some piece near the bottom of the spectrum. This however has been proven in essentially only one example [25]:

$$-\frac{d^2u}{dx^2}(x) \; + \; \frac{1}{\varepsilon}[\cos(x\omega_1) + \cos(\theta + x\omega_2)]u(x). \tag{13}$$

If we summarize some of the results we see that for any analytic non-constant potential $E : \mathbf{T} \to \mathbf{R}$ the discrete operator

$$\mathcal{L}_\theta^d : \quad -(u(x+1) + u(x-1)) \; + \; \varepsilon E(\theta + x\omega)u(x)$$

is absolutely continuous for small ε and pure point for large . What happens for values of ε between these extremes? The continuous operator (13) is known to be pure point on a piece near the bottom of the spectrum but absolutely continuous on the upper part. What happens in the middle of the spectrum? It is only for the almost Mathieu equation $E(\theta) = \cos(\theta)$ that this question has received some answer [29].

References

[1] R. A. Johnsson, G. R. Sell, *Smoothness of spectral subbundles and reducibility of quasi-periodic linear differential systems*, J. Diff. Eq. **41** (1981), 262–288

[2] R. A. Johnsson, J. Moser, *The rotation number for almost periodic potentials*, Commun. Math. Phys. **84** (1982), 403–438.

[3] M. Herman, *Une méthode pour minorer les exposants de Lyapunov*, Comment. Math. Helvetici **58** (1983), 453–502.

[4] F. Delyon, B. Souillard, *The rotation number for finite difference operators and its properties*, Commun. Math. Phys. **89** (1983), 418–427.

[5] H. L. Cycon, R. G. Froese, W. Kirsch, B. Simon, *Schrödinger operators with applications to quantum mechanics and global geometry*, Springer-Verlag, New York, (1987).

[6] T. Spencer, *The Schrödinger equation with a random potential — a mathematical review*, K. Osterwalder, R. Stora (eds.): Critical phenomena, random systems, gauge theories, North-Holland Publishing Co, Amsterdam-New York, (1986).

[7] B. Simon, *Bounded eigenfunctions and absolutely continuous spectrum for one-dimensional Schrödinger operators*, submitted to Proc. Amer. Math. Soc. (1995).

[8] R. Carmona, J. Lacroix, *Spectral Theory of Random Schrödinger Operators*, Birkhäuser, Boston, (1990).

[9] J. Avron, B Simon, *Singular continuous spectrum for a class of almost periodic Jacobi matrices*, Bull. Amer. Math. Soc. **6** (1982), 81–85.

[10] F. Delyon, D. Petritis, *Absence of Localization in a Class of Schrödinger Operators with Quasi-Periodic Potentials*, Commun. Math. Phys. **103** (1986), 441–443.

[11] S. Kotani, *Jacobi Matrices with Random Potentials taking Finitely many Values*, Rev. Math. Phys. **1** (1989), 129–133.

[12] E. I. Dinaburg, Ya. G. Sinai, *The one-dimensional Schrödinger equation with quasi-periodic potential*, Funkt. Anal. i. Priloz. **9** (1975), 8–21.

[13] J.Moser, J. Pöschel, *An extension of a result by Dinaburg and Sinai on quasi-periodic potentials*, Comment. Math. Helvetici **59** (1984), 39–85.

[14] L. H. Eliasson, *Floquet solutions for the one-dimensional quasi-periodic Schrödinger equation*, Commun. Math. Phys. **146** (1992), 447–482.

[15] J. Moser, *Convergent series expansions for quasiperiodic motions*, Math. Ann. **169** (1967), 136–176.

[16] L. H. Eliasson, *Perturbations of stable invariant tori for Hamiltonian systems*, Ann. Sc. Norm. Sup. Pisa Cl. Sci. **15** (1988), 115–147.

[17] A. Jorba, C Simo, *On the reducibility of linear differential equations with quasi-periodic coefficients*, J. Diff. Eq. **98** (1992), 111–124.

[18] L. H. Eliasson, *Ergodic skew systems on* $SO(3, \mathbf{R})$, preprint ETH-Zürich (1991).

[19] R. Krikorian, *Réductibilité presque partout des systèmes quasi périodiques analytiques dans le cas* $SO(3)$, C. R. Acad. Sci. Paris **321, Série I** (1995), 1039–1044.

[20] R. Krikorian, *Réductibilité des systèmes produits croisés quasi-périodiques à valeurs dans des groupes compacts*, Thesis École Polytechnique, Paris, (1996).

[21] B. Simon, *Almost periodic Schrödinger operators: a review*, Adv. Appl. Math. **3** (1982), 463–490.

[22] W. Craig, *Pure point spectrum for discrete almost periodic Schrödinger operators*, Commun. Math. Phys. **88** (1983), 113–131.

[23] J Pöschel, *Examples of discrete Schrödinger operators with pure point spectrum*, Commun. Math. Phys. **88** (1983), 447–463.

[24] Ya. G. Sinai, *Anderson localization for the one-dimensional difference Schrödinger operator with a quasi-periodic potential*, J. Stat. Phys. **46** (1987), 861–909.

[25] J. Fröhlich, T. Spencer, P. Wittver, *Loclization for a class of one-dimensional quasi-periodic Schrödinger operators*, Commun. Math. Phys. **132** (1990), 5–25.

[26] L. H. Eliasson, *Discrete one-dimensional quasi-periodic Schrödinger operators with pure point spectrum*, manuscript (1995).

[27] S. Jitomirskaya, B. Simon, *Operators with Singular Continuous Spectrum: III. Almost Periodic Schrödinger Operators*, Commun. Math. Phys. **165** (1994), 201–205.

[28] V. A. Chualevsky, Ya. G. Sinai, *Anderson localization and KAM-theory*, in P. Rabinowitz, E. Zehnder (eds.): Analysis etcetera, Academic Press, New York, 1989.

[29] S. Ya. Jitomirskaya, *Anderson Localization for the Almost Mathieu Equation: A Non-Perturbative Proof*, Commun. Math. Phys. **165** (1994), 49–57.

Progress in Mathematics, Vol. 168, © 1998 Birkhäuser Verlag Basel/Switzerland

Banach Spaces with Few Operators

W. T. Gowers

Department of Pure Mathematics
and Mathematical Statistics
16 Mill Lane, Cambridge CB2 1SB, England

Introduction

In this paper we shall survey certain results in infinite-dimensional Banach space theory. However, in order to avoid substantial overlap with another survey paper [G], we have not made the most obvious selection of results, and the selection does not exactly reflect the results mentioned in the author's lecture at the European Congress. This paper is designed rather as a companion to [G]. Although it can be read on its own by a non-expert, it is better read in conjunction with the other survey, which also contains a much more comprehensive set of references.

Briefly, there will be two main topics discussed here. Many interesting properties of so-called hereditarily indecomposable Banach spaces were mentioned in [G]. In this paper, we shall give some idea of how to construct them. It was also mentioned in [G] that there is a close connection between such constructions and a phenomenon known as the distortion of Banach spaces. We shall explain the connection here in some detail. In addition to these two topics, there will be a section about results that have appeared since the earlier paper was written.

A *hereditarily indecomposable* Banach space is a space X such that no subspace of X can be written as a sum $Y+Z$ with Y and Z infinite-dimensional and the projections to Y and Z continuous. Equivalently, on any subspace Y of X the only continuous projections have finite rank or corank. It is a straightforward exercise to prove another equivalence: X is hereditarily indecomposable if and only if for every $\epsilon > 0$ and every pair Y, Z of infinite-dimensional subspaces of X there exist unit vectors $y \in Y$ and $z \in Z$ such that $\|y-z\| < \epsilon$. It can be shown that every operator on such a space is the sum of a multiple of the identity and a strictly singular operator, which implies (by well known results) that it is either strictly singular or Fredholm with index zero. In either case, it cannot, for example, be an isomorphism to a proper subspace. For more detail about these results see [G,GM].

Unconditional bases and distortion

The first example of a hereditarily indecomposable space was constructed by Maurey and the author [GM]. We constructed it in order to do something weaker, and it was Johnson who pointed out to us that our space had the stronger property of hereditary indecomposability. Two definitions are needed in order to explain our original purpose. First, a *Schauder basis* of a Banach space X is a sequence (x_n) such that every element of X can be written in a unique way in the form $\sum a_n x_n$, where the sum converges in norm. Such a basis is called *unconditional* if there is some constant C such that, given any element $\sum a_n x_n$ of X and any sequence (ϵ_n) with $\epsilon_n = \pm 1$, the inequality

$$\left\|\sum \epsilon_n a_n x_n\right\| \leq C\left\|\sum a_n x_n\right\|$$

holds. If the inequality holds with constant C, one says that the basis is *C-unconditional.* (Note that this definition is non-trivial even for finite sequences.) A sequence which is an unconditional basis of its closed linear span is called an *unconditional basic sequence.* Our aim was to construct a Banach space containing no unconditional basic sequence, or equivalently, such that no subspace had an unconditional basis.

Let c_{00} stand for the vector space of all sequences of scalars that are eventually zero. Thus, given $x = (x_n) \in c_{00}$, the *support* of x, that is, the set of all n such that $x_n \neq 0$, is finite. We write $x < y$ if the largest element of the support of x is less than the smallest element of the support of y. (In other words, all non-zero coefficients of x come before all non-zero coefficients of y.) A *block subspace* of c_{00} is a subspace with a basis $y_1, y_2, \ldots$ such that $y_1 < y_2 < \ldots$. The basis is called a *block basis.* For every n, let P_n be the projection on c_{00} which kills all coordinates beyond n. It is not hard to show that constructing a Banach space such that no subspace has an unconditional basic sequence is equivalent to constructing a norm $\|.\|$ on c_{00} such that $\|P_n x\| \leq \|x\|$ for every $n \in \mathbb{N}$ and $x \in c_{00}$, and every block subspace Y of c_{00} contains for every C a sequence of vectors $x_1 < \ldots < x_n$ (n may depend on Y and C) such that

$$\|x_1 + x_2 + \ldots + x_n\| > C\|x_1 - x_2 + \ldots - (-1)^n x_n\| . \tag{1}$$

The second condition is of course the important one.

Even finding such a norm for a single value of $C > 1$ is not easy. The reasons for this are intimately connected with the idea of distortion of Banach spaces. We shall now give several arguments for this connection.

Suppose we try to use sequences of some *fixed* length in (1). That is, suppose we try to find some $C > 1$, some positive integer n and a norm $\|.\|$ on c_{00} such that every block subspace contains $x_1 < \ldots < x_n$ satisfying (1). Given such a norm, we can define another norm $|||.|||$ by the formula

$$|||x||| = \sup\left\|\sum_{i=1}^{n} \epsilon_i x_i\right\| ,$$

where the supremum is over all sequences $x_1 < \ldots < x_n$ such that $\sum_{i=1}^n x_i = x$ and over all sequences of signs $\epsilon_i = \pm 1$. By the triangle inequality and the assumption that the projections P_n had norm at most 1, one sees easily that $\|x\| \leq |||x||| \leq 2n\,\|x\|$ for every $x \in c_{00}$. Our assumption about the norm $\|.\|$ implies that every block subspace of c_{00} contains a vector x such that $|||x||| > C\,\|x\|$.

On the other hand it is also possible to find for every $N \in \mathbb{N}$, every $\epsilon > 0$ and every block subspace Y a sequence $x_1 < \ldots < x_N$ of unit vectors in Y forming a $(1+\epsilon)$-unconditional basis of its linear span. This is a consequence of Ramsey's theorem. Very roughly, for any $k \in \mathbb{N}$ and $\delta > 0$ one can pass to a subsequence (z_n) of the block basis (y_n) such that the norm of any linear combination $\sum_{i=1}^k a_i z_{n_i}$ such that $n_1 < \ldots < n_k$ depends, up to a factor of $(1+\delta)$, only on the sequence $a_1, \ldots, a_k$. If this is done for large enough k, then the vectors x_i can be chosen to be of the form $z_{m+1} - z_{m+2} + \ldots - (-1)^p z_{m+p}$. This is because $-x_i$ is then well-approximated by $z_{m+2} - z_{m+3} + \ldots - (-1)^{p+1} z_{m+p+1}$, and hence sign changes in a linear combination of the x_i do not have a large effect on the norm of the combination. This argument does not work if the z_n span a space too close to ℓ_∞, since we are not then able to regard $z_m + (-1)^{p+1} z_{m+p+1}$ as small, but this is a technicality which can be dealt with. See [MiS Chapter 11] for the argument in detail.

It is a straightforward exercise now to show that the vector $x = x_1 + \ldots + x_N$ satisfies $|||x||| \leq (1+2\epsilon)\,\|x\|$, at least if $\|x\|$ is significantly bigger than the $\|x_i\|$. We have more or less proved it already, except that when calculating $|||x|||$ by decomposing it we are allowed to split up the individual x_i. Again, other arguments can be used if the growth condition is never satisfied.

To summarize, we can find in every block subspace vectors x and y such that $|||x||| \leq (1+\epsilon)\,\|x\|$ and $|||y||| > C\,\|y\|$. It follows that on no block subspace are the norms $\|.\|$ and $|||.|||$ any better than C-equivalent. Standard arguments show that this is the same for an arbitrary (infinite-dimensional) subspace. If X is any normed space with equivalent norms $\|.\|$ and $|||.|||$ which are not C-equivalent on any subspace, then $|||.|||$ is called a *C-distortion* of $\|.\|$, and X is said to be *C-distortable.* If X is C-distortable for some $C > 1$, then X is called *distortable*; if for every $C > 1$ then it is *arbitrarily distortable.*

Going back to the space above, define A to be the set of all x such that $\|x\| = 1$ and $|||x||| \leq (1+\epsilon)$ and let B be the set of all y such that $\|y\| = 1$ and $|||y||| > C$, and recall that $|||z||| \leq 2n\,\|z\|$ for every z in the space. It is an easy exercise to show that $\|x - y\| \geq (C - 1 - \epsilon)/2n$ for every $x \in A$ and every $y \in B$. In general, it is possible to show that every distortable Banach space contains two subsets A and B of the unit sphere with the following two properties that are hard to reconcile. On the one hand, both A and B are *asymptotic*, which means that they intersect every subspace of X. On the other, they are *separated*, in the sense that there is a positive lower bound on the distance between any pair of vectors $x \in A$ and $y \in B$.

This can be regarded as an "anti-Ramsey" phenomenon. Suppose X is a Banach space such that the unit sphere of X contains separated asymptotic sets A and B, with the lower bound on the distance being at least ϵ. Let us colour points of the unit sphere of X according to whether they are closer to A or to B. Then the properties of A and B imply that no subspace of X has a monochromatic unit sphere, even up to an error of $\epsilon/2$.

By contrast, a theorem of Hindman [H] states the following. If the finite subsets of $\mathbb{N}$ are coloured with finitely many colours, then there is a collection $A_1, A_2, \ldots$ of finite subsets of $\mathbb{N}$ with $\max A_i < \min A_{i+1}$ for every i, such that all non-empty finite unions of the A_i have the same colour. If we reinterpret this in terms of the characteristic functions of the A_i, finite unions become finite sums, and the collection of sets guaranteed to have the same colour can be thought of as a sort of discrete subspace generated by the functions. In general, the existence of Ramsey-type theorems in combinatorics, some of which can even be applied to Banach spaces, makes it far from obvious whether distortable Banach spaces exist, and this goes some way towards explaining why they were not known to exist until comparatively recently, and why the solution by Odell and Schlumprecht [OS] of the so-called distortion problem, whether ℓ_2 is distortable, was a major breakthrough.

Let us now give two further arguments for a link between the construction of certain kinds of Banach spaces and the phenomenon of distortion. First, a result of Milman [Mi] states that if no subspace of a Banach space X is distortable, then that space must contain some ℓ_p-space or c_0 as a subspace. Since ℓ_p and c_0 have unconditional bases, we can deduce that if no subspace of X has an unconditional basis, then X has a distortable subspace.

The second argument brings us closer to the question of how to construct a Banach space such that equation (1) can be satisfied in every block subspace. If one were to try to prove that no such space existed, then a very useful first step would be to find for any $\epsilon > 0$ a vector x and a block subspace Y such that for every unit vector $y \in Y$ we have $x < y$ and $\|x + y\| \leq (1+\epsilon)\,\|x - y\|$. But can even this be guaranteed? If not, then in the language introduced above we can find for every x an asymptotic set A_x such that for every $y \in A_x$ we have the inequality $\|x + y\| > (1+\epsilon)\,\|x - y\|$. Notice that $-A_x$ is also asymptotic and separated from A_x, so that once again distortion seems to be relevant.

Turning this argument round, one can wonder whether the existence of an asymptotic set A separated from $-A$ may be used to construct a space with no C-unconditional basic sequence (for some $C > 1$). One soon finds that one such set is not enough, although this has not been proved as a formal mathematical result. For example, if x and y are very close, A_x cannot be close to $-A_y$ above. Considering a path from x to y one then sees that it would be useful to have two asymptotic sets A and B, separated from each other and from $-A$ and $-B$. More generally, the Borsuk-Ulam theorem suggests that no finite collection of separated asymptotic sets will be enough. These remarks provide a partial motivation for the following definition.

Definition. Let X be a Banach space. An *asymptotic biorthogonal system* on X is a pair of sequences $(A_n)_1^\infty$ and $(A_n^*)_1^\infty$ of subsets of X and X^* respectively, with the following properties.

(i) Elements of A_n have norm one and elements of A_n^* have norm at most one.
(ii) The sets A_n are all asymptotic.
(iii) For every $x \in A_n$ there exists $x^* \in A_n^*$ such that $|x^*(x)| \geq 1/2$.
(iv) There exists a real sequence $\delta_n \to 0$ such that $|x^*(x)| \leq \delta_k$ whenever $x \in A_n$, $x^* \in A_m^*$, $n \neq m$, $k = \min\{n, m\}$.

Passing to an appropriate subsequence one can of course ensure that the δ_n converge to zero at any desired speed. In particular we may assume that no δ_n is more than 1/4. Properties (iii) and (iv) above then give easily that the sets A_n are all separated from one another. Thus, the existence of an asymptotic biorthogonal system is a considerable strengthening of the property of distortability. The next theorem shows how we may use such a system.

Theorem 1. *Let X be a separable Banach space with an asymptotic biorthogonal system. For every C there is an equivalent norm $\|\!|.|\!\|$ on X such that the space $(X, \|\!|.|\!\|)$ has no subspace with a C-unconditional basis.*

We shall sketch the proof, which appears in full in [GM]. There is a small (and unimportant) choice in the definition of an asymptotic set, which we did not specify earlier, of exactly which subspaces it must intersect. For this proof it is convenient to insist that an asymptotic set intersects all infinite-dimensional subspaces, whether or not they are closed.

Since X is separable, we can find a countable subset Q of X^* such that for every $\epsilon > 0$, every $n \in \mathbb{N}$ and every $x \in A_n$ there exists $x^* \in Q \cap A_n^*$ such that $x^*(x) \geq 1/4$. Let $\delta > 0$ be sufficiently small, let k be such that $\delta_n \leq \delta$ for every $n \geq k$ and let σ be an injection from the set of all finite sequences of elements of Q, including the null sequence, into $\{k, k+1, k+2, \dots\}$. Define a *special sequence* of length k to be a sequence $(x_1^*, x_2^*, \dots, x_k^*)$ such that $x_i^* \in A_{\sigma(x_1^*,\dots,x_{i-1}^*)}$ for every $i \leq k$.

The main reason for the definition of special functionals is the following property that they have. If $(x_1^*, \dots, x_k^*)$ and $(y_1^*, \dots, y_k^*)$ are two special functionals then we can find t such that $x_i^* = y_i^*$ if $i < t$ while x_i^* and y_i^* lie in different A_n^* if $i > t$. It will soon become clear why this is useful.

The equivalent norm $\|\!|.|\!\|$ on X is defined by the formula

$$\|\!|x|\!\| = \|x\| \vee r \sup\Big\{|y^*(x)| : y \in S_r\Big\}$$

where r is an integer greater than $100C$ and S_r is the set of all special functionals of length r. We now wish to show that no sequence $z_1, z_2, \dots$ of linearly independent vectors in X is a C-unconditional basic sequence in the norm $\|\!|.|\!\|$. To do this, we shall construct a sequence $x_1, x_2, \dots, x_r$ of vectors in the (algebraic) linear span of the z_i, such that they are disjointly supported with respect to the z_i and such that

$$C\Big\|\!\Big|\sum_{i=1}^{r}(-1)^i x_i\Big|\!\Big\| < \Big\|\!\Big|\sum_{i=1}^{r} x_i\Big|\!\Big\| .$$

To do this, let Z be the linear span of the z_i. Since $A_{\sigma()}$ is asymptotic, we can find $y_1 \in A_{\sigma()} \cap Z$ and we can then find $x_1^* \in Q \cap A_{\sigma()}^*$ such that $x_1^*(y_1) \geq 1/4$. Suppose we have chosen $y_1, \dots, y_k$ and $x_1^*, \dots, x_k^*$ such that $x_i^* \in Q$ for each i. Since each y_i is finitely supported and the A_n are asymptotic, we can choose $y_{k+1} \in A_{\sigma(x_1^*,\dots,x_k)}$ supported disjointly from the y_i with $i \leq k$, and $x_{k+1}^* \in Q \cap A_{\sigma(x_1^*,\dots,x_k)}^*$ such that $x_{k+1}^*(y_{k+1})$ is at least 1/4. Finally, set $x_i = y_i / x_i^*(y_i)$, so that $\|x_i\| \leq 4$ and $x_i^*(x_i) = 1$ for every i.

The sequences $x_1, \dots, x_r$ and $x_1^*, \dots, x_r^*$ are constructed so that $x_1^* + \dots + x_r^*$ is a special functional which is large on the vector $x_1 + \dots + x_r$. It is easy to use property (iv) of asymptotic biorthogonal systems to check that this is so: $x_i^*(x_j)$ is 1 when $i=j$ and at most 4δ otherwise. One can choose δ so that

$$(x_1^*+\ldots+x_r^*)(x_1+\ldots+x_r)\geq r/2\ ,$$

which shows that $|||x_1+\ldots+x_r|||\geq r^2/2$. A slightly subtler argument gives an upper bound for the norm of the alternating sum $-x_1+x_2-\ldots+(-1)^r x_r$. By the triangle inequality we know that it is at most $4r$ in the old norm, so the only chance for it to be large in the new norm is if we can find a suitable special functional $y_1^*+\ldots+y_r^*$ to norm it. However, as we commented earlier, $y_i^*=x_i^*$ for i less than some t and for $i>t$ y_i^* lies in the wrong A_m^* for any $y_i^*(x_j)$ to be more than δ. Thus $y_i^*\left((-1)^j x_j\right)$ is $(-1)^i$ if $i=j<t$, and at most δ if either $i=j>t$ or $i\neq j$. The first contributions cancel (or almost all cancel if t is even) and for suitable δ we have shown that no special functional is large at $-x_1+x_2-\ldots+(-1)^r x_r$, which completes the proof of Theorem 1.

We have now seen that distortability is necessary for constructing a space with no unconditional basic sequence, and that a strengthening of distortability is sufficient. However, the existence of an asymptotic biorthogonal system is such a strong assumption that it is not surprising that it has interesting consequences, and we must now turn our attention to how such a space and system can be constructed.

Inductive constructions of Banach spaces

First, let us remark that in the paper of Odell and Schlumprecht showing that ℓ_2 is distortable [OS] it is shown also that ℓ_2 contains an asymptotic biorthogonal system. In this case the statement can be reformulated as follows: there are asymptotic subsets $A_1, A_2, \ldots$ of ℓ_2 such that for any $x\in A_m$ and any $y\in A_n$ the size of the inner product $\langle x,y\rangle$ is at most $\delta_{\min(m,n)}$ when $m\neq n$. Thus for any C there is an equivalent norm on ℓ_2 such that the resulting space contains no C-unconditional basic sequence.

In this paper we shall discuss a different example, essentially due to Schlumprecht [S]. The space is more complicated to define than ℓ_2, but it is easier to find an asymptotic biorthogonal system. He defines his space by modifying a famous space of Tsirelson [T]. We shall give the definitions of both spaces. It will be necessary to introduce a little more notation. Given finite subsets $E,F\subset\mathbb{N}$, let us write $E<F$ if $\max E<\min F$. Let us also write $k<E$ if $k<\min E$. If $x\in c_{00}$, then let Ex stand for the projection to the coordinates in E. That is, $(Ex)_i=x_i$ if $i\in E$ and 0 otherwise. Tsirelson's space is the completion of the normed space $(c_{00},\|.\|)$, where $\|.\|$ is the unique norm satisfying

$$\|x\|=\|x\|_\infty\vee(1/2)\sup\Big\{\sum_{i=1}^k\|E_i x\|:k<E_1<E_2<\ldots<E_k\Big\}$$

for every $x\in c_{00}$. It is an easy exercise to show the existence and uniqueness of this norm. One way of thinking about it is that to calculate the norm of x one takes sets $k < E_1 < \ldots < E_k$ and adds half the norms of the vectors $E_i x$. These norms we already know by induction on the size of their supports. When the support has size one the norm is the ℓ_∞-norm.

It is not as obvious how to analyse Tsirelson's space and prove the result for which it was defined: there exists a Banach space not containing c_0 or ℓ_p $(1 \le p < \infty)$. It seems to be essential that the norm is described in an implicit (or inductive) way rather than by a more direct formula. Certainly the idea of such a description was a fundamental breakthrough of Tsirelson, opening up a whole range of possibilities and leading to the solution of many problems for which more conventional constructions have so far been inadequate.

In Schlumprecht's space, the condition $k < E_1$ is dropped and the factor of $1/2$ is replaced by one of $(\log_2(k+1))^{-1}$. Thus his space is the completion of $(c_{00}, \|.\|)$ with $\|.\|$ defined by the equation

$$\|x\| = \|x\|_\infty \vee \sup\Big\{\frac{1}{\log_2(k+1)} \sum_{i=1}^{k} \|E_i x\| : k \ge 2, E_1 < E_2 < \ldots < E_k\Big\} .$$

It is important in both definitions that there should be something to discourage splitting x into too many pieces with the projections E_i. In Tsirelson's space it is the condition $k < E_1$: if we use too many sets E_i then we lose the beginning of the vector x. In Schlumprecht's space we divide by a larger number the more sets we use. This may seem a relatively small change, but it turned out to be crucial. We shall now define an asymptotic biorthogonal system in Schlumprecht's space, but it is an open problem whether Tsirelson's space is even arbitrarily distortable.

Notice that if $x_1, x_2, \ldots, x_N$ is a sequence in c_{00} such that $x_1 < \ldots < x_N$, then it follows from the definition of the norm that

$$\|x_1 + \ldots + x_N\| \ge (\log_2(N+1))^{-1}(\|x_1\| + \ldots + \|x_N\|) . \qquad (2)$$

The sets A_n^* are now easy to define. If $f_1, f_2, \ldots, f_n$ is a sequence of vectors in c_{00}, such that $f_1 < \ldots < f_N$ and each f_i has norm at most 1 when considered as a linear functional on Schlumprecht's space, then equation (2) above implies that the sum $(\log_2(n+1))^{-1}(f_1 + \ldots + f_n)$ is also a functional of norm at most 1. The set of all functionals that arise in this way is the set A_n^*.

The sets A_n are a little more complicated. First, an argument of James [J] can be used to deduce from (2) that every block subspace of Schlumprecht's space contains for every M and every $\epsilon > 0$ a sequence $x_1 < \ldots < x_N$ with $x_i = 1$ for every i, such that $\|x_1 + \ldots + x_M\| \ge (1-\epsilon)M$. Given an integer k and $\delta > 0$ it is not hard to choose M and ϵ such that the vector $x = x_1 + \ldots + x_M$ has the following property: for any choice of sets $E_1 < \ldots < E_k$ the sum $\sum_{i=1}^k E_i x$ has norm at most $(1+\delta)\|x\|$. That is, cutting x into k pieces and adding their norms does not give an answer much bigger than $\|x\|$. (By the triangle inequality it cannot be smaller.) Let us call such a vector x a (k,δ)-*vector*.

These (k,δ)-vectors are the building blocks for elements of A_n. The precise definition of A_n involves various arbitrary choices, and it is perhaps clearer not to be precise. A typical element of A_n is defined as follows. Let $f(n)$ be an integer much greater than $f(n-1)$ and let $\delta > 0$ be sufficiently small. For k_1 large enough choose a (k_1, δ)-vector x_1 of norm one. Then, for k_2 large enough (depending on δ and the maximum of the support of x_1) choose a (k_2, δ)-vector x_2 of norm one with $x_1 < x_2$. Continue up to $x_{f(n)}$. Finally, let $y = x_1 + \ldots + x_{f(n)}$ and let $x = y/\|y\|$. Basically, x is a sum of (k_i, δ)-vectors x_i with the k_i increasing very rapidly, but it is important that the rapidity depends on the x_i as well as the k_i. Any vector x of this form is an element of A_n.

Proving that the sequences (A_n) and (A_n^*) form an asymptotic biorthogonal system is somewhat technical. The main idea is that the the typical element x of A_n described above was carefully designed so that $\|x\|$ is not much larger than $(\log_2(f(n)+1))^{-1}\sum_{i=1}^{f(n)}\|x_i\|$, which is a trivial lower bound, by (2). The proof of this fact involves showing that if k is either much less or much greater than $f(n)$ and $E_1 < \ldots < E_k$, then $(\log_2(k+1))^{-1}\sum_{i=1}^{k}\|E_i x\|$ is very much less than $\|x\|$, which shows that (A_n) and (A_n^*) form an asymptotic biorthogonal system.

We have now shown, for any $C > 1$, how to construct a space with no C-unconditional basic sequence. However, it is substantially harder to find a single space that works for every C. The most obvious idea would be to adapt the proof of Theorem 1 but use special sequences of unbounded length. The resulting space would not be isomorphic to the space with which one began (necessarily) but that would not matter. It turns out, however, that a space constructed in such a way contains c_0. (To see this, consider a fast-growing sequence $k_1, k_2, \ldots$ of integers and a normalized block basis $x_1 < x_2 < \ldots$ with x_n constructed as in the proof of Theorem 1 to be normed by a special functional arising from a special sequence of length k_n.)

The construction in [GM] *does* use special sequences of unbounded length, but they are defined intrinsically. Roughly speaking, the definition of the sets A_n^* in Schlumprecht's space makes sense in any space $X = (c_{00}, \|.\|)$, and these sets can then be used in a manner related to the proof of Theorem 1 to define functionals that are *special in* X. Although we have not explained exactly what it means, the following expression for the norm on the space in [GM] gives some idea of how it is defined. The sets $E_1 < \ldots < E_N$ appearing in the definition are restricted to being *intervals*, that is, sets of the form $\{m, m+1, \ldots, n\}$ (so that the resulting space will not trivially have an unconditional basis).

$$\|x\| = \|x\|_\infty \vee \sup\Big\{\frac{1}{\log_2(N+1)}\sum_{i=1}^{N}\|E_i x\| : N \geq 2, E_1 < \ldots < E_N\Big\}$$
$$\vee \sup\{|x^*(x)| : x^* \text{ is special in } X\}$$

The space X in the definition is the space we are defining! This is another example of apparent circularity, and again one can show that there is a unique norm satisfying the above equation for every x.

The most important idea above is to include special functionals as part of an inductive definition. Doing so raises several technical difficulties, but these can be overcome and we will not discuss them here.

We conclude this section with the observation, due to Johnson, that the space above is actually hereditarily indecomposable. The proof that it contains no unconditional basic sequence involves constructing in every block subspace and for every C a sequence $x_1 < \ldots < x_n$ satisfying (1). If Z and W are two (block) subspaces of X, the same argument can be used to construct such a sequence but this time with $x_i \in Z$ for i even and $x_i \in W$ for i odd. Setting $z = x_1 + x_3 + x_5 + \ldots$ and $w = x_2 + x_4 + x_6 + \ldots$ we find that $\|z + w\| > C\|z - w\|$, from which it is easy to deduce that $z/\|z\|$ and $w/\|w\|$ are close. This argument should be compared with the deduction of Theorem 3 from Theorem 4 in [G].

Further results and open problems

Since the paper [G] was written there have been two main sources of progress. First, there have been constructions with stronger properties, and second the properties of hereditary indecomposability and distortability are now better understood. We shall discuss these two directions in turn.

Although it was never likely that any mild condition on a Banach space would stop it from being hereditarily indecomposable, it was surprisingly hard to modify the construction of [GM] to rid it of some of its seemingly accidental features. For example, the space in [GM] contains almost isometric copies of ℓ_∞^n for every n and is therefore not uniformly convex. Ferenczi [F1] has constructed a uniformly convex hereditarily indecomposable space, but the proof is significantly more difficult than that of [GM].

It was also interesting for various reasons to know whether a hereditarily indecomposable space could be built which resembled Tsirelson's space more than Schlumprecht's. The main property of interest here is that if $x_1 < x_2 < \ldots < x_n$ is a block basis in Tsirelson's space and if the minimum of the support of x_1 is greater than n, then $(x_1, \ldots, x_n)$ is 2-equivalent to the unit vector basis of ℓ_1^n. This property makes Tsirelson's space an example of an *asymptotic* ℓ_1 *space* (cf. [MMT]). It was not clear whether such a space could be hereditarily indecomposable until Argyros and Delyanni produced an example [AD], which is, again, complicated.

A third example is due to Habala [H]. He constructs a space such that no subspace has the Gordon-Lewis property, a property considerably weaker than having an unconditional basis. His space is also hereditarily indecomposable, and is of interest in the context of the homogeneous spaces problem (see [G] for a statement of this).

We now turn to results about hereditary indecomposablity and distortion. We shall use certain terms that are defined in [G] without repeating the definitions here. Ferenczi [F2] has shown the following strengthening of [G Theorem 1].

Theorem 2. *Let X be a complex hereditarily indecomposable Banach space and let Y be a subspace of X. Every operator from Y to X is the sum of a strictly singular operator and a multiple of the inclusion map.*

He has also studied what happens in the real case [F3].

There are some interesting results concerning hereditarily indecomposable spaces and duality, also due to Ferenczi. It turns out that the dual of a hereditarily indecomposable space need not be hereditarily indecomposable. This motivates the following definition. A Banach space X is *quotient hereditarily indecomposable* if no subspace of any quotient of X can be written as a topological direct sum $Y+Z$. See [F4] for results in this direction.

Despite all the evidence outlined earlier in this paper for a connection between distortion and spaces with no unconditional basic sequence, the following beautiful result of Tomczak-Jaegermann [T-J] was for a surprisingly long time an open problem.

Theorem 3. *Let X be a Banach space which has no C-distortable subspace. Then X contains a $4C$-unconditional basic sequence.*

It follows easily that if X contains no unconditional basic sequence, then X has an arbitrarily distortable subspace, and that if X is hereditarily indecomposable, then X is itself arbitrarily distortable. Although these results are very interesting and satisfactory, the discussion earlier in this paper suggests that more should be true. The following question is open.

Problem 1. *Suppose X is a Banach space containing no unconditional basic sequence. Does X have a subspace which contains an asymptotic biorthogonal system?*

Another open problem is the following.

Problem 2. *Does every distortable Banach space have an arbitrarily distortable subspace?*

One can ask more: no example is known of a distortable space not containing an asymptotic biorthogonal system. The most likely candidate is Tsirelson's space, which is distortable but is not known to be arbitrarily distortable. This last problem is fascinating because it is so concrete, and somehow encapsulates the difficulties that still exist in the area.

Alternatively, one can weaken Problem 1. For example, it would be very interesting to show that X must have a subspace containing a collection of asymptotic sets, all separated by a uniform amount.

It is possible that the answers to some of the questions above (especially Problem 2) may be positive for uninteresting reasons. This would be the case, for example, if one could somehow produce a relevant classification of Banach spaces and observe by examining each class in turn that all spaces happened either not to be distortable or to contain an asymptotic biorthogonal system. Some arguments like this already exist. For example, Maurey [M] has shown that a uniformly convex space with an unconditional basis has an arbitrarily distortable subspace. Combined with Theorem 3 above this shows that every uniformly convex space is arbitrarily distortable. (Of course, such classifications are themselves interesting – it is just the deduction of one form of distortability from another that would not be "genuine".)

References

[F1] V. Ferenczi, *A uniformly convex and hereditarily indecomposable Banach spaces*, Israel J. Math. (to appear).

[F2] V. Ferenczi, *Operators on subspaces of hereditarily indecomposable Banach spaces*, Bull. L. M. S. **29** (1997), 338–344.

[F3] V. Ferenczi, *Hereditarily finitely decomposable Banach spaces*, Studia Math. (to appear).

[F4] V. Ferenczi, PhD. Thesis, Université de Marne La Vallée, France.

[G] W. T. Gowers, *Recent results in the theory of Banach spaces*, Proceedings of the International Congress of Mathematicians 1994, Birkhäuser Verlag 1995, 933–942.

[GM] W. T. Gowers and B. Maurey, *The unconditional basic sequence problem*, J. Amer. Math. Soc. **6** (1993), 851–874.

[H] P. Habala, *Banach spaces all of whose subspaces fail the Gordon-Lewis property*, (submitted).

[H] N. Hindman, *Finite sums from sequences within cells of a partition of* $\mathbb{N}$, J. Comb. Th. (A) **17** (1974), 1–11.

[J] R. C. James, *Uniformly non-square Banach spaces*, Ann. of Math. **80** (1964), 738–743.

[M] B. Maurey, *A remark about distortion*, in Operator Theory: Advances and Applications **77** Birkhäuser Verlag 1995, 131–142.

[MMT-J] B. Maurey, V.D. Milman and N. Tomczak-Jaegermann, *Asymptotic infinite-dimensional theory of Banach spaces*, in Operator Theory: Advances and Applications **77** Birkhäuser Verlag 1995, 149–175.

[Mi] V. D. Milman, *Spectrum of continuous bounded functions on the unit sphere of a Banach space*, Funct. Anal. Appl. **3** (1969), 67–79 (translated from Russian).

[MiS] V. D. Milman and G. Schechtman, Asymptotic Theory of Finite Dimensional Normed Spaces, Lecture Notes in Mathematics 1200, Springer Verlag 1986.

[OS] E. Odell and T. Schlumprecht, *The distortion problem*, Acta Math. **173** (1994), 259–281.

[S] T. Schlumprecht, *An arbitrarily distortable Banach space*, Israel J. Math. **76** (1991), 81–95.

[T-J] N. Tomczak-Jaegermann, *Banach spaces of type p have arbitrarily distortable subspaces*, GAFA (to appear).

[T] B. Tsirelson, *Not every Banach space contains* ℓ_p *or* c_0, Funct. Anal. Appl. **8** (1974), 139–141.

Progress in Mathematics, Vol. 168, © 1998 Birkhäuser Verlag Basel/Switzerland

Recent Developments in the Function Theory of the Bergman Space

HÅKAN HEDENMALM

Department of Mathematics
Lund University, Box 118
S–221 00 Lund, Sweden
E-mail: haakan@maths.lth.se

ABSTRACT. The recent developments in the function theory of the Bergman space are reviewed. Key ingredients are: (1) factorization based on extremal divisors, (2) an analog of Beurling's invariant subspace theorem, (3) concrete examples of invariant subspaces of index higher than 1, (4) a partial description of zero sequences, (5) characterizations of interpolating and sampling sequences, and (6) some remarks on weighted Bergman spaces.

1. The Hardy and Bergman spaces – a comparison

The Hardy space H^2 consists of all holomorphic functions on the open unit disk $\mathbb{D}$ such that

$$\|f\|_{H^2} = \sup_{0<r<1} \left(\int_{\mathbb{T}} |f(rz)|^2 \, ds(z) \right)^{\frac{1}{2}} < +\infty, \tag{1-1}$$

where $\mathbb{T}$ stands for the unit circle, and ds is arc-length measure, normalized so that the mass of $\mathbb{T}$ equals 1. In terms of Taylor coefficients, the norm takes a more appealing form: if $f(z) = \sum_n a_n z^n$, then

$$\|f\|_{H^2} = \left(\sum_n |a_n|^2 \right)^{\frac{1}{2}}.$$

The Bergman space L^2_a, on the other hand, consists of all holomorphic functions on $\mathbb{D}$ such that

$$\|f\|_{L^2_a} = \left(\int_{\mathbb{D}} |f(z)|^2 dS(z) \right)^{\frac{1}{2}} < +\infty,$$

where dS is area measure, normalized so that the mass of $\mathbb{D}$ equals 1. Though the integral expression of the norm is more straightforward than for the Hardy space, it is more complicated in terms of Taylor coefficients: if $f(z) = \sum_n a_n z^n$, then

$$\|f\|_{L^2_a} = \left(\sum_n \frac{|a_n|^2}{n+1} \right)^{\frac{1}{2}}.$$

*) This research was supported in part by the Swedish Natural Science Research Council.

The Bergman space L^2_a contains H^2 as a dense subspace. It is intuitively clear from the definition of the norm in H^2 that functions in it have well-defined boundary values in $L^2(\mathbb{T})$. This is however not the case for L^2_a. In fact, there is a function in it which fails to have radial limits at every point of $\mathbb{T}$. This is a consequence of a more general statement due to Mac-Lane [**23**]; see also [**22**] and [**6**]. Apparently the two spaces H^2 and L^2_a are rather different from a function-theoretic perspective.

The Hardy space theory. The classical factorization theory for the Hardy spaces (these are the spaces H^p, with $0 < p \le +\infty$, which are defined by property (1-1), with 2 replaced by p), which relies on work by Blaschke, Riesz, and Szegö, requires some familiarity with the concepts of a Blaschke product, a singular inner function, an inner function, and an outer function. Let H^∞ stand for the space of bounded analytic functions in $\mathbb{D}$, supplied with the supremum norm. Given a (finite or infinite) sequence $A = \{a_j\}_j$ of points in $\mathbb{D}$, one considers the product

$$B_A(z) = \prod_j \frac{\bar{a}_j}{|a_j|}\frac{a_j - z}{1 - \bar{a}_j z}, \qquad z \in \mathbb{D},$$

which converges to a function in H^∞ of norm 1 if and only if the Blaschke condition $\sum_j 1 - |a_j| < +\infty$ is fulfilled, in which case A is said to be a Blaschke sequence, and B_A is said to be a Blaschke product. We note that for Blaschke sequences A, B_A vanishes precisely on the A in $\mathbb{D}$, with appropriate multiplicities, depending on how many times a point is repeated in the sequence. Moreover, the function B_A has boundary values of modulus 1 almost everywhere, provided that the limits are taken in nontangential approach regions. We also note that if the sequence A fails to be Blaschke, then the above product B_A collapses to 0. Given a finite positive Borel measure μ on the unit circle $\mathbb{T}$, which is singular to arc-length Lebesgue measure, one associates a singular inner function

$$S_\mu(z) = \exp\left(-\int_{\mathbb{T}} \frac{\zeta + z}{\zeta - z}\, d\mu(\zeta)\right), \qquad z \in \mathbb{D},$$

which is in H^∞, and has norm 1 there. Also, S_μ has no zeros in $\mathbb{D}$, and its nontangential boundary values are almost all 1 in modulus. This is the general criterion for a function in H^∞ to be inner: to have boundary values of modulus 1 almost everywhere. A product of a unimodular constant, a Blaschke product, and a singular inner function, is still inner, and all inner functions are obtained this way. If h is a real-valued L^1 function on $\mathbb{T}$, the associated outer function is

$$O_h(z) = \exp\left(\int_{\mathbb{T}} \frac{\zeta + z}{\zeta - z}\, h(\zeta)\, ds(\zeta)\right), \qquad z \in \mathbb{D},$$

which is an analytic function in $\mathbb{D}$ with $|O_h(z)| = \exp\big(h(z)\big)$ almost everywhere on the circle, the boundary values of O_h being thought of in the non-tangential sense. The function O_h is in H^2 if and only if $\exp(h)$ is in $L^2(\mathbb{T})$. The factorization theorem in H^2 then states that every nonidentically vanishing f in H^2 has the form

$$f(z) = \gamma B_A(z) S_\mu(z) O_h(z), \qquad z \in \mathbb{D},$$

where γ is a unimodular constant, and $\exp(h) \in L^2(\mathbb{T})$. The natural setting for the factorization theory is a larger class of functions, known as the Nevanlinna class. To cut a long story short, it consists of all functions of the above type, where no additional requirement is made on h, and where the singular measure μ is allowed to take negative values as well. We denote the Nevanlinna class by N. It is well known that $f \in N$ if and only if the function f is holomorphic in $\mathbb{D}$, and

$$\sup_{0<r<1} \int_{\mathbb{T}} \log^+ |f(rz)|\, ds(z) < +\infty.$$

The Bergman space case: inner functions. The Bergman space L^2_a contains H^2. How then does it relate to N? It turns out that there are functions in N that are not in L^2_a, and that there are functions in L^2_a which are not in N. The latter statement follows from the fact alluded to above that there is a function in L^2_a lacking nontangential boundary values altogether. The functions in N, on the other hand, all do have finite nontangential boundary values almost everywhere. The former statement follows from a much simpler example: take μ equal to a point mass at, say 1, and consider the function $1/S_\mu$. It is in the Nevanlinna class, but it is much too big near 1 to be in L^2_a.

The classical Nevanlinna factorization theory is ill-suited for the Bergman space. This is particularly apparent from the fact that there are zero sequences for L^2_a that are not Blaschke. The question is which functions can replace the Blaschke products or more general inner functions in the Bergman space setting. There may be several ways to do this, but only one is canonical from the point of view of operator theory.

A subspace M of H^2 is invariant, if it is closed and $zM \subset M$. It is well known that inner functions in H^2 are characterized as elements of unit norm in some $M \ominus zM$, where M is a nonzero invariant subspace. Following Halmos, we call $M \ominus zM$ the wandering subspace for M. For a collection L of functions in H^2, we let $[L]$ stand for the smallest invariant subspace containing L. We note that $u \in H^2$ is an inner function if and only if

$$h(0) = \int_{\mathbb{T}} h(z)|u(z)|^2 ds(z), \qquad h \in L^\infty_h(\mathbb{D}), \tag{1-2}$$

$L^\infty_h(\mathbb{D})$ being the Banach space of bounded harmonic functions on $\mathbb{D}$.

We take (1-2) as the starting point in our search for analogues of inner functions for the Bergman space L^2_a. We say that a function $G \in L^2_a$ is L^2_a-inner provided that

$$h(0) = \int_{\mathbb{D}} h(z)|G(z)|^2 dS(z), \qquad h \in L^\infty_h(\mathbb{D}). \tag{1-3}$$

A function G of unit norm in L^2_a is L^2_a-inner if and only if it is in a wandering subspace $M \ominus zM$ for some nonzero invariant subspace M of L^2_a. In contrast with the H^2 case, where $M \ominus zM$ always has dimension 1 (unless M is the zero subspace), this time the dimension may take any value in the range $1, 2, 3, \dots, +\infty$. This follows from the dilation theory developed by Apostol, Bercovici, Foiaş, and

Pearcy [**2**]. The dimension of $M \ominus zM$ will be referred to as the index of the invariant subspace M.

For the space H^2, Beurling's invariant subspace theorem yields a complete description [**3**].

THEOREM 1.1. (Beurling) *Let M be an invariant subspace of H^2, and let $M \ominus zM$ be the associated wandering subspace. Then $M = [M \ominus zM]$. If M is not the zero subspace, then $M \ominus zM$ is one-dimensional and spanned by an inner function, call it φ. It follows that $M = [\varphi] = \varphi H^2$.*

A natural question is whether the analogous statement $M = [M \ominus zM]$ (with the brackets referring to the invariant subspace lattice of L^2_a this time) holds for general invariant subspaces M of L^2_a. Pleasantly, and perhaps surprisingly, this turns out to be true [**1**]. We shall return to this matter in Section 3.

2. Factorization of zeros in the Bergman space

The treatment of the subject matter of this section is taken from [**9**], [**11**], and [**7, 8**]. It should be mentioned that the first results on factorization in Bergman spaces were obtained by Horowitz [**17**], and slightly later, but independently, by Korenblum [**19**].

An example. Let us begin with a simple but illuminating example: a multiple zero at the origin of multiplicity n. Let M_n be the subspace of L^2_a of all such functions, which is clearly invariant. The associated wandering subspace $M_n \ominus zM_n$ is one-dimensional, and spanned by the unit vector $G_n(z) = \sqrt{n+1}\, z^n$. According to the terminology introduced in the previous section, the function G_n is an L^2_a-inner function. Since it comes from a zero-based invariant subspace, it is a Bergman space analog of a (finite) Blaschke product. Let f be an arbitrary element of M_n, which then has a Taylor expansion $f(z) = \sum_{j=n}^{\infty} a_j z^j$. It can be factored $f = G_n g$, where $g(z) = (n+1)^{-1/2} \sum_{j=n}^{\infty} a_j z^{j-n}$. Let us compare the norms of f and g,

$$\|g\|^2_{L^2_a} = \frac{1}{n+1} \sum_{j=n}^{\infty} \frac{|a_j|^2}{j-n+1} \le \sum_{j=n}^{\infty} \frac{|a_j|^2}{j+1} = \|f\|^2_{L^2_a};$$

here, we used that

$$j+1 \le (n+1)(j-n+1), \qquad j = n, n+1, \ldots .$$

Since $g = f/G_n$, we see that division by the unit element G_n of the wandering subspace $M_n \ominus zM_n$ is contractive on M_n; in other words, multiplication by G_n is norm expansive on L^2_a.

General zero sets. Now let A be a zero sequence for the space L^2_a, counting multiplicities, and let M_A be the subspace of all functions in L^2_a that vanish on A, with at least the given multiplicity at each point. It is an invariant subspace, and its wandering subspace $M_A \ominus zM_A$ is one-dimensional. Let G_A denote a unit element of the wandering subspace. Let j be the multiplicity of the origin in the sequence A (which is 0 if the origin is not in A). By multiplying G_A by an appropriate unimodular constant, we may suppose that it solves the extremal problem

$$\sup\left\{ \operatorname{Re} G^{(j)}(0) : \ G \in M_A, \ \|G\|_{L^2_a} = 1 \right\}. \tag{2-1}$$

The above example with a multiple zero at the origin suggests that the function G_A may be a contractive divisor on M_A. This turns out to be the case. To begin with, we must rule out the possibility that the function G_A may have extraneous zeros.

More general invariant subspaces. We consider a more general situation with an invariant subspace M of index 1, and let G_M be a unit element of the one-dimensional wandering subspace $M \ominus zM$. We shall show that for $f \in H^2$, $G_M f$ is in L^2_a, and that

$$\|f\|_{L^2_a} \le \|G_M f\|_{L^2_a} \le \|f\|_{H^2}. \tag{2-2}$$

One of the inequalities states that multiplication by G_M expands the L^2_a norm of functions in H^2; this is what entails, after some work, that division by G_M is well-defined and norm contractive $M \to L^2_a$. As in the above case $M = M_A$, we may assume, by multiplying G_M by an appropriate unimodular constant, that it solves the extremal problem

$$\sup\left\{ \operatorname{Re} G^{(j)}(0) : \ G \in M, \ \|G\|_{L^2_a} = 1 \right\}, \tag{2-3}$$

where j is the multiplicity of the common zero at the origin of all the functions in M. For this reason, we shall refer to G_M as the extremal function for M. Since G_M is an L^2_a-inner function,

$$h(0) = \int_{\mathbb{D}} h(z)|G_M(z)|^2 dS(z), \qquad h \in L^\infty_h(\mathbb{D}),$$

and so

$$\int_{\mathbb{D}} h(z)\big(|G_M(z)|^2 - 1\big) dS(z) = 0, \qquad h \in L^\infty_h(\mathbb{D}). \tag{2-4}$$

Equation (2-4) has the interpretation that $|G_M|^2 - 1$ annihilates the bounded harmonic functions on $\mathbb{D}$.

Potential theory. Consider the function Φ_M that solves the boundary value problem

$$\begin{cases} \Delta \Phi_M = |G_M|^2 - 1 & \text{on } \mathbb{D}, \\ \Phi_M = 0 & \text{on } \mathbb{T}, \end{cases} \tag{2-5}$$

where Δ is one quarter of the usual Laplacian (this is not so important, really, one can use the standard Laplacian, only later would we then have to use slightly different Green functions). If we play around with Green's formula, in the form

$$\int_{\mathbb{D}} (v\Delta u - u\Delta v)\, dS = \int_{\mathbb{T}} \left(v\frac{\partial u}{\partial n} - u\frac{\partial v}{\partial n} \right) \tfrac{1}{2} ds, \tag{2-6}$$

where the normal derivatives are taken in the outward direction, and forget about regularity requirements, then (2-4) can be reformulated as saying that the normal derivative of Φ_M vanishes on $\mathbb{T}$. If we add this condition to (2-5), this system becomes overdetermined. Elliptic equations of order $2m$ are determined by m boundary data, so we may raise the order of the partial differential equation to 4 and keep a unique solution. This is accomplished by applying a Laplacian to both sides, and we get, in view of $\Delta\big(|G_M|^2 - 1\big) = |G'_M|^2$, that

$$\begin{cases} \Delta^2 \Phi_M = |G'_M|^2 \quad \text{on } \mathbb{D}, \\ \Phi_M = 0 \quad \text{on } \mathbb{T}, \\ \frac{\partial}{\partial n}\Phi_M = 0 \quad \text{on } \mathbb{T}. \end{cases} \tag{2-7}$$

The Green function for Δ^2 is the function $U(z,\zeta)$ on $\mathbb{D}\times\mathbb{D}$ that solves for fixed $\zeta \in \mathbb{D}$

$$\begin{cases} \Delta^2 U(\cdot,\zeta) = \delta_\zeta \quad \text{on } \mathbb{D}, \\ U(\cdot,\zeta) = 0 \quad \text{on } \mathbb{T}, \\ \frac{\partial}{\partial n}U(\cdot,\zeta) = 0 \quad \text{on } \mathbb{T}, \end{cases}$$

and it is given explicitly as

$$U(z,\zeta) = |z-\zeta|^2\Gamma(z,\zeta) + (1-|z|^2)(1-|\zeta|^2),$$

where

$$\Gamma(z,\zeta) = 2\log\left|\frac{z-\zeta}{1-\bar{\zeta}z}\right|$$

is the Green function for Δ. These are the expressions obtained when it is agreed to identify locally integrable functions φ with the corresponding measures φdS (recall that dS involved some normalization) to interpret the functions as distributions.

By now it should not require too much of a leap of faith to believe that

$$\Phi_M(z) = \int_{\mathbb{D}} U(z,\zeta)|G'_M(\zeta)|^2 dS(\zeta),$$

so that in view of the fact that $0 < U(z,\zeta)$ on $\mathbb{D}\times\mathbb{D}$, $0 < \Phi_M$ on $\mathbb{D}$, unless G_M is constant, in which case $\Phi_M = 0$. If we apply Green's formula (2-6) and recall the definition of Φ_M, it follows that

$$\int_{\mathbb{D}} \big(|G_M(z)|^2 - 1\big)|f(z)|^2 dS(z) = \int_{\mathbb{D}} \Phi_M(z)|f'(z)|^2 dS(z),$$

for, say, polynomials f. We rewrite this as

$$\begin{aligned} \int_{\mathbb{D}} |G_M(z)f(z)|^2 dS(z) &= \int_{D} |f(z)|^2 dS(z) + \int_{\mathbb{D}} \Phi_M(z)|f'(z)|^2 dS(z) \\ &= \int_{D} |f(z)|^2 dS(z) + \int_{\mathbb{D}}\int_{\mathbb{D}} U(z,\zeta)|f'(z)|^2|G'_M(\zeta)|^2 dS(z)dS(\zeta). \end{aligned} \tag{2-8}$$

Let Ψ solve

$$\begin{cases} \Delta\Psi = -1 & \text{on } \mathbb{D}, \\ \Psi = 0 & \text{on } \mathbb{T}; \end{cases}$$

the solution comes out to be $\Psi(z) = 1-|z|^2$. The function $\Phi_M - \Psi$ is subharmonic, and has 0 boundary values, so inside $\mathbb{D}$ it must be ≤ 0. In view of what we have already shown, it follows that $0 \leq \Phi_M(z) \leq \Psi(z) = 1-|z|^2$. It is well known that

$$\|f\|^2_{H^2} = \|f\|^2_{L^2_a} + \int_{\mathbb{D}} (1-|z|^2)|f'(z)|^2 dS(z), \qquad f \in H^2,$$

so that by continuity, identity (2-8) extends to all $f \in H^2$, and (2-2) holds. Let $\mathcal{A}(G_M)$ be the space of all functions $f \in L^2_a$ with

$$\|f\|^2_{\mathcal{A}(G_M)} = \|f\|^2_{L^2_a} + \int_{\mathbb{D}}\int_{\mathbb{D}} U(z,\zeta)|f'(z)|^2|G'_M(\zeta)|^2 dS(z)dS(\zeta) < +\infty,$$

and let $\mathcal{A}_0(G_M)$ be the closure of the polynomials in $\mathcal{A}(G_M)$. Then multiplication by G_M is an isometry $\mathcal{A}_0(G_M) \to M \subset L^2_a$, and $H^2 \subset \mathcal{A}_0(G_M) \subset \mathcal{A}(G_M) \subset L^2_a$. Moreover, the injection mappings $H^2 \to \mathcal{A}_0(G_M)$ and $\mathcal{A}(G_M) \to L^2_a$ are contractions. It follows that the invariant subspace generated by G_M, $[G_M]$, equals $G_M\mathcal{A}_0(G_M)$. A number of questions appear:

- Is $\mathcal{A}_0(G_M) = \mathcal{A}(G_M)$?
- Is $[G_M] = M$?
- Is $\mathcal{A}(G_M) = \{f \in L^2_a : G_M f \in L^2_a\}$?

The answer to the first two questions is yes [**1**] (see Section 3). The answer to the third question is no [**4, 5**].

Extraneous zeros. Let λ be a point of $\mathbb{D}$, and let G_λ be the extremal function associated with M_λ, the invariant subspace of all functions vanishing at λ. In terms of the Bergman kernel function $k(z,\zeta) = (1-\bar{\zeta}z)^{-2}$, it has the form

$$G_\lambda(z) = \left(1 - \frac{1}{k(\lambda,\lambda)}\right)^{-\frac{1}{2}} \left(1 - \frac{k(z,\lambda)}{k(\lambda,\lambda)}\right),$$

and one quickly verifies that on $\mathbb{T}$, it has modulus bigger than 1, and in $\mathbb{D}$, it has a simple zero at λ, and nowhere else. This means that if $f \in L^2_a$ vanishes at λ, then $f/G_\lambda \in L^2_a$, and since multiplication by G_λ is norm expansive on L^2_a (see (2-2) and (2-8) for $M = M_\lambda$), $\|f/G_\lambda\|_{L^2_a} \leq \|f\|_{L^2_a}$. Now suppose that G_M has an extraneous zero at λ, by which we mean that G_M vanishes at λ with a multiplicity higher than that of some element of M. By inspection of the extremal problem (2-3), to which G_M is the unique solution, we see that λ cannot be 0. If we divide G_M by G_λ, we get an element of L^2_a. If G_M/G_λ is in fact in M, then the function $\widetilde{G} = \gamma G_M/G_\lambda$, with $\gamma = \|G_M/G_\lambda\|^{-1}_{L^2_a}$, is a competing function with G_M in the extremal problem (2-3). It has norm 1, belongs to M, and the j-th derivative at 0 is $\widetilde{G}^{(j)}(0) = \gamma G^{(j)}_M(0)/G_\lambda(0)$, which is bigger than $G^{(j)}_M(0)$, as $1 < \gamma$ and $G_\lambda(0) < 1$. Hence $\widetilde{G}$ is more extremal than G_M itself, which leads to a contradiction. So, the assumption that G_M had an extraneous zero must be false. The weak point thus far is that we have not explained why G_M/G_λ was in M in the first place. Recall that $M \ominus zM$ was one-dimensional: from a perturbation argument it follows that $M \ominus (z-\lambda)M$ is one-dimensional for each $\lambda \in \mathbb{D}$. The subspace $(z-\lambda)M$ having codimension one in M means that it consists of all functions in M having an extra zero (or a zero of multiplicity one higher) at λ, so that G_M, having an extraneous zero at λ, must be in $(z-\lambda)M$. The conclusion that G_M/G_λ is in M follows.

Factorization of zeros. Let us see what kinds of conclusions we can draw from the above. For a finite zero sequence A, G_A has no extraneous zeros in $\mathbb{D}$, extends analytically across $\mathbb{T}$ to a rational function with poles at the reflected points in $\mathbb{T}$ of A, and the expansive multiplier property (2-2) implies that $1 \le |G_A|$ on $\mathbb{T}$. One shows that $\mathcal{A}_0(G_A) = L^2_a$ (the norms are different, though equivalent), so that G_A is an expansive multiplier on all of L^2_a. It follows that

$$\|f\|^2_{L^2_a} = \|f/G_A\|^2_{L^2_a} + \int_{\mathbb{D}}\int_{D} U(z,\zeta)|(f/G_A)'(z)|^2|G'_A(\zeta)|^2 dS(z)dS(\zeta), \quad f \in M_A,$$

and if we go to the limit as A approaches an infinite zero sequence, Fatou's lemma yields a $\ge$ in place of the $=$ sign. In particular, $f/G_A \in \mathcal{A}(G_A) \subset L^2_a$, and division by G_A is norm contractive $M_A \to L^2_a$.

3. General invariant subspaces

Most of the results mentioned here are from the paper [**1**] by Aleman, Richter, and Sundberg.

The following version of (2-8) will prove useful:

$$\|f\|^2_{L^2_a} = \|f/G_M\|^2_{L^2_a} + \int_{\mathbb{D}}\int_{\mathbb{D}} U(z,\zeta)|(f/G_M)'(z)|^2|G'_M(\zeta)|^2 dS(z)dS(\zeta), \qquad f \in [G_M]. \tag{3-1}$$

A skewed projection operator. Let M be an invariant subspace in L^2_a, and suppose for the moment that it has index 1; G_M denotes the associated extremal function. For $\lambda \in \mathbb{D}$,

$$f = \frac{f(\lambda)}{G_M(\lambda)}G_M + \left(f - \frac{f(\lambda)}{G_M(\lambda)}G_M\right), \qquad f \in M$$

offers a unique decomposition of M as a sum; $M = (M \ominus zM) + (z-\lambda)M$. As the two summands are obtained by bounded (skewed) projection operators, the subspaces $M \ominus zM$ and $(z-\lambda)M$ are at a positive angle. Note that the first projection,

$$Q_\lambda f = \frac{f(\lambda)}{G_M(\lambda)}G_M, \qquad f \in M,$$

is well-defined for all $\lambda \in \mathbb{D}$, as we know that G_M has no extraneous zeros. In terms of Q_λ, identity (3-1) can be written as

$$\begin{aligned}\|f\|_{L^2_a}^2 &= \int_{\mathbb{D}} \|Q_\lambda f\|_{L^2_a}^2 dS(\lambda) \\ &+ \int_{\mathbb{D}}\int_{\mathbb{D}} U(z,\zeta)\Delta_z\Delta_\zeta |Q_z f(\zeta)|^2 dS(z)dS(\zeta), \qquad f \in [G_M].\end{aligned} \tag{3-2}$$

This form lends itself to generalization to general invariant subspaces M, not necessarily of index 1. Namely, one shows that a skewed decomposition of the type $M = (M \ominus zM) + (z-\lambda)M$ holds in general, so that a corresponding projection $Q_\lambda : M \to M \ominus zM$ can be defined, and it depends analytically on $\lambda \in \mathbb{D}$. Moreover, (3-2) carries over, almost letter by letter:

$$\begin{aligned}\|f\|_{L^2_a}^2 &= \int_{\mathbb{D}} \|Q_\lambda f\|_{L^2_a}^2 dS(\lambda) \\ &+ \int_{\mathbb{D}}\int_{\mathbb{D}} U(z,\zeta)\Delta_z\Delta_\zeta |Q_z f(\zeta)|^2 dS(z)dS(\zeta), \qquad f \in [M \ominus zM].\end{aligned} \tag{3-3}$$

Abel summation. Define the bounded linear operator $L : M \to M$ by declaring that $Lf = f/z$ for $f \in zM$, and $Lf = 0$ for $f \in M \ominus zM$. Also, let P be the orthogonal projection $M \to M \ominus zM$. If f is in M, we can decompose f as a sum of an element of $M \ominus zM$ and a "remainder term" by the formula $f = Pf + zLf$. Repeating this for Lf, we obtain $f = Pf + zPLf + z^2L^2f$. Continuing this process, we get the formal series

$$f = Pf + zPLf + z^2PL^2f + z^3PL^3f + \dots,$$

each term of which is in $[M \ominus zM]$. If the series were to converge to f for each given $f \in M$, the assertion $M = [M \ominus zM]$ would be immediate. However, this is probably not the case in general, so we choose the second best thing: we form the Abel series

$$R_t f = Pf + tzPLf + t^2z^2PL^2f + t^3z^3PL^3f + \dots, \qquad 0 \le t < 1,$$

which does converge to an element of $[M \ominus zM]$, in the hope that $R_t f \to f$ as $t \to 1$. The skewed projection Q_λ has a similar series expansion,

$$Q_\lambda f = Pf + \lambda PLf + \lambda^2 PL^2f + \lambda^3PL^3f + \dots, \qquad \lambda \in \mathbb{D},$$

which one can use to show that $Q_\lambda R_t = Q_{t\lambda}$. The operator R_t may also be thought of as given by $R_t f(z) = Q_{tz}f(z)$. As $t \to 1$, $Q_{tz}f(z) \to Q_z f(z)$, and $Q_z f(z) = f(z)$, because by the definition of Q_λ, $f(z) - Q_\lambda f(z)$ is zero when $z = \lambda$. If follows that $R_t f(z) \to f(z)$ as $t \to 1$ pointwise in $\mathbb{D}$. It remains to show that the convergence holds in norm, too.

Controlling the norm of the Abel sum. General functional analysis arguments show that we do not really need to show that $R_t f$ tends to f in norm, weak convergence would suffice. Moreover, weak convergence would follow if we only had a uniform bound of the norm of $R_t f$ as $t \to 1$. This is the crux of the problem. By (3-3) and the identity $Q_\lambda R_t = Q_{t\lambda}$,

$$\begin{aligned} \|R_t f\|^2_{L^2_a} &= \int_{\mathbb{D}} \|Q_{t\lambda} f\|^2_{L^2_a} dS(\lambda) \\ &\quad + \int_{\mathbb{D}}\int_{\mathbb{D}} U(z,\zeta)\Delta_z\Delta_\zeta |Q_{tz} f(\zeta)|^2 dS(z) dS(\zeta), \qquad f \in M. \end{aligned} \tag{3-4}$$

Certain regularity properties of the biharmonic Green function $U(z,\zeta)$ can be used to show that, as $t \to 1$, the right-hand side of (3-4) tends to

$$\int_{\mathbb{D}} \|Q_\lambda f\|^2_{L^2_a} dS(\lambda) + \int_{\mathbb{D}}\int_{\mathbb{D}} U(z,\zeta)\Delta_z\Delta_\zeta |Q_{tz} f(\zeta)|^2 dS(z) dS(\zeta), \tag{3-5}$$

so if we can only bound this expression, we are done. A bound that works is $\|f\|^2_{L^2_a}$. The approach in [**1**] is based on the identity

$$\|f\|^2_{L^2_a} = \int_{\mathbb{T}} \|Q_{r\lambda} f\|^2_{L^2_a} ds(\lambda) + \int_{\mathbb{D}}\int_{\mathbb{T}} (|z|^2 - r^2) \left| \frac{f(z) - Q_{r\lambda} f(z)}{z - r\lambda} \right|^2 ds(\lambda) dS(z),$$

for $0 < r < 1$. By cleverly applying Green's theorem to the above right-hand side expression and obtaining estimates of "remainder terms" as $r \to 1$, Aleman, Richter, and Sundberg were able to show that the expression (3-5) is no bigger than $\|f\|^2_{L^2_a}$, whence the following analogue of Beurling's theorem follows.

THEOREM 3.1. *Let M be an invariant subspace of L^2_a. Then $M = [M \ominus zM]$.*

Invertibility and cyclicity. A function $f \in L^2_a$ is said to be cyclic if $[f] = L^2_a$. It has been a long standing problem whether there are noncyclic functions $f \in L^2_a$ that are invertible, that is, $1/f \in L^2_a$. A complicated construction of such functions was recently found by Borichev and Hedenmalm [**4**, **5**]. The idea is that if the given function f grows maximally fast on a "big" set, then an H^∞ function which is small there is small everywhere, and hence cannot lift the small values of f as required by cyclicity. This example has certain consequences for the uniqueness of inner-outer factorization in L^2_a [**1**]. It is not clear whether the invariant subspace associated to the constructed invertible noncyclic function is in the closure of the collection of zero-based invariant subspaces, with respect to any of the topologies suggested by Korenblum [**21**].

4. Zero sequences

The first results on zero sequences for Bergman space functions were obtained by Horowitz [**17**, **18**]. For instance, he showed that the union of two zero sequences need not be a zero sequence. If we consider the corresponding zero-based invariant subspaces, call them M_A and M_B, then $M_A \cap M_B = \{0\}$. Actually, this behaviour of index one invariant subspaces is the raison d'être for the invariant subspaces of higher index [**24**]. Also, look at the explicit constructions in Section 6, and the papers [**10**, **16**].

The sharpest results so far were obtained by Seip in [**26**, **27**]. The tools were borrowed from the fundamental work of Korenblum on the topological algebra $A^{-\infty}$ [**19**, **20**], which consists of all functions f holomorphic in $\mathbb{D}$ that meet the growth condition $|f(z)| \le C(1-|z|)^{-N}$ for some positive constants C, N.

For a finite subset F of $\mathbb{T}$, let $\mathbb{T} \setminus F = \cup_k I_k$ be the complementary arcs, and consider the Beurling-Carleson entropy of F,

$$\widehat{\varkappa}(F) = \sum_k \frac{|I_k|}{2\pi} \left(\log \frac{2\pi}{|I_k|} + 1 \right).$$

Let $d(\cdot,\cdot)$ be the curvilinear metric $d(e^{it}, e^{is}) = \pi^{-1}|t-s|$ on $\mathbb{T}$, where it is assumed that $|t-s| \le \pi$. The Korenblum star associated with the finite set F is

$$G(F) = \left\{ z \in \overline{\mathbb{D}} \setminus \{0\} : d(z/|z|, F) \le 1 - |z| \right\} \cup \{0\},$$

and for a sequence A of points in $\mathbb{D}$, let

$$\sigma(A, F) = \sum_{z \in A \cap G(F)} \log \frac{1}{|z|}$$

be the local "Blaschke sum". The Korenblum density $\delta(A)$ of the sequence A is the infimum over all β such that

$$\sup_F \left(\sigma(A,F) - \beta \widehat{\varkappa}(F) \right) < +\infty,$$

the supremum being taken over all finite subsets of $\mathbb{T}$.

The following result was obtained by Seip [**26**].

THEOREM 4.1. *Let A be a sequence of points in $\mathbb{D} \setminus \{0\}$. If A is the zero sequence of some function in L^2_a, then $\delta(A) \le \frac{1}{2}$. On the other hand, if $\delta(A) < \frac{1}{2}$, then A is the zero sequence of some L^2_a function.*

5. Interpolating and sampling sequences

A sequence $A = \{a_j\}_j$ of distinct points in $\mathbb{D}$ is said to be a sampling sequence for L^2_a if

$$\sum_j (1-|a_j|^2)^2 |f(a_j)|^2 \asymp \|f\|^2_{L^2}, \qquad f \in L^2_a,$$

where the $\asymp$ sign means that the left-hand side is bounded from above and below by positive constant multiples of the right-hand side. The reason why the factor

$(1-|a_j|^2)^2$ is needed is that in a more general setting, one should use the reciprocal of the reproducing kernel, $k(a_j,a_j)^{-1}$. Similarly, A is interpolating for L^2_a provided that to every l^2 sequence $\{w_j\}_j$, there exists a function $f \in L^2_a$ such that

$$(1-|a_j|^2)f(a_j) = w_j \quad \text{for all } j.$$

Generally, sampling sequences are fat, and interpolating sequences are thin. Clearly, a sampling sequence cannot be a zero sequence for L^2_a. However, every interpolating sequence for L^2_a is also a zero sequence, for the following reason. Take an interpolant for the sequence $w_1 = 1$ and $w_j = 0$ for all other j, and multiply this function by $z - a_1$ to get a nonidentically vanishing function that vanishes on the sequence A. This actually only shows that A must be a subsequence of an L^2_a zero sequence, but it is well known that every subsequence of a zero sequence for L^2_a is itself a zero sequence for L^2_a [**17**, **9**]. In [**25**], Seip obtains a complete description of the sampling and interpolating sequences for L^2_a. We shall try to describe the result, but in order to do so, we need some notation.

A sequence $A = \{a_j\}_j$ of points in $\mathbb{D}$ is said to be uniformly discrete if for some $\varepsilon > 0$,

$$\varepsilon \le \left|\frac{a_j - a_k}{1-\overline{a_k}a_j}\right|, \qquad j \ne k.$$

For $\lambda \in \mathbb{D}$, let A_λ be the image of A under the conformal automorphism of the unit disk

$$\varphi_\lambda(z) = \frac{z-\lambda}{1-\overline{\lambda}z}, \qquad z \in \mathbb{D}.$$

Associate with A_λ the function $n(r, A_\lambda)$, which counts the number of points of A_λ contained within the disk $r\mathbb{D}$ $(0 < r < 1)$. We shall need the definite integral

$$N(r, A_\lambda) = \int_0^r n(t, A_\lambda)dt, \qquad 0 < r < 1.$$

If $A(r)$ now stands for the function

$$A(r) = \log\frac{1+r}{1-r}, \qquad 0 < r < 1,$$

Seip defines the upper density of A as

$$D^+(A) = \limsup_{r\to 1}\sup_{\lambda\in\mathbb{D}}\frac{N(r,A_\lambda)}{A(r)},$$

and the lower density as

$$D^-(A) = \liminf_{r\to 1}\inf_{\lambda\in\mathbb{D}}\frac{N(r,A_\lambda)}{A(r)}.$$

For the standard Bergman space, his result is as follows.

THEOREM 5.1. *A sequence A of distinct points in $\mathbb{D}$ is sampling for L^2_a if and only if it can be expressed as a finite union of uniformly discrete sets and it contains a uniformly discrete subsequence A' for which $D^-(A') > \frac{1}{2}$.*

THEOREM 5.2. *A sequence A of distinct points in $\mathbb{D}$ is interpolating for L^2_a if and only if it is uniformly discrete and $D^+(A') < \frac{1}{2}$.*

6. Invariant subspaces of index two or higher

Back in Section 1 it was mentioned that an invariant subspace M of the Bergman space L^2_a may have index $1, 2, 3, \ldots$, whereas the most obvious examples have index 1. For instance, every zero based invariant subspace M_A has index 1, and so does every singly generated one. Invariant subspaces based on singular masses at the boundary have index 1, too [**15**]. It is therefore a natural question to ask what these invariant subspaces of higher index look like. A simple constuction was found in [**10**].

Let A and B be two disjoint zero sequences for L^2_a, and let $C = A \cup B$. Let $M = M_A \vee M_B$, the smallest invariant subspace containing both M_A and M_B; it is obtained as the closure of the sum $M_A + M_B$. It turns out that in this situation, either $M = L^2_a$ or M has index 2. Moreover, what determines which of these alternatives occurs is the fatness of the sequence C. If one of the sequences A and B fails to accumulate on an arc of $\mathbb{T}$, then C is not fat enough, and so $M = L^2_a$. On the other hand, if C is sampling, then M has index 2, because M_A and M_B are at a positive angle. To see this, let $C = \{c_j\}_j$, and use the sampling property

$$\|f\|^2_{L^2_a} \asymp \sum_j (1 - |c_j|^2)^2 |f(c_j)|^2, \qquad f \in L^2_a.$$

Let $f \in M_A$ and $g \in M_B$ be arbitrary. Then for every point c_j of C, we have

$$|f(c_j) + g(c_j)|^2 = |f(c_j)|^2 + |g(c_j)|^2,$$

so that

$$\begin{aligned}\|f + g\|^2_{L^2_a} &\asymp \sum_j (1 - |c_j|^2)^2 |f(c_j) + g(c_j)|^2 \\ &= \sum_j (1 - |c_j|^2)^2 \left(|f(c_j)|^2 + |g(c_j)|^2\right) \asymp \|f\|^2_{L^2_a} + \|g\|^2_{L^2_a}.\end{aligned}$$

Consequently, the sum $M_A + M_B$ is closed, and it is easy to show that M has index 2: $zM = zM_A + zM_B$ [**9**].

7. Green functions for weights and factorization

The results related to in this section are mostly from [**14**].

It is clear from Section 2 that Green functions for certain elliptic operators of order 4 play an important role in the study of factorization in Bergman spaces. Let ω be a nonnegative sufficiently smooth function in $\mathbb{D}$, and let $L^2_a(\omega)$ be the corresponding weighted Bergman space of all holomorphic functions f in $\mathbb{D}$ with

$$\|f\|^2_{L^2_a(\omega)} = \int_{\mathbb{D}} |f(z)|^2 \omega(z) dS(z) < +\infty.$$

It is a Hilbert space if ω is not equal to 0 too frequently near $\mathbb{T}$; this is not precise, but it is enough here. For the extremal functions to have a chance to be good divisors, we need to ask of ω that

$$h(0) = \int_{\mathbb{D}} h(z)\omega(z)dS(z), \qquad h \in L_h^\infty(\mathbb{D}).$$

The relevant Green function is that of the operator $\Delta\omega^{-1}\Delta$ on $\mathbb{D}$, and issue is whether it is positive (or at least nonnegative). It was shown in [**12**] that it is positive for the weights $\omega_\alpha(z) = (\alpha+1)|z|^{2\alpha}$, with $\alpha > -1$, and the issue at hand is whether this information can be used to tell us anything about weights that are convex combinations of these. For instance, is it true in general that if ω and μ are two weights, with associated Green functions U_ω and U_μ, that we have, with $\omega[t] = (1-t)\omega + t\mu$,

$$(1-t)U_\omega(z,\zeta) + tU_\mu(z,\zeta) \le U_{\omega[t]}(z,\zeta), \qquad (z,\zeta) \in \mathbb{D}\times\mathbb{D},\ 0<t<1? \quad \text{(7-1)}$$

This is probably not so, although I cannot supply an immediate counterexample. However, with some additional information given in terms of the Green functions, it is true. If we apply a Laplacian to $U_\omega(z,\zeta)$, we get

$$\Delta_z U_\omega(z,\zeta) = \omega(z)\big(\Gamma(z,\zeta) + H_\omega(z,\zeta)\big),$$

where $H_\omega(z,\zeta)$ is harmonic in z. We call H_ω the harmonic compensator. If $H_\mu(z,\zeta) \le H_\omega(z,\zeta)$ holds pointwise, then (7-1) holds, and $H_\mu(z,\zeta) \le H_{\omega_t}(z,\zeta)$ also holds pointwise. A consequence of this result is that if

$$\omega(z) = \int_{]-1,+\infty[} (\alpha+1)|z|^{2\alpha} d\rho(\alpha),$$

where ρ is a probability measure, then

$$0 \le \int_{]-1,+\infty[} U_{\omega_\alpha}(z,\zeta)d\rho(\alpha) \le U_\omega(z,\zeta), \qquad (z,\zeta) \in \mathbb{D}\times\mathbb{D}.$$

Related work has been carried out by Shimorin; see, for instance, [**28**, **29**].

There are questions concerning the usual Bergman space, that, after reformulation become questions about weighted Bergman spaces. For instance, it is an open problem whether in general multiplication by G_A is less norm expansive than multiplication by G_B whenever the sequence A is contained in the finite zero sequence B. This is so for a fixed A if and only if multiplication by extremal functions is norm expansive on $L_a^2(\omega)$, with $\omega = |G_A|^2$). Moreover, the latter statement follows if we know that the Green function U_ω is positive, with $\omega = |G_A|^2$. However, no general methods for checking whether a Green function U_ω is positive are known. We conjecture that $U_\omega \ge 0$ provided that

- $\log\omega$ is subharmonic, and
- ω represents the origin, which means that

$$h(0) = \int_{\mathbb{D}} h(z)\omega(z)dS(z), \qquad h \in L_h^\infty(\mathbb{D}).$$

References

1. A. Aleman, S. Richter, C. Sundberg, *Beurling's theorem for the Bergman space*, preprint 1995.
2. C. Apostol, H. Bercovici, C. Foiaş, and C. Pearcy, *Invariant subspaces, dilation theory, and the structure of the predual of a dual algebra*, J. Funct. Anal. **63** (1985), 369–404.
3. A. Beurling, *On two problems concerning linear transformations in Hilbert space*, Acta Math. **81** (1949), 239–255.
4. A. A. Borichev, P. J. H. Hedenmalm, *Cyclicity in Bergman-type spaces*, Internat. Math. Res. Notices (1995), no. no. 5, 253–262.
5. A. Borichev, H. Hedenmalm, *Harmonic functions of maximal growth: invertibility and cyclicity in the Bergman spaces*, J. Amer. Math. Soc. **10** (1997), 761–796.
6. D. G. Cantor, *A simple construction of analytic functions without radial limits*, Proc. Amer. Math. Soc. **15** (1964), 335–336.
7. P. Duren, D. Khavinson, H. S. Shapiro, C. Sundberg, *Contractive zero-divisors in Bergman spaces*, Pacific J. of Math. **157** (1993), 37–56.
8. P. Duren, D. Khavinson, H. S. Shapiro, C. Sundberg, *Invariant subspaces in Bergman spaces and the biharmonic equation*, Michigan Math. J. **41** (1994), 247–259.
9. H. Hedenmalm, *A factorization theorem for square area-integrable analytic functions*, J. Reine Angew. Math. **422** (1991), 45–68.
10. P. J. H. Hedenmalm, *An invariant subspace of the Bergman space having the codimension two property*, J. Reine Angew. Math. **443** (1993), 1–9.
11. H. Hedenmalm, *A factoring theorem for the Bergman space*, Bull. London Math. Soc. **26** (1994), 113–126.
12. H. Hedenmalm, *A computation of Green functions for the weighted biharmonic operators* $\Delta|z|^{-2\alpha}\Delta$, *with* $\alpha > -1$, Duke Math. J. **75** (1994), 51–78.
13. H. Hedenmalm, *Open problems in the function theory of the Bergman space*, Festschrift in honour of Lennart Carleson and Yngve Domar (Uppsala, 1993), Acta Univ. Upsaliensis C58, Uppsala, Sweden (1995), pp. 153–169.
14. H. Hedenmalm, *Boundary value problems for weighted biharmonic operators*, St Petersburg Math. J. **8** (1997), 661–674.
15. H. Hedenmalm, B. Korenblum, K. Zhu, *Beurling type invariant subspaces of the Bergman spaces*, J. London Math. Soc. **53** (1996), 601–614.
16. H. Hedenmalm, S. Richter, K. Seip, *Interpolating sequences and invariant subspaces of given index in the Bergman spaces*, J. Reine Angew. Math. (to appear).
17. C. Horowitz, *Zeros of functions in the Bergman spaces*, Duke Math. J. **41** (1974), 693–710.
18. C. Horowitz, *Factorization theorems for functions in the Bergman spaces*, Duke Math. J. **44** (1977), 201–213.
19. B. Korenblum, *An extension of the Nevanlinna theory*, Acta Math. **135** (1975), 187–219.
20. B. Korenblum, *A Beurling-type theorem*, Acta Math. **138** (1977), 265–293.
21. B. Korenblum, *Outer functions and cyclic elements in Bergman spaces*, J. Funct. Anal. **115** (1993), 104–118.
22. N. N. Luzin, I. I. Privalov, *Sur l'unicité et la multiplicité des fonctions analytiques*, Ann. Sci. Ecole Norm. Sup. **42** (1925), 143–191.
23. G. R. Mac-Lane, *Holomorphic functions, of arbitrarily slow growth without radial limits*, Mich. Math. J. **9** (1962), 21–24.

24. S. Richter, *Invariant subspaces in Banach spaces of analytic functions*, Trans. Amer. Math. Soc. **304** (1987), 585–616.
25. K. Seip, *Beurling type density theorems in the unit disk*, Invent. Math. **113** (1993), 21–39.
26. K. Seip, *On a theorem of Korenblum*, Ark. Mat. **32** (1994), 237–243.
27. K. Seip, *On Korenblum's density condition for the zero sequences of* $A^{-\alpha}$, J. Analyse Math. **67** (1995), 307–322.
28. S. M. Shimorin, *Factorization of analytic functions in weighted Bergman spaces*, St Petersburg Math. J. **5** (1994), 1005–1022.
29. S. M. Shimorin, *On a family of conformally invariant operators*, St Petersburg Math. J. **7** (1996), 287–306.

Progress in Mathematics, Vol. 168,

Extensions of Motives

Annette Huber

Mathematisches Institut, Einsteinstr. 62
D-48149 Muenster, Germany
huber@math.uni-muenster.de

Introduction

These pages grew from my talk at the ECM in Budapest. It was aimed at a general audience who wanted to get an idea of what my work was about. The present article is written in the same spirit. The written form allows some of the definitions to be made more precise and to give exact references. It can be read as an introduction to the field of motives for non-specialists, or to be honest to some aspects of it. Specialists might get a quick guide to the main results of [H1] and even an idea of the line of proof.

I would like to take this opportunity to thank the European Mathematical Society for honoring me with their prize. I can only hope to justify their confidence in my work. I would also like to thank warmly my "Doktorvater", Christopher Deninger, for all the wonderful mathematics I learned from him and his constant support. Finally, I am grateful to E. Landvogt for his comments on the text.

We will work over the base field $\mathbb{Q}$ for simplicity. Everything can be done for number fields or even be extended to arbitrary fields of characteristic 0.

1. Motives

Let us start with an example. Consider the elliptic curve E over $\mathbb{Q}$ given by the equation

$$E : y^2 = x(x+a)(x+b) \qquad \text{where } a, b \in \mathbb{Q} \smallsetminus \{0\}; a \neq b \ .$$

By $E(\mathbb{C})$ we denote the solutions of E in $\mathbb{C}^2$. They form a curve over the complex numbers, i.e., a Riemann surface. It is compact if we add a point ∞. We know that this Riemann surface is of genus 1 ([Si] II Ex. 5.7), i.e., it has the topological type of a torus. This is of course a very interesting geometric object.

On the other hand, note that the coefficients of E are in $\mathbb{Q}$. Hence the rational solutions $E(\mathbb{Q})$ are a very interesting object in number theory.

There are of course many ways to study elliptic curves. A very successful set of methods is led by the idea of linearization. We want to attach objects of linear algebra to E. These could be numbers or vector spaces – we will see more elaborate examples later. The main point is that we can make calculations with these invariants. What I call "linearization" usually goes by the name of cohomology.

Example 1: It is known ([Si] VI 5.1.1) that $E(\mathbb{C}) \cong \mathbb{C}/\Gamma$ where Γ is a lattice in $\mathbb{C}$. This is an isomorphism of Riemann surfaces. We can rewrite it like this:

$$\Gamma \longrightarrow V \longleftarrow V_{\mathbb{R}} \quad ; \tag{$*$}$$

where Γ is considered as a free abelian group of rank 2, V is a one-dimensional $\mathbb{C}$-vector space and $V_{\mathbb{R}}$ is a one-dimensional $\mathbb{R}$-subvector space of V. This means we do not fix a basis of V as a $\mathbb{C}$- (or even $\mathbb{R}$-) vector space but we do want to fix the complex structure given by the embedding of the real into the complex numbers. The triple $(*)$ is purely an object of linear algebra. Of course it can tell us a lot about E. We can fully recover the Riemann surface.

Example 2: Fix a prime l. (Traditionally l denotes a prime different from the characteristic of the base field which would be p.) Let $E[l^n]$ be the l^n-torsion of E, i.e., $\frac{1}{l^n}\Gamma \subset \mathbb{C}/\Gamma$. It is one of the facts about elliptic curves that the group law on $E(\mathbb{C})$ can be defined by polynomial equations with coefficients in $\mathbb{Q}$ ([Si] III §2). Hence the points in $E[l^n]$ have coordinates in $\bar{\mathbb{Q}}$ and are permuted by $\mathrm{Gal}(\bar{\mathbb{Q}}/\mathbb{Q})$. We define the Tate-module ([Si] III §7)

$$T_l(E) := \lim_{\leftarrow} E[l^n] \quad . \tag{$**_l$}$$

It is a free $\mathbb{Z}_l$-module of rank 2 with a continuous operation of the absolute Galois group ([Si] III 7.1), again an object of linear algebra. The Tate module is an extremely useful invariant from number theory, but only as far as the prime l is concerned. Note that all other primes are invertible in $\mathbb{Z}_l$.

Let us summarize: we have found different linearizations of E that reflect different aspects of E. $(*)$ reflects the complex analytic structure whereas $(**_l)$ reflects number-theoretic properties. Note also that free modules of rank 2 occur in both examples. Now we would like to have one object, in terms of linear algebra, that reflects all these properties at the same time. And this would be the *motive* of the elliptic curve.

Before we make this more precise, let us generalize. Let X be a general variety over $\mathbb{Q}$, i.e., locally given as the zero locus of finitely many polynomial equations over $\mathbb{Q}$ with all transition morphisms given by polynomials over $\mathbb{Q}$ (for a precise definition see [Mu] I.5). These are fairly general objects. We allow singular varieties or non-complete ones. We want to see how the above examples generalize.

Example 1 (Hodge structures):

DEFINITION 1.1 (DELIGNE [D1] 2.3.1, [H1] 8.1.1)
A (mixed) $\mathbb{Q}$-Hodge structure A consists of a tuple

$$((A_{\mathbb{Q}}, W_*), (A_{\mathbb{C}}, W_*, F^*), \alpha_{\mathbb{Q}})$$

where $A_{\mathbb{Q}}$ is a $\mathbb{Q}$-vector space with an increasing filtration W_* (called weight filtration*), $A_{\mathbb{C}}$ is a $\mathbb{C}$-vector space with weight filtration W_* and a decreasing filtration F^* (called* Hodge filtration*),*

$$\alpha_{\mathbb{Q}} : A_{\mathbb{Q}} \otimes \mathbb{C} \to A_{\mathbb{C}}$$

is a filtered isomorphism. Let $(\bar{\cdot})$ *be the complex conjugation on* $A_{\mathbb{C}}$ *with respect to the real structure induced by* $A_{\mathbb{Q}}$. *We assume the following property:*

$$W_n/W_{n-1}A_{\mathbb{C}} = \bigoplus_{p+q=n} F^p(W_n/W_{n-1}A_{\mathbb{C}}) \cap \bar{F}^q(W_n/W_{n-1}A_{\mathbb{C}}) \quad \text{for all } n \in \mathbb{Z}.$$

A Hodge structure is called pure if it is direct sum of Hodge structures which have only one non-trivial step in the weight filtration.

A Hodge-de Rham structure *(or* Hodge structure defined over $\mathbb{Q}$*) is a* $\mathbb{Q}$-*Hodge structure* A *plus a* $\mathbb{Q}$-*vector space* $(A_{\mathrm{DR}}, W_*, F^*)$ *with weight and Hodge filtration and an isomorphism of bifiltered vector spaces*

$$\alpha_{\mathrm{DR}} : A_{\mathrm{DR}} \otimes \mathbb{C} \to A_{\mathbb{C}} \ .$$

You should think of a Hodge structure as a dressed up $\mathbb{C}$-vector space. The most important thing to know about Hodge structures is that they form an abelian category ([D1] Thm 2.3.5). There is a linearization of the category of varieties with values in this abelian category.

THEOREM 1.2 (DELIGNE) *For any variety* X *over* $\mathbb{Q}$ *there is functorially a mixed Hodge-de Rham structure*

$$X \mapsto \underline{H}_{\mathcal{H}}(X) = \bigoplus_{n=0}^{2\dim(X)} \underline{H}^n_{\mathcal{H}}(X) \ .$$

The underlying $\mathbb{C}$-*vector space of* $\underline{H}^n_{\mathcal{H}}(X)$ *is the* n-*th singular or Betti-cohomology of the complex manifold (complex space in the singular case)* $X(\mathbb{C})$, *i.e.,* $H^n_B(X(\mathbb{C}), \mathbb{C})$. *If* X *is smooth and complete, then* $\underline{H}^n_{\mathcal{H}}(X)$ *is pure of weight* n *and the direct sum decomposition induced by the Hodge filtration and its complex conjugate is given by the theorem of Hodge for compact Kähler manifolds.*

Proof. The pure case is [D1] 2.3.1, for smooth varieties it is [D1] 3.2.5 (iii) and in general [D2] 8.2.2. □

We want to see how this relates to our original example.

LEMMA 1.3 *The triple* $(*)$ *for the elliptic curve* E *determines* $\underline{H}^1_{\mathcal{H}}(X)$ *as a mixed Hodge structure.*

Proof. We put $A_{\mathbb{Q}} = \{a : \Gamma \to \mathbb{Q}\}$, $A_{\mathbb{C}} = \{a : \Gamma \to \mathbb{C}\}$ ($\mathbb{Z}$-linear maps) and $\alpha_{\mathbb{Q}}$ induced by the natural inclusion. The weight filtration is concentrated in degree 1, i.e., $W_n/W_{n-1} = 0$ for $n \neq 1$. The Hodge filtration is given by

$$F^p A_{\mathbb{C}} = \begin{cases} A_{\mathbb{C}} & p \leq 0 \\ V^* & p = 1 \\ 0 & p \geq 2 \end{cases}$$

where the embedding of V^* into $A_{\mathbb{C}}$ is the dual of the inclusion $\Gamma \to V$. □

Example 2 (l-adic cohomology): This cohomology theory was developed by Grothendieck and many collaborators ([SGA 4], [SGA 4 1/2], [SGA 5]). Even to give a sketch of the construction is beyond the scope of this overview. The cohomology theory takes values in the category of finite dimensional $\mathbb{Q}_l$-vector spaces equipped with a continuous operation of $\mathrm{Gal}(\bar{\mathbb{Q}}/\mathbb{Q})$.

THEOREM 1.4 (GROTHENDIECK ET AL.) *For any variety X over $\mathbb{Q}$ there is functorially a $\mathbb{Q}_l$-$\mathrm{Gal}(\bar{\mathbb{Q}}/\mathbb{Q})$-module*

$$X \mapsto \underline{H}_{\mathcal{H}}(X) = \bigoplus_{n=0}^{2\dim(X)} \underline{H}_l^n(X) \ .$$

The dimension of the underlying $\mathbb{Q}_l$-vector space of $\underline{H}_l^n(X)$ is the n-th Betti number of $X(\mathbb{C})$, i.e., the dimension of Betti cohomology.

Proof. For the definition, see e.g. [FK] §12. The comparison with Betti-cohomology follows from [FK] §11. □

Again we check our first example:

LEMMA 1.5 *In the case of an elliptic curve E we have*

$$T_l(E) \otimes \mathbb{Q}_l = \underline{H}_l^1(E)^\vee$$

where the dual on the right is taken in the category of Galois modules.

Proof. [Mi2] Thm 15.1. □

In the light of lemmas 1.3 and 1.5 we see that the rank 2 in the introductory example was simply the first Betti-number of E which turns up in both cohomology theories. The case of elliptic curves is particularly nice insofar as the curve can be recovered from the cohomological data. This is not true for general X.

Note that again the theory in the first example reflects the complex structure of $X(\mathbb{C})$. In the second example number-theoretic properties are reflected – again one prime at a time. On the other hand the examples are not independent of each other. The dimensions of the underlying vector spaces agree. More precisely, after choice of $\bar{\mathbb{Q}} \subset \mathbb{C}$, there is canonical isomorphism of $\mathbb{Q}_l$-vector spaces

$$\alpha : (\underline{H}_{\mathcal{H}}^n(X))_{\mathbb{Q}} \otimes_{\mathbb{Q}} \mathbb{Q}_l \longrightarrow \underline{H}_l^n(X)$$

where the term on the left denotes the $\mathbb{Q}$-vector space which is part of the data of a Hodge structure. But the compatibility is even better. We have:

THEOREM 1.6 (DELIGNE) *Let X be a smooth and complete variety over $\mathbb{Q}$. Then $\underline{H}_l^n(X)$ is pure of weight n, i.e., the operation of $\mathrm{Gal}(\bar{\mathbb{Q}}/\mathbb{Q})$ is unramified for almost all primes s and the eigenvalues of the operation of Fr_s on $\underline{H}_l^n(X)$ have the absolute value $s^{n/2}$.*

If X is a general variety over $\mathbb{Q}$, then $\underline{H}_l^n(X)$ carries a (unique) weight filtration W_* such that $\mathrm{Gr}_k^W \underline{H}_l^n(X)$ is pure of weight k.

Proof. The pure case is [D3] Thm 1.6. For the general case [D5] and [D4] 14. □

Moreover, the weight filtrations on Hodge-cohomology and l-adic cohomology correspond to each other under comparison of the underlying vector spaces. The above comparison morphism α is an isomorphism of filtered vector spaces. All these effects lead to the idea (or perhaps they were found because of the conjecture) that there should be one cohomology theory that reflects all properties of X at a time.

Conjecture 1.7 (Grothendieck) *There is an abelian category $\mathcal{MM}$ of* (mixed) motives *and a universal linearization*

$$\begin{array}{rcl} \textit{varieties over } \mathbb{Q} & \to & \mathcal{MM} \\ X & \mapsto & h(X) = \bigoplus_{n=0}^{2\dim(X)} h^n(X) \end{array}$$

of the category of varieties with values in $\mathcal{MM}$. It has the universal property that any linearization (with an explicit set of properties) factors through h.

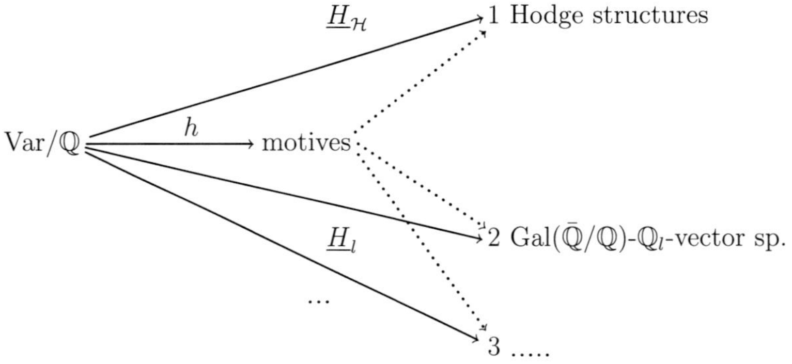

- We have a good idea what pure motives are (= corresponding to smooth, complete varieties). In this case there is an explicit construction of a category, called Grothendieck-motives ([Kl1]). If a set of explicit conjectures, called the *standard conjectures* are true, then this construction yields a semi-simple abelian category which does the job in this pure case ([Kl2]). Each variety would give rise to a motive $h(X)$. The morphisms from $h(X)$ to $h(Y)$ in the category of motives are essentially given by subvarieties of $X \times Y$ (generalizing the graph of a morphism of varieties). The standard conjectures mostly ask for existence of certain subvarieties. The details are not important for our considerations.

- We think that in general a motive M has a weight filtration by submotives

$$0 = W_k M \subset W_{k+1} M \subset \ldots \subset W_K M = M$$

such that all $W_m M / W_{m-1} M$ are pure motives ([D8] 1.1). Recall that these filtrations exist in our two examples for linearizations.

For the purposes of this article let us make the (somewhat exaggerated) assumption that we understand pure motives. In order to understand mixed motives we have to solve the following problem:

Question: *Given "easy" motives M_1 and M_2, what are the possible extensions of M_2 by M_1? I.e., what are the possible short exact sequences*

$$0 \to M_1 \to ? \to M_2 \to 0 \ ?$$

Let $\mathrm{Ext}^1_{\mathcal{MM}}(M_2, M_1)$ be the set of these sequences up to isomorphism. It is even an abelian group by Yoneda's addition of sequences ([Y], Cor. in 3.3).

2. Mixed realizations

Let us try to find (partial) answers to the question posed in the last section. How can we possibly talk about extensions of motives if we do not know what the category of mixed motives is in the first place? Strictly speaking this is of course impossible. However, Deligne [D6] Ch. 0 and Jannsen [Ja1] independently suggested the following approach.

Let us replace the conjectural object $h^n(X)$ by the tuple

$$\underline{H}^n_{\mathcal{MR}}(X) = \begin{pmatrix} \underline{H}^n_{\mathcal{H}}(X); \underline{H}^n_l(X) \text{ for each prime } l \ ; \\ \text{all comparison isomorphisms} \end{pmatrix} ,$$

a well-defined object in the category $\mathcal{MR}$ of *mixed realizations* ([Ja1] 2.1, or [H1] 11.1.1) (which is perfectly well-defined in terms of linear algebra). $\mathcal{MR}$ is a $\mathbb{Q}$-linear abelian category. There is a $\otimes$-product and a notion of duals. The functor $\underline{H}_{\mathcal{MR}}$ is of course not universal but at least it is universal for those cohomology theories we have considered so far. This looks like cheating. However, if either the Hodge or the Tate conjecture is true, then there is an equivalence of categories ([D7] Ex. 2.1)

$$\{\text{pure motives}\} \cong \left\{ \begin{array}{c} \text{objects in } \mathcal{MR} \text{ which are in the image} \\ \text{of smooth, complete varieties under } \underline{H}_{\mathcal{MR}} \end{array} \right\}$$

As we have agreed to assume that everything on pure motives is known, we can view $\mathcal{MM}$ as a subcategory of $\mathcal{MR}$. It makes sense to study extensions of motives in $\mathcal{MR}$. More precisely we can construct extensions of mixed realizations which (we think) are in fact extensions of motives.

Example: Let us do this in the example of our elliptic curve $E/\mathbb{Q}$. (The following construction is a special case of the "Abel-Jacobi map" considered in [Ja1] §9.) Let $D = \sum_{i=1}^k a_i P_i$ be a divisor of degree 0 on E, i.e., a formal linear combination with $a_i \in \mathbb{Q}$ and $P_i \in E(\mathbb{Q})$ with $\sum a_i = 0$. Let $U = E \smallsetminus \{P_1, \dots, P_k\}$ be the open complement of these points. This is a smooth but non-complete variety over $\mathbb{Q}$. Before we start our construction, we need a little preparation.

For $n \in \mathbb{Z}$ the *Tate-motive* $\mathbb{Q}(n) \in \mathcal{MR}$ is the tuple corresponding to a one-dimensional object in each component (see [Ja1] Def. 2.17). The weight filtration is pure of weight $-2n$.

The Hodge-filtration in the Hodge-component is concentrated in degree $-n$. The l-adic component $\mathbb{Q}_l(n)$ is a one-dimensional vector space but the operation of $\mathrm{Gal}(\bar{\mathbb{Q}}/\mathbb{Q})$ is given by the n-th power of the cyclotomic character. E.g. the Frobenius elements at primes $s \neq p$ operate by multiplication with s^n. For the precise definition see [Ja1] 2.17 b). The Tate-motive appears in geometry as the dual of the Lefschetz-motive $h^2(\mathbb{P}^1)$ (which even exists as a Grothendieck motive [Sch2] 1.9).

If M is a mixed realization, then we abbreviate $M(n) = M \otimes \mathbb{Q}(n)$.

LEMMA 2.1

$$\underline{H}_{\mathcal{MR}}(\coprod_{i=1}^{k} P_i) = \bigoplus_{i=1}^{k} \underline{H}_{\mathcal{MR}}(P_i) = \bigoplus_{i=1}^{k} \mathbb{Q}(0)$$

Proof. The variety $\coprod_{i=1}^{k} P_i$ corresponds to the complex manifold consisting of k points. Its singular cohomology is k-dimensional. The assertions on the extra structures in Hodge or l-adic cohomology are easy exercises with the definitions. □

We consider the geometric situation of our example

$$\coprod_{i=1}^{k} P_i \subset E \supset U \quad .$$

The long exact sequence familiar from singular cohomology also exists in terms of $\mathcal{MR}$. It reads

$$0 \to \underline{H}^1_{\mathcal{MR}}(E) \to \underline{H}^1_{\mathcal{MR}}(U) \to \underline{H}^0_{\mathcal{MR}}(\coprod_{i=1}^{k} P_i)(-1) \to \underline{H}^2_{\mathcal{MR}}(E) \to \cdots \quad .$$

The zero on the left comes from the vanishing of a $\underline{H}^{-1}_{\mathcal{MR}}$. The twist (-1) is an effect that is invisible on the level of singular cohomology. The divisor D induces a map

$$(a_1, \cdots, a_k) : \mathbb{Q}(-1) \longrightarrow \bigoplus_{i=1}^{k} \mathbb{Q}(-1) \quad .$$

Hence we get a commutative diagram

$$\begin{array}{ccccccccc}
0 & \longrightarrow & \underline{H}^1_{\mathcal{MR}}(E) & \longrightarrow & \underline{H}^1_{\mathcal{MR}}(U) & \xrightarrow{\delta} & \bigoplus_{i=1}^{k} \mathbb{Q}(-1) & \longrightarrow & \underline{H}^2_{\mathcal{MR}}(E) \\
\uparrow{=} & & \uparrow & & \uparrow & & \uparrow{D} & & \uparrow \\
0 & \longrightarrow & \underline{H}^1_{\mathcal{MR}}(E) & \longrightarrow & M & \longrightarrow & \mathbb{Q}(-1) & \longrightarrow & 0
\end{array}$$

The commutativity of the right-hand square depends on the degree zero of the divisor ([GH] Ex. 1 p. 144). M is defined as pull-back of $D(\mathbb{Q}(-1))$ under δ.

This second line is an extension class in $\mathcal{MR}$. But if there is a theory of motives at all, then this must also be an extension class in $\mathcal{MM}$. We think that M is a motive.

This construction is a special case of a general method. We have to generalize from divisors to elements in K-groups. For each variety X and $m \in \mathbb{Z}$, there is a group $K_m(X)$ ([Q], [TT]). It carries a filtration γ. We will need the graded parts of the γ-filtration up to torsion. The definition of K-theory is very formal and it is hard to give explicit examples. But e.g. if X is affine or smooth $\mathrm{Gr}_1^\gamma K_1(X)_\mathbb{Q} = \mathcal{O}(X)^* \otimes_\mathbb{Z} \mathbb{Q}$ (global invertible algebraic functions on X up to torsion, [So] 2.9 Prop. 1). Or if X is a smooth curve, then $\mathrm{Gr}_1^\gamma K_0(X)_\mathbb{Q}$ is the divisor class group of X. More generally for a smooth variety X, the group $\mathrm{Gr}_i^\gamma K_0(X)_\mathbb{Q}$ is isomorphic to $\mathrm{Ch}^i(X)$ ([SGA 6], Exp. XIV, §4).

CONJECTURE 2.2 (BEILINSON, MACPHERSON, SCHECHTMAN, [BMS] 0.2)

1. *For each $n, j \in \mathbb{Z}$ there is (functorially in X) a* motivic cohomology group *$H^n_{\mathcal{MM}}(X, j)$ ([BMS]) which stands canonically in the short exact sequence*

$$0 \to \mathrm{Ext}^1_{\mathcal{MM}}\left(\mathbb{Q}(-j), h^{n-1}(X)\right) \to H^n_{\mathcal{MM}}(X,j) \to \mathrm{Hom}_{\mathcal{MM}}\left(\mathbb{Q}(-j), h^n(X)\right) \to 0 \ .$$

2. *Let $m = 2j - n$. There are* Chern class maps

$$c_j^m : K_m(X) \longrightarrow H^n_{\mathcal{M}}(X, j) \ .$$

3. *If X is smooth, then c_j^m induces an isomorphism*

$$c_j^m : \mathrm{Gr}^j_\gamma K_m(X)_\mathbb{Q} \longrightarrow H^n_{\mathcal{M}}(X,j) \ .$$

Remark: In part of the literature, motivic cohomology of smooth varieties is *defined* naively by the formula conjectured in 2.2.3. By simplicial methods, this definition can be extended to singular varieties (see [HW] Appendix B). Voevodsky has defined directly a cohomology theory for varieties over fields called *motivic cohomology* as well. There are good reasons to believe that this theory is the conjectured one. For example, Conjecture 2.2.3 holds. However, the definition of the abelian category of mixed motives is still open.

The existence of motivic cohomology is itself an expression of a yet bigger set of conjectures which we will discuss in the next section. But first let us see how our example E fits in with the conjecture. $D \in \mathrm{Gr}^1_\gamma K_0(E)$ is mapped to $H^2_{\mathcal{MM}}(E, 1)$ by c_1. As the degree is zero, the image of $c_1(D)$ in the Hom-group vanishes and hence it is an extension.

Note that we have fallen back to a fully conjectural theory. Now we want to see what we can do in terms of $\mathcal{MR}$.

THEOREM 2.3 *Analogues of the above conjecture hold for $\mathcal{MR}$. In particular:*

1. *For $n, j \in \mathbb{Z}$, there is (functorially in X) a* mixed realization cohomology *$H^n_{\mathcal{MR}}(X, j)$.*

2. There is a spectral sequence

$$\mathrm{Ext}^p_{\mathcal{MR}}(\mathbb{Q}(-j), \underline{H}^q_{\mathcal{MR}}(X)) \Rightarrow H^{p+q}_{\mathcal{MR}}(X,j)$$

where the $\mathrm{Ext}^p_{\mathcal{MR}}$ *vanish for* $p \neq 0,1,2$.

3. Let $m = 2j - n$. *There are* Chern class maps

$$c_j^m : K_m(X) \longrightarrow H^n_{\mathcal{MR}}(X,j) \quad .$$

Proof. [H1] 11.3, 22.3.5, 18.2.6. □

We will say more about the method of proof in the section 4. Note that the spectral sequence in 2. replaces the short exact sequence in the conjecture. These additional terms are non-trivial in general. This shows that $\mathcal{MR}$ is definitely different from the conjectural $\mathcal{MM}$.

Remark: We are cheating a little with our notations. Our Ext^2 is not the group of Yoneda-2-extensions of objects in $\mathcal{MR}$ but something very closely related, cf. section 4.

The theorem translates into a construction of extension of mixed realizations by the general yoga of spectral sequences. From the spectral sequence we have

$$\begin{array}{ccccc} \mathrm{Ker}\, d & \longrightarrow & H^n_{\mathcal{MR}}(X,j) & \xrightarrow{d} & \mathrm{Hom}_{\mathcal{MR}}(\mathbb{Q}(-j), \underline{H}^n_{\mathcal{MR}}(X)) \\ \downarrow & & & & \\ \mathrm{Ext}^1_{\mathcal{MR}}(\mathbb{Q}(-j), \underline{H}^{n-1}_{\mathcal{MR}}(X)) & & & & \end{array} \quad .$$

An element k of $\mathrm{Gr}^j_\gamma K_m(X)_\mathbb{Q}$ is called *cohomologically trivial* if the image of $c_j(k)$ under d vanishes. This is equivalent to the vanishing of the Chern class of k in singular cohomology. Let $V^n(X,j)$ be the space of these elements.

Corollary 2.4 ([H1] 18.2.8) *The Chern class of mixed realization cohomology induces functorially in* X *a map*

$$V^n(X,j) \to \mathrm{Ext}^1_{\mathcal{MR}}(\mathbb{Q}(-j), \underline{H}^{n-1}_{\mathcal{MR}}(X)) \quad .$$

Similar extensions had been constructed before by Scholl [Sch1] in terms of Bloch's higher Chow groups. The advantage of the above method is that the extensions fit in the formalism of a cohomology theory that has all properties of a Bloch-Ogus cohomology theory ([H1] 15.3.8).

We can also show that the extensions in the corollary are "motives". By "motive" people mean different things in different circumstances. It might be

- an object in the conjectural category of motives;
- an object in the well-defined category of (pure) Grothendieck-motives, which has not yet been shown to be abelian;
- a mixed realization M which is subobject of an object in the image of the functor $\underline{H}_{\mathcal{MR}}$.

We take the third point of view. More precisely we use the following notion that is a little more general than Jannsen's [Ja1] Def. 4.1:

DEFINITION 2.5 ([H1] 22.1.3) *Let* $\mathbb{Q}\otimes\underline{Var}$ *be the* $\mathbb{Q}$*-linear additive category generated by* $\underline{Var}$*. Its objects are varieties but its morphisms are formal rational linear combinations of morphisms of varieties. Let* $C^b(\mathbb{Q}\otimes\underline{Var})$ *be the category of bounded complexes. It is not obvious but true that* $\underline{H}_{\mathcal{MR}}$ *extends to a functor on this category.*

An object $M \in \mathcal{MR}$ *is called a* motive *if it is subobject of some* $\underline{H}^n_{\mathcal{MR}}(X_.)$ *where* $X_.$ *is some complex of objects in* $\mathbb{Q}\otimes\underline{Var}$*.*

E.g., a variety X can be considered as a complex concentrated in degree 0 and then all its $\underline{H}^n_{\mathcal{MR}}(X)$ are motives in this sense. An extension class in $\mathcal{MR}$ is called *motivic* if the extension can be represented by a motive.

PROPOSITION 2.6 ([H1] 22.1.5) *The extension in* $\mathcal{MR}$ *associated to a cohomologically trivial element in* $\mathrm{Gr}^\gamma_j K_m(X)_\mathbb{Q}$ *is motivic.*

There are infinitely many open questions with respect to motives. But even in the more restricted context of mixed realizations there are many things to do.

1. Is it really necessary to use complexes of varieties in definition 2.5 or do we get the same category from smooth varieties alone? (Once we allow complexes it does not matter whether we consider all varieties or only smooth quasi-projective ones.)

2. We have $\mathrm{Ext}^n_{\mathcal{MR}}(M_1, M_2) = \mathrm{Ext}^n_{\mathcal{MR}}(\mathbb{Q}, M_1^\vee \otimes M_2)$. Is every $M_1^\vee \otimes M_2$ (M_i motives) subquotient of some $\underline{H}^n_{\mathcal{MR}}(X_.)(j)$ where $X_.$ is a complex of varieties? If so, are all motivic extensions in the image of the Chern class map?

3. The above construction makes no use of crystalline conditions. It is possible to refine the definition of $\mathcal{MR}$ by including such conditions. They are definitely necessary if we want $\mathcal{MR}$ to have a chance to contain $\mathcal{MM}$ as a full subcategory which is closed under extensions. It is not obvious if the above main theorem also extends to these refined versions. In principle it would need a well-developed theory of cohomology with coefficients in p-adic cohomology. However, in the light of recent developments ([Jo]) it seems plausible that such a program could be carried out.

4. If this works, do we get cohomological dimension one?

3. Motivic sheaves

Suppose we have a family of elliptic curves $p : \mathcal{E} \to S$ over some other variety S. This is a smooth proper map such that each fibre $\mathcal{E}_s = p^{-1}(s)$ for $s \in S$ is an elliptic curve. We then have for each point $s \in S$ the first Hodge structure $\underline{H}^1_\mathcal{H}(\mathcal{E}_s)$ of the fibre, its l-adic cohomology $\underline{H}^1_l(\mathcal{E}_s)$, its mixed realization $\underline{H}^1_{\mathcal{MR}}(\mathcal{E}_s)$ or even the motive $h^1(\mathcal{E}_s)$. The right setting to consider these objects is in terms of sheaves. This will also lead to a more conceptual interpretation of conjecture 2.2. Everything in this section is essentially due to Beilinson.

Let us look at the main example. Let X be a smooth complex manifold. There is an equivalence of categories between locally constant sheaves of $\mathbb{Q}$-vector spaces and finite dimensional representations of the fundamental group $\pi_1(X)$. We call them *smooth analytic sheaves.* If the manifold consists only of point, then this is just a vector space.

In the l-adic situation we use the algebraic fundamental group: if X is a smooth variety over $\mathbb{Q}$ and $X(\mathbb{C})$ the manifold of its complex valued points, then there is a short exact sequence ([SGA 1] XIII 4.3 and [SGA 1] XII 5.1)

$$1 \to \pi_1(\widehat{X(\mathbb{C})}) \to \pi_1^{alg}(X) \to \mathrm{Gal}(\bar{\mathbb{Q}}/\mathbb{Q}) \to 1$$

where $\widehat{(\,\cdot\,)}$ means the pro-finite completion. A *smooth l-adic sheaf* is a continuous representation of $\pi_1^{alg}(X)$ on a finite dimensional $\mathbb{Q}_l$-vector space. It corresponds to a pro-sheaf on the étale site of X ([Jou] 1.2.3.1). Note that a smooth l-adic sheaf on a point is a continuous representation of $\mathrm{Gal}(\bar{\mathbb{Q}}/\mathbb{Q})$.

In the Hodge setting, the corresponding notion is that of an admissible variation of Hodge structure. It is a smooth analytic sheaf (i.e. locally constant) sheaf of $\mathbb{C}$-vector spaces on $X(\mathbb{C})$ which is equipped with extra structures ([Ka] 1.6–1.9). In the case of a point we get back a Hodge structure. They do not have an interpretation as sheaves in the strict sense. But we still think of them as sheaves because they admit the same formalism of functors as sheaves do.

In each of these cases smooth sheaves are used to build a general theory of "sheaves". The term is a little misleading. We mean *perverse sheaves* which are not sheaves at all. If X is a variety, then a simple perverse sheaf consists of a simple smooth sheaf on a smooth (locally closed) subvariety of X which is extended canonically (using a functor called the middle direct image, [BBD] 1.4.22) to all of X. A general perverse sheaf is finite extension of simple ones. Of course the non-trivial part of the definition is to specify the possible extensions of simple sheaves. In the analytic case this is carried out in [BBD] (perverse sheaves are certain *complexes of sheaves* with morphisms taken in the derived category). They also consider the l-adic case with finite or algebraically closed ground field. In the l-adic case over a number field, a little modification is necessary, cf. [H2]. In the Hodge case, the corresponding theory of *mixed algebraic Hodge modules* was introduced by Saito ([Sa1], [Sa2]).

We expect that such categories of sheaves also exist for motives.

CONJECTURE 3.1 (BEILINSON [B2] APP. A) *On each variety X there should be a category of* motivic sheaves $\mathcal{MM}(X)$. *The category of motivic sheaves on the base point* Spec $\mathbb{Q}$ *is* $\mathcal{MM}(\mathrm{Spec}\,\mathbb{Q}) = \mathcal{MM}$. *The categories of sheaves satisfy a formalism of Grothendieck functors, i.e., pull-backs, push-forward, duals etc.*

Even more so than with motives we do not know these categories. But we know candidates for some motivic sheaves.

For each smooth variety X there should be at least the Tate-sheaves $\mathbb{Q}(n)_X$, i.e., the ones corresponding to the constant sheaf with the motive $\mathbb{Q}(n)$ above each $x \in X$. As we assume existence of Grothendieck functors, many other motivic sheaves can be derived from these.

Example: Let us look at our family of elliptic curves $p : \mathcal{E} \to S$. There should be a motivic sheaf on S corresponding to the motive $h^1(\mathcal{E}_s)$ above $s \in S$. In fact this would be the motivic sheaf $R^1 p_* \mathbb{Q}(0)_{\mathcal{E}}$.

CONJECTURE 3.2 ([JA2] 4.8 (4)) *If $a : X \to \operatorname{Spec} \mathbb{Q}$ is a variety, the functor h^n should be given as the higher direct image $R^n a_*$ of the motivic sheaf $\mathbb{Q}(0)_X$ on X.*

Global sections of a motivic sheaf $\mathcal{F}$ are defined as $\operatorname{Hom}(\mathbb{Q}(0)_X, \mathcal{F})$. (If we use the same definition with respect to smooth analytic or l-adic sheaves, we really get the global section functor in the usual sense.)

DEFINITION 3.3 Motivic cohomology *(cf. conjecture 2.2) is defined as derived functor of the global section functor, i.e.,*

$$H^n_{\mathcal{MM}}(X, j) := R\Gamma^n(X, \mathbb{Q}(j)_X) = \operatorname{Ext}^n_{\mathcal{MM}(X)}(\mathbb{Q}(-j)_X, \mathbb{Q}(0)_X) \quad .$$

We now explain how the conjectures on motivic cohomology as stated in 2.2 fit in with the ideas on motivic sheaves.

1. Note that $\Gamma(X, \cdot\,) = \Gamma(\operatorname{Spec} \mathbb{Q}, a_*(\,\cdot\,))$. On the level of derived functors this induces a Leray spectral sequence. In our special case (over a number field) the cohomological dimension of motivic cohomology is expected to be one ([Ja2] 4.12 c). All that remains of the spectral sequence is the short exact sequence used in conjecture 2.2 1.

2. As mentioned above we expect that the categories of motivic sheaves on different varieties satisfy a nice formalism of six Grothendieck functors ([B2] App. A). We have already used some of them above. This would imply that motivic cohomology is part of a Bloch-Ogus cohomology theory. c_j exists for all such (but cf. the remark below). This explains 2.2 2.

3. The Gr^γ_j themselves form a Bloch-Ogus cohomology theory, at least on the category of smooth varieties ([So] Thm 9). On the other hand we think that motivic cohomology is universal in the same way that motives were universal. Hence it makes sense to expect the isomorphism of 2.2 3.

c_j only seems to generate extensions of a special type. However, using duality for motives, all extensions can be reduced to these. Beilinson's conjecture 2.2 really predicts all extensions of motives.

Under the assumption of conjectures 3.1–3.2 and the isomorphism of conjecture 2.2.3 in the smooth case, the Chern class morphism of conjecture 2.2.2 cannot be expected to be an isomorphism in the singular case.

Counterexample: Let $X = \operatorname{Spec} k[s,t]/st$ (k a field), i.e., two affine lines crossing in a simple double point. It is easy to see by a sheaf-theoretic computation and using the homotopy property of K-theory that

$$H^n_{\mathcal{M}}(X,j) = H^n_{\mathcal{M}}(\operatorname{Spec} k, j) \text{ for all } n,j \in \mathbb{Z}.$$

But for a field of characteristic zero

$$K_2(X)_{\mathbb{Q}} \neq K_2(\operatorname{Spec} k)_{\mathbb{Q}}.$$

The difference between the two K-groups is measured by multi-relative K-theory. It is non-torsion, e.g. [DK], Remark 4.10.

Remark: For the existing constructions of higher Chern classes on higher K-groups, we need more than the axioms of a Bloch-Ogus cohomology (respectively more than the formalism of six Grothendieck functors). This is true for both Gillet's sheaf theoretic construction ([G]) and the simplicial variant in [H1] Part IV. In order to obtain Chern classes the cohomology groups have to be given as cohomology groups of complexes of abelian groups which are strictly functorial (rather than up to homotopy or in the derived category), i.e., form a presheaf of complexes on the big Zariski-site. It seems natural to expect such a property for motivic cohomology as well.

4. Realization cohomology

In the light of the ideas of the previous section, the category $\mathcal{MR}$ should be the special case on a point of some category of sheaves. In fact:

DEFINITION 4.1 ([H2] 5.3) *An object $\mathcal{F}$ in the category $\mathcal{MR}(X)$ of* mixed realization sheaves *on a variety X is given by the tuple consisting of*

- *a mixed algebraic Hodge module $\mathcal{F}_{\mathcal{H}}$ on $X(\mathbb{C})$;*
- *for each prime l a horizontal perverse l-adic sheaf $\mathcal{F}_l$;*
- *for each inclusion $\bar{\mathbb{Q}} \to \mathbb{C}$ a comparison isomorphism between the perverse sheaf underlying $\mathcal{F}_{\mathcal{H}}$ and the pull-back of $\mathcal{F}_l$ to $\mathbb{C}$.*

In $\mathcal{MR}(X)$ let $\mathcal{MR}(X)^w$ be the subcategory of those objects where all l-adic components carry a weight filtration (this is part of the data in the case of the Hodge module anyway) and the comparison isomorphisms are isomorphisms of filtered objects.

There is a slight problem with this definition. $\mathcal{MR}$ agrees with

$$\mathcal{MR}(\operatorname{Spec} \mathbb{Q})^w \neq \mathcal{MR}(\operatorname{Spec} \mathbb{Q}).$$

However, it is not known if push-forward respects the categories $\mathcal{MR}(X)^w$ in general. The bigger categories $\mathcal{MR}(X)$ allow the full formalism of Grothendieck functors.

In principle it is possible to define a cohomology theory using the categories $\mathcal{MR}(X)$, even so the program has not yet been carried out. We use a different approach which also allows the use of the stronger category $\mathcal{MR}$. The key is the following observation of Beilinson [B1] 0.1 (in terms of Hodge structures there): we expect the existence of a Leray spectral sequence

$$\mathrm{Ext}^p_{\mathcal{MR}}(\mathbb{Q}(0)_{\mathrm{Spec}\,\mathbb{Q}}, R^q a_* \mathbb{Q}(j)) \Rightarrow \mathrm{Ext}^{p+q}_{\mathcal{MR}(X)^w}(\mathbb{Q}(0)_X, \mathbb{Q}(j)_X) \quad .$$

Recall that the right-hand side should be $H^{p+q}_{\mathcal{MR}}(X,j)$ whereas the $R^q a_* \mathbb{Q}(j) = \underline{H}^q_{\mathcal{MR}}(X)(j)$. In order to define the terms on the right-hand side it is enough to define the terms on the left. Even though those have an interpretation in terms of sheaves this is not necessary for the definition.

Remark: Specialists will notice that the indices in the above Leray spectral sequence are those of ordinary sheaves rather than perverse sheaves.

In order to give the precise definition, we have to introduce derived categories ([V]). Let $\mathcal{A}$ be an abelian category. Let $C^b(\mathcal{A})$ be the category of bounded cohomological complexes in $\mathcal{A}$. A morphism $f : K \to L$ in $C^b(\mathcal{A})$ is called quasi-isomorphism if the induced morphisms $H^i(f) : H^i(K) \to H^i(L)$ are isomorphisms. The derived category $D^b(X)$ is constructed from $C^b(\mathcal{A})$ by formally inverting all quasi-isomorphisms ([V] 1.1). Many cohomology theories are defined by constructing certain complexes and then taking their cohomology objects. In a typical case (e.g. singular cohomology of topological spaces) the complexes are not well-defined but their cohomology objects are. By considering the complex in the derived category it becomes well-defined itself. We are interested in a few main features of derived categories:

1. If M and N are objects in $\mathcal{A}$, then ([V] §1 no. 2)

$$\mathrm{Hom}_{D^b(\mathcal{A})}(M, N[n]) = \mathrm{Ext}^n_{\mathcal{A}}(M,N)$$

   where $N[n]$ denotes the complex concentrated in degree $-n$.

2. If $F : \mathcal{A} \to \mathcal{B}$ is a left exact functor between abelian categories, then we denote the induced functor by $RF : D^b(\mathcal{A}) \to D^b(\mathcal{B})$ (if it exists). We then have $R^n F(\,\cdot\,) = H^n(RF(\,\cdot\,))$. (See [V] §2 no 2)

3. If $F : \mathcal{A} \to \mathcal{B}$ and $G : \mathcal{B} \to \mathcal{C}$ are functors between abelian categories, then the Grothendieck spectral sequence for their composition (if it exists) can be expressed as ([V] §2 3.1)

$$R(G \circ F) = RG \circ RF \quad .$$

In particular, the above Leray spectral sequence would read

$$\mathrm{Hom}_{\mathcal{MR}}(\mathbb{Q}_{\mathrm{Spec}\,\mathbb{Q}}, Ra_*(\mathbb{Q}(j)_X)[n]) = \mathrm{Ext}^n_{\mathcal{MR}(X)^w}(\mathbb{Q}_X, \mathbb{Q}_X(j)) \quad .$$

$\mathcal{MR}$ is of course an abelian category, hence it has a derived category which we could use. It turns out that this is not practical. We do not know how to define the functor $R_{\mathcal{MR}}$, which we have to construct in a minute, with values in $D(\mathcal{MR})$. Instead we replace it by another triangulated category with t-structure (this notion formalizes most of the properties of the derived category of an abelian category, [BBD] 1.3) which is easier to handle. We carry out the following program in [H1]:

First Step: We construct a category $D_{\mathcal{MR}}$ such that the cohomology objects of its objects are mixed realizations ([H1] 11.1.3). If A and B are mixed realizations, then ([H1] 11.1.6, 11.1.8)

$$\mathrm{Hom}_{D_{\mathcal{MR}}}(A, B[n]) = \begin{cases} 0 & \text{if } n \neq 0,1,2 \\ \mathrm{Hom}_{\mathcal{MR}}(A,B) & n=0 \\ \mathrm{Ext}^1_{\mathcal{MR}}(A,B) & n=1 \\ \prod_l W_0 \mathrm{Hom}_{D_l}(A_l, B_l[2]) & n=2 \end{cases}$$

(Here D_l is the triangulated category used as a stand-in for the derived category of $\mathrm{Gal}(\bar{\mathbb{Q}}/\mathbb{Q})$-$\mathbb{Q}_l$-modules. The Hom-groups carry a weight filtration W_*, [H1] 9.4.2).

It is not clear whether $\mathrm{Hom}_{D_{\mathcal{MR}}}(A, B[2]) = \mathrm{Ext}^2_{\mathcal{MR}}(A,B)$ but it is probably wrong. This means we do not know whether $D_{\mathcal{MR}} = D(\mathcal{MR})$ and expect it to be different.

Let us give a hint as to the construction of $D_{\mathcal{MR}}$. An object in $D_{\mathcal{MR}}$ is a tuple of filtered complexes in the different components appearing in $\mathcal{MR}$. We do not assume comparison isomorphisms but only comparison quasi-isomorphisms. Morphisms in $D_{\mathcal{MR}}$ are morphisms of such tuples with quasi-isomorphisms formally inverted.

Second Step: We then construct a functor ([H1] 11.2)

$$R_{\mathcal{MR}} : C^+(\mathbb{Q} \otimes \underline{Var}^s) \to D_{\mathcal{MR}}$$

By $\underline{Var}^s$ we denote the category of smooth varieties. As in 2.5, the notation $\mathbb{Q} \otimes \underline{Var}^s$ means the $\mathbb{Q}$-linear category spanned by $\underline{Var}^s$. The key property is the equality ([H1] 11.3.1)

$$H^i(R_{\mathcal{MR}}(X)) = \underline{H}^i_{\mathcal{MR}}(X) \tag{$*$}$$

for each smooth variety X. The components of $R_{\mathcal{MR}}(X)$ are given by those complexes that compute the components of $\underline{H}^i_{\mathcal{MR}}(X)$. We have to do this very carefully in order to extend the definition to complexes of smooth varieties. We think of $R_{\mathcal{MR}}(X)$ as $Ra_*\mathbb{Q}(0)_X$.

Third Step: The next step is to extend $R_{\mathcal{MR}}$ to all of $C^+(\mathbb{Q} \otimes \underline{Var})$. For an arbitrary variety X, there is always a complex $X^s_\cdot \to X$ of smooth varieties with the same cohomology groups. We define $R_{\mathcal{MR}}(X) = R_{\mathcal{MR}}(X^s_\cdot)$. (The resolution is not unique but the functor is well-defined up to canonical quasi-isomorphism). The property $*$ now holds for arbitrary X. This extends easily to complexes of arbitrary varieties.

Fourth Step: We define the global section functor

$$\Gamma : \mathcal{MR} \to \underline{ab} \quad \text{as} \quad \Gamma = \mathrm{Hom}_{\mathcal{MR}}(\mathbb{Q}(0), \,\cdot\,).$$

Instead of Ext-groups we use morphisms in $D_{\mathcal{MR}}$. Taking the Leray spectral sequence as a guideline, we put for a complex $X_.$ of varieties:

$$H^n_{\mathcal{MR}}(X_., j) := \operatorname{Hom}_{D_{\mathcal{MR}}}(\mathbb{Q}(0), R_{\mathcal{MR}}(X_.)(j)[n])$$

This is the group whose existence was claimed in Theorem 2.3.1. The spectral sequence in 2.3 is a direct consequence of our definition.

We then check all properties of a Bloch-Ogus-cohomology theory on the category $\underline{Var}$ ([H1] Ch. 15). This concerns certain long exact sequences, the definition of cohomology with support, homology (as the dual of cohomology with compact support), Poincaré duality between cohomology with support and homology, etc.

E.g., if $i : Y \to X$ is closed subvariety with open complement $j : U \to X$, then we define cohomology with support in Y as

$$H^n_{\mathcal{MR},Y}(X, k) = H^n_{\mathcal{MR}}(C(j), k)$$

where $C(j)$ is the cone of j in the category $C^b(\mathbb{Q}\otimes\underline{Var})$. It is not hard to construct the long exact sequence ([H1] 15.1)

$$\to H^n_{\mathcal{MR},Y}(X, k) \to H^n_{\mathcal{MR}}(X, k) \to H^n_{\mathcal{MR}}(U, k) \to H^{n+1}_{\mathcal{MR},Y}(X, k)$$

from the triple $C(j) \to X \to U$. The extra freedom we get from working with whole complexes of varieties instead of varieties alone replaces the usual sheaf-theoretic methods.

We then construct Chern classes (see 2.3.3)

$$c_j : K_m(X_.) \to H^{2j-m}_{\mathcal{MR}}(X_., j)$$

for all objects $X_.$ in $C^b(\mathbb{Q} \otimes \underline{Var})$ ([H1] 18.2.6). Let us sketch the method of loc. cit. in the special case of an affine variety $X = \operatorname{Spec} A$. We need to introduce the simplicial pro-variety $B_.Gl$. It is defined as the projective limit over $n \in \mathbb{N}$ of the simplicial varieties

$$B_.Gl_n : \quad \operatorname{Spec}(k) \;\substack{\longleftarrow \\ \longleftarrow}\; Gl_n \;\substack{\longleftarrow \\ \longleftarrow \\ \longleftarrow}\; Gl_n \times Gl_n \;\substack{\longleftarrow \\ \longleftarrow \\ \longleftarrow \\ \longleftarrow}\; \cdots \; .$$

We consider it as a complex by taking alternating sums of the differentials.

There is a natural injective map ([So] 2.10)

$$K_m(X)_{\mathbb{Q}} \to H_m(Gl(A), \mathbb{Q}) \qquad \text{(group cohomology)}$$

The latter is computed as H_m of the homological complex

$$\begin{aligned}\mathbb{Q}[B_kGl(A)]_{k\in\mathbb{Z}} &= \mathbb{Q}\left[\operatorname{Hom}_{\underline{Var}}(\operatorname{Spec} A, B_kGl)\right]_{k\in\mathbb{Z}} \\ &= \left[\operatorname{Hom}_{\mathbb{Q}\underline{Var}}(\operatorname{Spec} A, B_kGl)\right]_{k\in\mathbb{Z}}\end{aligned}$$

By functoriality it is mapped to the cohomological complex

$$\mathrm{Hom}_{D_{\mathcal{MR}}}(R_{\mathcal{MR}}(B_k Gl), R_{\mathcal{MR}}(\mathrm{Spec}\, A))_{k\in\mathbb{Z}}$$

An element $k \in K_m(X)$ induces an element $\alpha(k)$ in H^m of this last complex. This is nothing but a morphism in $D_{\mathcal{MR}}$

$$\alpha(k) : R_{\mathcal{MR}}(B_. Gl)[m] \to R_{\mathcal{MR}}(X) \quad .$$

We can show the following lemma:

LEMMA 4.2 (SPLITTING PRINCIPLE, [H1] 17.4.1)

$$R_{\mathcal{MR}}(B_. Gl) = \bigoplus_{j\in\mathbb{N}_0} \underline{H}^{2j}_{\mathcal{MR}}(B_. Gl)[-2j]$$

$$\underline{H}^{2j}_{\mathcal{MR}}(B_. Gl) = \bigoplus_{\substack{j_1,\dots\leq 0;\\ \sum j_l=-j}} \mathbb{Q}(j_1)\otimes\cdots\otimes\mathbb{Q}(j_m)$$

In particular, there are canonical maps $c_j^{univ} : \mathbb{Q}(-j) \to R_{\mathcal{MR}}(B_. Gl)[2j]$.

Finally we can use this to define:

DEFINITION 4.3 *For* $k \in K_m(X)$ *let*

$$c_j(k) : \mathbb{Q}(-j) \xrightarrow{c_j^{univ}} R_{\mathcal{MR}}(B_. Gl)[2j] \xrightarrow{\alpha(k)} R_{\mathcal{MR}}(X)[2j-m]$$

By definition this is an element of $H^{2j-m}_{\mathcal{MR}}(X, j)$.

For the general case, we first have to generalize this method to complexes of affine varieties. We then replace an arbitrary complex by an affine resolution. This will give Theorem 2.3 c) in full generality.

Finally, in proving Proposition 2.6, we have to produce a complex of varieties which defines the extension. This complex is the cone of k where we use the above identification of elements in K-groups with morphism of complexes $\mathrm{Spec}\, A \to B_. Gl[m]$.

References

[B1] A.A. Beilinson, "Notes on absolute Hodge cohomology", in "Applications of Algebraic K-theory to Algebraic Geometry and Number Theory", Proceedings of a Summer Research Conference held June 12–18, 1983, in Boulder, Colorado, Contemp. Math., vol. 55, Part I, AMS, Providence, pp. 35–68.

[B2] Beilinson, "Height Pairings between algebraic cycles", in K-theory, Arithmetic and Geometry (Y. Manin ed.), LNM 1289, Springer 1987.

[BBD] A.A. Beilinson, J. Bernstein, P. Deligne, "Faisceaux pervers", in B. Teissier, J.L. Verdier, "Analyse et Topologie sur les Espaces singuliers" (I), Astérisque 100, Soc. Math. France 1982.

[BMS] A. Beilinson, R. Macpherson, V. Schechtman, "Notes on Motivic Cohomology", Duke Math. Journ. 54,2 (1987), 679-710.

[D1] P. Deligne, "Théorie de Hodge, II", Publ. Math. IHES 40 (1971), pp. 5–57.

[D2] P. Deligne, "Théorie de Hodge, III", Publ. Math. IHES 44 (1974), pp. 5–77.

[D3] P. Deligne, "La Conjecture de Weil. I", Publ. Math. IHES 43 (1974), pp. 273–308.

[D4] P. Deligne, " Poids dans la cohomologie", Actes du Congrès International des Mathématiciens, Vancouver 19974, pp. 79–85.

[D5] P. Deligne, "La Conjecture de Weil. II", Publ. Math. IHES 52 (1981), pp. 313–428.

[D6] P. Deligne, "Valeurs de fonctions L et périodes d'integrales", Proc. Symp. Pure Math. 33,2 (79), 313-342.

[D7] P. Deligne, "Hodge cycles on abelian varieties", in "Hodge Cycles, Motives, and Shimura Varieties", LNM 900, Springer 1982, pp. 9–100.

[D8] P. Deligne, "A quoi servent les motifs?", Proc. Symp. Pure Math., Vol55, Part I, (1994), pp. 143–162.

[DK] R.K. Dennis, M.I. Krusemeyer: "$K_2(A[X,Y]/XY)$, a problem of Swan, and related computations", J. Pure Appl. Algebra 15, 125–148 (1979).

[FK] E. Freitag, R. Kiehl, "Etale cohomology and the Weil Conjecture", Ergebnisse der Math. 3. Folge Band 13, Springer 1988.

[G] H. Gillet, "Riemann-Roch theorems for higher algebraic K-theory", Adv. in Math. 40 (1981), pp. 203–289.

[GH] P. Griffith, J. Harris, "Principles of Algebraic Geometry", John Wiley & Sons, New York 1978.

[H1] A. Huber, "Mixed Motives and Their Realization in Derived Categories", LNM 1604, Springer 1995.

[H2] A. Huber, "Mixed perverse sheaves for schemes over number fields", Comp. Math. 108 (1997), 107–121.

[HW] A. Huber, J. Wildeshaus: "Classical motivic polylogarithm according to Beilinson and Deligne", preprint 1996.

[Ja1] U. Jannsen, "Mixed Motives and Algebraic K-Theory", LNM 1400, Springer-Verlag 1990.

[Ja2] U. Jannsen, "Motivic Sheaves and Filtrations on Chow Groups", Proc. Symp. Pure Math., Vol55, Part I, (1994), pp. 245–302.

[Jo] A.J. de Jong, "Smoothness, semi-stability and alterations", Preprint nr. 916, University Utrecht, Dept. of Mathematics, 1995.

[Jou] J.-P. Jouanolou, "Cohomologie l-adique", Exp. VI, pp. 251–281, in SGA 5, A. Grothendieck et al., "Cohomologie l-adique et Fonctions L", Springer LNM 589, Springer 1977.

[Ka] M. Kashiwara, "A study of variation of mixed Hodge structure", Publ. RIMS, Kyoto Univ. 22 (1986), pp. 991–1024.

[Kl1] S. Kleiman, "Motives", Proc. Alg. Geometry, Oslo 1970 (F. Oort, ed.), Walters-Noordhoff, Groningen, 1972, pp.53–82.

[Kl2] S. Kleiman, "The standard conjectures", Proc. Symp. Pure Math., Vol55, Part I, (1994), pp. 3–20.

[Mi1] J. Milne, "Etale Cohomology", Princeton Math. Series 33, Princeton Univ. Press 1980.

[Mi2] J.S. Milne, "Abelian varieties", in "Arithmetic Geometry", G. Cornell, J.H. Silverman ed., Springer 1986, pp. 103–150.

[Mu] D. Mumford, "The Red Book of Varieties and Schemes", LNM 1358, Springer 1988.

[Q] D. Quillen, "Higher Algebraic K-Theory", in "Alg. K-Theory I", LNM 341, Springer 1973 (1988), pp. 207–245.

[Sa1] Morihiko Saito, "Modules de Hodge Polarisables", Publ. RIMS, Kyoto Univ. 24 (1988), pp. 849–995.

[Sa2] Morihiko Saito, "Mixed Hodge Modules", Publ. RIMS, Kyoto Univ. 26 (1990), pp. 221–333.

[Sch1] A.J. Scholl, "Extensions of motives, higher Chow groups and special values of L-functions", Sém. de Théorie de Nombres, Paris 1991/92, 279-292.

[Sch2] A. Scholl, "Classical Motives", Symp. Pure Math., Vol55, Part I, (1994), pp. 163–188.

[SGA 1] A. Grothendieck et al., "Revètements Etales et Groupe Fondamental", LNM 224, Springer-Verlag 1971.

[SGA 4] M. Artin, A. Grothendieck, J.L. Verdier, et al., "Théorie des Topos et Cohomologie Etale des Schémas", Springer LNM 269, 270, 305, Springer 1972, 1973.

[SGA 4 1/2] P. Deligne et al., "Cohomologie Etale", Springer LNM 569, Springer 1977.

[SGA 5] A. Grothendieck et al., "Cohomologie l-adique et Fonctions L", Springer LNM 589, Springer 1977.

[SGA 6] P. Berthelot, A. Grothendieck, L. Illusie, "Théorie des Intersections et Théorème de Riemann-Roch", Springer LNM 225, Springer 1971.

[Si] J.H. Silverman, "The Arithmetic of Elliptic Curves", Graduate Texts in Math. 106, Springer 1986.

[So] C. Soulé, "Opérations en K-théorie algébrique", Can. J. Math., vol 28, no 3 (1985), pp. 488–550.

[TT] R.W. Thomason, T. Trobaugh, "Higher Algebraic K-Theory of Schemes and of Derived Categories", in The Grothendieck Festschrift Vol III, 247-436, Birkhäuser 1990.

[V] J.L. Verdier, "Catégories dérivées, état 0", in SGA 4 1/2, LNM 569, 262-311, Springer 1977.

[Y] N. Yoneda, "On Ext and Exact Sequences", J. Fac. Sci. Tokyo, Sec. I 8 (1960) 507-526.

Progress in Mathematics, Vol. 168, © 1998 Birkhäuser Verlag Basel/Switzerland

Boundary Values of Dirichlet Series and the Distribution of Primes

JERZY KACZOROWSKI*

Adam Mickiewicz University
Faculty of Mathematics and Computer Science
ul. Matejki 48/49, Poznań, Poland

1. Introduction

There exist many interesting and difficult problems concerning irregularities in the distribution of primes. As early as 1853 P.L. Chebyshev investigated differences between numbers of primes in arithmetic progressions (mod 4), and B. Riemann in his famous memoir from 1859 conjectured that (in usual notation) $\pi(x) < \mathrm{li}x$ for $x > 2$. Questions of this type appear in a very natural way and attracted the attention of many leading mathematicians including J.E. Littlewood, E. Landau, E. Phragmén, A.E. Ingham, G. Pólya, N. Levinson and many others. S. Knapowski and P. Turán in the early 60's for the first time considered *comparative prime number theory* as a separate branch of the theory of numbers, cf. [20]. We refer the interested reader to [20] for historical remarks.

In abstract form the basic problem of this theory can be formulated as follows. Suppose N functions $A_1(x), A_2(x), \ldots, A_N(x)$ are given which are asymptotically equivalent to $M(x)$ as x tends to infinity. The object is the study of the A_j's or equivalently of the associated error terms $E_j(x) = A_j(x) - M(x)$, $j = 1, 2, \ldots, N$, from the point of view of discrepancies. For instance one can ask about the size of exceptionally large values of errors and their locations, the sign of the E_j's or the differences $A_j(x) - A_k(x)$, $j \neq k$, etc. In concrete cases the A_j's depend on arithmetic properties, usually the counting functions of some algebraic objects, sometimes taken with appropriate weights. In most cases they are difficult to study but there exist explicit formulae, expressing them in terms of zeros and poles of the associated zeta functions.

We have a variety of interesting problems of this type. The basic example is of course the distribution of primes in various arithmetic progressions $a(\mathrm{mod}\, q)$, $(a, q) = 1$, belonging to the same modulus q. In this case the A_j's are equal to $\pi(x, q, a)$ or $\psi(x, q, a)$ with $(a, q) = 1$. We basically restrict our attention to this particular case, but it is fair to stress that it is also very interesting to study other weighted sums over primes, or summatory functions of arithmetic functions depending on factorization properties like e.g. Möbius or Euler's totient functions.

*Partially supported by KBN Grant nr.2 PO3A 028 09.

Every such particular case has its own nature and usually involves some specific technical problems.

The content of this paper is the following. After introducing some technical tools in Section 2, we discuss limiting distributions related to primes (Section 3), Chebyshev's problem (Section 4), the Shanks-Rényi race problem (Section 5), and finally changes of sign of the remainder term in Dirichlet's prime number theorem (Sections 6 and 7). A short list of open problems ends the paper. Some of the presented results are conditional (e.g. proved under the assumption of the Generalized Riemann Hypothesis). This is due to the complexity of the problems discussed. For instance until recently no results were known on the general race problem of Shanks-Rényi (see Section 5 below), and it was believed that this problem was "intractable at present", cf. P. Erdős, [5].

It is worth mentioning that most of the results have their counterparts in other contexts, for example in algebraic number fields (distribution of prime ideals in ideal classes, Chebotarev-type questions). Many of them can be proved in arithmetic semigroups with a finite classgroup and sufficiently regular L-functions, for instance belonging to the Selberg class. Prime counting functions appear also in prime geodesic problems in hyperbolic geometry. Here, as in the arithmetic case, prime geodesics are equidistributed amongst the homology classes (up to the main terms), and one can study finer distribution problems in this context (cf. [27], [28]).

2. Explicit formulae

Explicit formulae provide the bridge between primes and zeta zeros. In our investigations we use k-functions to interpret number-theoretic errors as the boundary values of some harmonic functions on the upper half-plane $H = \{z \in \mathbf{C} : \Im m(z) > 0\}$. Let $\chi(\mathrm{mod}\, q)$, $q \geq 1$, be a primitive Dirichlet character. For $z \in H$ we write

$$k(z,\chi) = \sum_{\gamma>0} e^{\rho z},$$
$$K(z,\chi) = \sum_{\gamma>0} \frac{1}{\rho} e^{\rho z},$$

where the summation is taken over all non-trivial $L(s,\chi)$ zeros $\rho = \beta + i\gamma$ with positive imaginary parts. Of course both functions are holomorphic on H. Functions of this type have been studied in [10], and in the special case (for the Riemann zeta function) in [3] and [6]. They appear in a natural way in the cohomological approach to analytic number theory, cf. [4]. In a sense analytic properties of k-functions can be viewed as the *complex explicit formulae*, cf. [10] for more details.

THEOREM 1 (cf. [10]) *The function k can be continued analytically to a meromorphic function on the Riemannian surface M of* $\log z$ *having infinitely many simple poles. We have*

$$2\pi i k(z,\chi) = N_0(z,\chi) + \frac{e^z E_1(\frac{3}{2}z) + E_1(\frac{5}{2}z) + H(z,d)}{1 - e^{2z}} + h(z,\chi),$$

where

$$\begin{aligned}
N_0(z,\chi) &= e^{3z/2}\sum_{n\geq 2}\frac{\Lambda(n)\chi(n)}{n^{3/2}(z-\log n)} \\
&+ e^{-z/2}\sum_{n\geq 2}\frac{\Lambda(n)\overline{\chi}(n)}{n^{3/2}(z+\log n)} - \frac{1}{z}e^{-z/2}\log\frac{q}{\pi}, \\
d &= d(\chi)\in\{0,1\} \quad \text{is such that} \quad \chi(-1)=(-1)^d, \\
H(z,0) &= e^{2z}\int_{l(-5/4,-1/4)}\frac{\Gamma'}{\Gamma}(w)e^{2wz}\,dw + e^{3z}\int_{3/4}^{7/4}\frac{\Gamma'}{\Gamma}(w)e^{-2wz}\,dw, \\
H(z,1) &= e^{z}\int_{l(-3/4,1/4)}\frac{\Gamma'}{\Gamma}(w)e^{2wz}\,dw + e^{4z}\int_{5/4}^{9/4}\frac{\Gamma'}{\Gamma}(w)e^{-2wz}\,dw, \\
h(z,\chi) &= \int_{l(-1/2,3/2)}\frac{L'}{L}(w,\chi)e^{wz}\,dw,
\end{aligned}$$

Γ being the Euler gamma function and $l(a,b)$, $a,b\in\mathbf{R}$, $a<b$, denoting a smooth curve

$$\tau:[0,1]\to\mathbf{C},$$

such that $\tau(0)=a$, $\tau(1)=b$, $\Im m\ \tau(t)>0$ for $0<t<1$ and such that there are no $L(s,\chi)$-zeros between $l(a,b)$ and the real axis.

Moreover, k satisfies two functional equations on M (cf. Theorems 3.1,3.2 and 3.3 in [10]). Since for $z\in H$ we have $K(z,\chi)=\int_{i\infty}^{z}k(s,\chi)\,ds$, where we integrate along the vertical half-line $s=z+it$, $\infty>t\geq 0$, the function $K(z,\chi)$ can be continued analytically along every path on M not passing through the singularities of k; it becomes therefore a multi-valued function on this surface. Every pole of k becomes a logarithmic branch point of K. The only singularities on the real axis can appear at $x=0$ or $x=\pm\log n$, $n\in\mathbf{N}$. For further details the reader is referred to [10].

Let $x\in\mathbf{R}$ and

$$F(x,\chi)=\lim_{y\to 0+}\left\{K(x+iy,\chi)+\overline{K(x+iy,\overline{\chi})}\right\}.$$

It can be proved that this limit exists for all real x (cf. [10], page 200).

THEOREM 2 (cf. [10]) *For $x>0$ we have*

$$F(x,\chi)=-\psi_0(e^x,\chi)-\sum_{\gamma=0}e^{\beta x}/\beta+e(\chi)e^x-e_1(\chi)x-R(x,d(\chi))+B(\chi), \quad (1)$$

where the summation is over all non-trivial, real $L(s,\chi)$ zeros (if there are any) and

$$\psi_0(x,\chi) = \frac{1}{2}\left(\psi(x+0,\chi)+\psi(x-0,\chi)\right),$$

$$\begin{aligned}
\psi(x,\chi) &= \sum_{n\le x} \Lambda(n)\chi(n),\\
e(\chi) &= \begin{cases} 1 & \text{if } \chi=\chi_0 \text{ (the principal character)}\\ 0 & \text{otherwise,}\end{cases}\\
d(\chi) &= \begin{cases} 1 & \text{if } \chi(-1)=-1\\ 0 & \text{otherwise,}\end{cases}\\
e_1(\chi) &= \begin{cases} 1 & \text{if } d(\chi)=0,\ \chi\ne\chi_0\\ 0 & \text{otherwise,}\end{cases}\\
R(x,d(\chi)) &= \begin{cases} \frac12\log(1-e^{-2x}) & \text{if } d(\chi)=0\\ \frac12\log\frac{e^x-1}{e^x+1} & \text{if } d(\chi)=1.\end{cases}
\end{aligned}$$

The constants $B(\chi)$ are given by

$$B(\chi)=\sum_{\gamma=0}\frac{1}{\beta}-e(\chi)-d(\chi)\log 2-C/2-\frac12\log\frac{\pi}{q}+F(0,\chi),$$

where $C=0.57721566\ldots$ denotes the Euler constant[1]*). For $x<0$ we have*

$$F(x,\chi)=\tilde\psi_0(e^{|x|},\chi)-\sum_{\gamma=0}e^{\beta x}/\beta+e(\chi)e^x+e(\chi)x-R(|x|,1-d(\chi))+C(\chi),$$

where

$$\begin{aligned}
\tilde\psi_0(x,\chi) &= \frac12(\tilde\psi(x+0,\chi)+\tilde\psi(x-0,\chi)),\\
\tilde\psi(x,\chi) &= \sum_{n\le x}\frac{\Lambda(n)\chi(n)}{n},
\end{aligned}$$

and

$$C(\chi)=B(\chi)+C+\log(2\pi/q).$$

Let $q\ge 1$, $0<a\le q$, $(a,q)=1$ and let

$$F(z,q,a)=-2e^{-z/2}\frac{1}{\varphi(q)}\sum_{\chi(\mathrm{mod}\ q)}\overline{\chi(a)}K(z,\chi')-\frac{2}{\varphi(q)}\sum_{\chi(\mathrm{mod}\ q)}\overline{\chi(a)}m(1/2,\chi),$$

where $m(1/2,\chi)$ denotes the multiplicity of $L(s,\chi)$ zero at $s=1/2$; we put $m(1/2,\chi)=0$ when $L(1/2,\chi)\ne 0$. Let us write for $x>0$

$$\begin{aligned}
\psi(x,q,a) &= \sum_{\substack{n\le x\\ n\equiv a(\mathrm{mod}\ q)}}\Lambda(n), \quad E(x,q,a)=(\psi(e^x,q,a)-\frac{1}{\varphi(q)}e^x)e^{-x/2},\\
\tilde\psi(x,q,a) &= \sum_{\substack{n\le x\\ n\equiv a(\mathrm{mod}\ q)}}\frac{\Lambda(n)}{n}, \quad \tilde E(x,q,a)=(\tilde\psi(e^x,q,a)-\frac{1}{\varphi(q)}x+b(q,a))e^{x/2},
\end{aligned}$$

[1]Note that there is a misprint in formula (4.6) of [10].

where $b(q,a)$ denotes a certain (complicated) constant depending on q and a alone (cf. [11]).

THEOREM 3 (cf. [11]) *For* $x \geq 1$ *we have*

$$\lim_{y\to 0+} \Re e F(x+iy,q,a) = E(x,q,a) + O(xe^{-x/2}),$$

and for $x \leq -1$ *we have*

$$\lim_{y\to 0+} \Re e F(x+iy,q,a) = \tilde{E}(|x|,q,\overline{a}) + O(e^{-|x|/2}),$$

where $\overline{a}(\bmod q)$ *denotes the inverse of* $a(\bmod q)$*:* $a\overline{a} \equiv 1(\bmod q)$.

The behaviour of $F(z,q,a)$ near $z=0$ is important. We have for $0<r<1$, $0 \leq \phi \leq \pi$

$$\Re e F(re^{i\phi},q,1) = \frac{1}{\pi}(\frac{\pi}{2} - \phi)\log r + O(1), \tag{2}$$

and

$$\Re e F(re^{i\phi},q,a) = O(1), \tag{3}$$

when $a \not\equiv 1(\bmod q)$; cf. [14], Lemma 2.

We need the following general result on boundary values of generalized Dirichlet's series. Let us denote by $\mathcal{B}$ the set of all functions

$$F(z) = \sum_{n=1}^{\infty} \alpha_n e^{iw_n z}, \quad z = x+iy, y>0, \tag{4}$$

satisfying the following conditions.

1. $0 \leq w_1 < w_2 < \ldots$ are real numbers.

2. $\alpha_n \in \mathbf{C}, n = 1,2,3,\ldots$

3. The series in (4) converges absolutely for all $y > 0$.

4. The limit

   $$P(x) = \lim_{y\to 0+} P(x+iy),$$

   where $P(x+iy) = \Re e F(x+iy), y>0$, exists for almost all real x. (Putting $P(x) = 0$ for the remaining x we get P well defined on the closed upper half-plane $\overline{H} = \{z \in \mathbf{C} : \Im m(z) \geq 0\}$.)

5. We have

   $$\lim_{y\to 0+} \sup_{x\in\mathbf{R}} \int_{-1/2}^{1/2} \mid P(x+t) - P(x+t+iy) \mid^2 \, dt = 0.$$

The functions $F(z,q,a)$ and $F(z,q,a) - F(z,q,b)$, where $q \geq 1$, $(a,q) = (b,q) = 1$ are members of the class $\mathcal{B}$.

THEOREM 4 (cf. [14]) *Let $F_j \in \mathcal{B}$ for $j = 1, 2, \ldots, k$ and let $x_0 \in \mathbf{R}$ be a continuity point of the mapping*

$$\begin{aligned} P : \overline{H} &\ni z \mapsto (P_1(z), P_2(z), \ldots, P_k(z)) \in \mathbf{R}^k, \\ P_j &= \Re e F_j, j = 1, 2, \ldots, k. \end{aligned}$$

Then for every open neighborhood $U \subset \mathbf{R}^k$ of $P(x_0)$ there exist constants $b_0 = b_0(U) > 0$ and $l_0 = l_0(U) > 0$ such that

$$\mu(P^{-1}(U) \cap \mathcal{J}) > b_0,$$

for arbitrary interval $\mathcal{J} \subset \mathbf{R}$ of length $\geq l_0$; μ being the Lebesgue measure on the real axis.

3. Limiting distributions

It was first observed by Wintner [30] that, under the RH, the remainder term in the prime number theorem after suitable normalization and change of variable has a limiting distribution. His method of proof can be extended to cover the similar situation in arithmetic progressions. So essentially Wintner's result asserts that under the GRH the distribution functions

$$\mathcal{F}_T(u,q,a) = \frac{1}{T}\mu\{0 \leq t \leq T : E(t,q,a) < u\}$$

weakly converge as $T \to \infty$ to a limit $\mathcal{F}(u,q,a)$. In general we do not know much about $\mathcal{F}(u,q,a)$. It would be of great interest to determine the support of it assuming the GRH. Probably it is equal to $(-\infty, \infty)$. The relevance of this result to the distribution of primes is obvious. It would immediately imply unconditionally that

$$\limsup_{x \to \infty} \frac{|\pi(x,q,a) - \frac{1}{\varphi(q)}\mathrm{li}x|}{\sqrt{x}/\log x} = \infty$$

for every $(a,q) = 1$, where, as usual, $\pi(x,q,a)$ denotes the number of primes $p \equiv a (\mathrm{mod}\, q)$, $p \leq x$. Under the additional condition that no L-function associated with a Dirichlet's character $(\mathrm{mod}\, q)$ vanishes on $(1/2, 1)$ we would even have that

$$\limsup_{x \to \infty} \frac{\pi(x,q,a) - \frac{1}{\varphi(q)}\mathrm{li}x}{\sqrt{x}/\log x} = \infty$$

and

$$\liminf_{x \to \infty} \frac{\pi(x,q,a) - \frac{1}{\varphi(q)}\mathrm{li}x}{\sqrt{x}/\log x} = -\infty.$$

In particular, it would imply that the difference $\pi(x,q,a) - \frac{1}{\varphi(q)}\mathrm{li}x$ changes sign infinitely many times as x tends to infinity. None of the above statements is as yet proved and without doubt they belong to the most important open problems in the theory.

One of the rare instances when we can prove something on limiting distributions assuming the GRH alone is the following result describing connections between primes in reciprocal arithmetic progressions

$$a (\operatorname{mod} q) \quad \text{and} \quad \overline{a} (\operatorname{mod} q),$$

where

$$a\overline{a} \equiv 1 (\operatorname{mod} q).$$

Suppose we consider the following general situation. Let the generalized Dirichlet series $F(z) \in \mathcal{B}$ be given. For a fixed $y > 0$ the function $F_y(x) = F(x + iy)$ is an almost periodic function of Bohr, so its real and imaginary parts have limiting distributions. The same is true for the boundary function $P(x)$. Suppose that we are given two functions f and g defined on the positive real axis, and that there exists a function $F \in \mathcal{B}$ such that for $x > 0$

$$f(x) = \Re e F(x)$$

and

$$g(x) = \Re e F(-x).$$

Since f and g are boundary values of the same almost periodic function their limiting distributions are closely related. Applying this principle to the function $F(z, q, a)$, and using Theorem 3 we obtain the following theorem.

THEOREM 5 *Assume the GRH for L-functions* (mod q)*. Then both remainders* $E(x, q, a)$ *and* $\tilde{E}(x, q, a)$ *have limiting distributions for every* $(a, q) = 1$*. Denoting them by* $\mathcal{F}(u, q, a)$ *and* $\tilde{\mathcal{F}}(u, q, a)$ *respectively we have*

$$\mathcal{F}(u, q, a) = 1 - \tilde{\mathcal{F}}(u, q, \overline{a})$$

for almost all $u \in \mathbf{R}$*.*

M. Rubinstein and P. Sarnak [28] considered limiting distributions of the vector-valued function ($x \geq 2$)

$$E_{q,a_1,\dots,a_r}(x) = \frac{\log x}{\sqrt{x}}(\varphi(q)\pi(x, q, a_1) - \pi(x), \dots, \varphi(q)\pi(x, q, a_r) - \pi(x)),$$

where $a_1, a_2, \dots, a_r$ are distinct classes coprime to q.

THEOREM 6 (cf. [28]) *Assume GRH. Then* $E_{q,a_1,\dots,a_r}$ *has a limiting distribution* $\mathcal{F}_{q,a_1,\dots,a_r}$ *on* $\mathbf{R}^r$*, that is,*

$$\lim_{T\to\infty} \frac{1}{\log T} \int_2^T f(E_{q,a_1,\dots,a_r}(x)) \,\frac{dx}{x} = \int_{\mathbf{R}^r} f(x)\, d\mathcal{F}_{q,a_1,\dots,a_r}$$

for all bounded continuous functions f *on* $\mathbf{R}^r$*.*

As before we do not know much about $\mathcal{F}_{q,a_1,\ldots,a_r}$, except that it exists. The situation changes when some more speculative conjectures on the non-trivial zeros of L-functions are assumed, (cf. [8], [25], [28], [30]).

Linear Independence Hypothesis (LIH). *The set of $\gamma \geq 0$ such that $L(\frac{1}{2}+i\gamma,\chi)=0$, for χ running over primitive Dirichlet characters, is linearly independent over* $\mathbf{Q}$.

This hypothesis implies in particular that all the zeros are simple and that $L(1/2,\chi)\neq 0$. The relevance of LIH to our subject is that assuming in addition GRH, the terms in explicit formulae corresponding to different non-trivial zeros can be treated as independent random variables and the Fourier transform of the limiting distribution can be computed explicitly (cf. [8], [28], [2]).

THEOREM 7 (cf. [28]) *Under GRH and LIH we have the following explicit formula for the Fourier transform of $\mathcal{F}_{q,a_1,\ldots,a_r}$*

$$\hat{\mathcal{F}}_{q,a_1,\ldots,a_r}(\xi_1,\ldots,\xi_r)=\exp\left(i\sum_{j=1}^{r}c(q,a_j)\xi_j\right)\prod_{\substack{\chi\neq\chi_0\\ \chi(\bmod q)}}\prod_{\gamma_\chi>0}J_0\left(\frac{2|\sum_{j=1}^{r}\chi(a_j)\xi_j|}{\sqrt{\frac{1}{4}+\gamma_\chi^2}}\right),$$

where

$$c(q,a)=-1+\sum_{\substack{b^2\equiv a(\bmod q)\\ 0\leq b\leq q-1}}1,$$

and $J_0(z)$ is the Bessel function

$$J_0(z)=\sum_{m=0}^{\infty}\frac{(-1)^m(\frac{1}{2}z)^{2m}}{(m!)^2}.$$

4. Chebyshev's problem

The classical conjecture of Chebyshev asserts that "there are more primes $\equiv 3(\bmod 4)$ than $\equiv 1(\bmod 4)$" (cf. [1], [13], [20]). Denoting by $N(T)$ the number of natural numbers $m\leq T$ for which

$$\pi(m,4,1)>\pi(m,4,3),$$

S. Knapowski and P. Turán [20] asserted that

$$\lim_{T\to\infty}\frac{N(T)}{T}=0. \tag{5}$$

This conjecture however is unlikely to be true. As an application of Theorem 4 in the case $k=1$ and

$$F_1(z)=F(z,4,1)-F(z,4,3)=-2e^{-z/2}K(z,\chi_1),$$

where χ_1 denotes the non-principal Dirichlet character $(\bmod 4)$, we have the following result.

THEOREM 8 (cf. [13]). *If the GRH holds for $L(s,\chi_1)$, then both sets*

$$P(4;1,3) := \{m \in \mathbf{N} : \pi(m;4,1) > \pi(m;4,3)\}$$

and

$$P(4;3,1) := \{m \in \mathbf{N} : \pi(m;4,3) > \pi(m;4,1)\}$$

have positive lower densities.

It would be of great interest to prove this statement without appealing to any unproved hypotheses. The best unconditional result is however significantly weaker (cf.[16])

THEOREM 9 (cf. [16]) *We have*

$$\#(P(4;1,3) \cap [2,T]) = \Omega(T^{1-\varepsilon})$$

and

$$\#(P(4;3,1) \cap [2,T]) = \Omega(T^{1-\varepsilon})$$

for every positive ε and T tending to infinity.

For real x we write

$$P_1(x) = \lim_{y \to 0+} \Re e F_1(x+iy).$$

We call $x_0 \in \mathbf{R}$ an *a-point* when there exists a positive ε such that

$$P(x_0 - t) < 1 \quad \text{and} \quad P(x_0 + t) > 1$$

for every $t \in (0,\varepsilon)$. Similarly we call $x_0 \in \mathbf{R}$ a *b-point* when there exists a positive ε such that

$$P(x_0 - t) > 1 \quad \text{and} \quad P(x_0 + t) < 1$$

for every $t \in (0,\varepsilon)$. There exist infinitely many a-points and infinitely many b-points. Let us denote by a_n and b_n, $n \in \mathbf{Z}$, consecutive a-points and b-points, numbered in increasing order so that

$$\cdots a_{-2} < b_{-2} < a_{-1} < b_{-1} < a_0 < b_0 < a_1 < b_1 < a_2 < b_2 < \cdots$$

Let us denote by $d^*(4;1,3)$ and $d^*(4;3,1)$ the upper densities of the sets $P(4;1,3)$ and $P(4;3,1)$ respectively. Then we have the following result.

THEOREM 10 (cf. [17]) *Let us assume the GRH for the L-function $L(s,\chi_1)$. Then we have*

$$d^*(4;1,3) = \sup_{N \in \mathbf{N}} e^{-b_N} \sum_{k=0}^{\infty} (e^{b_{N-k}} - e^{a_{N-k}})$$

and

$$d^*(4;3,1) = \sup_{N \in \mathbf{N}} e^{-a_N} \sum_{k=0}^{\infty} (e^{a_{N-k}} - e^{b_{N-k-1}}).$$

It is not difficult to find numerically a-points and b-points with small absolute values. The interval $[-15.42, 15.42]$ contains exactly 139 a-points and the same number of b-points. Using these numerical data we obtain the following estimates.

COROLLARY 1 (cf. [17]) *Let us assume the GRH for* $L(s,\chi_1)$. *Then we have*

$$\begin{aligned} d^*(4;1,3) &\geq 0.040540454\ldots \\ d^*(4;3,1) &\geq 0.999989360\ldots \end{aligned}$$

It can be observed that these estimates indicate a preponderance of primes $\equiv 3 (\bmod 4)$ (*Chebyshev's bias* in the terminology of Rubinstein and Sarnak [28]), but not so sharp as conjectured in (5). It follows also that neither of the sets $P(4;1,3)$ and$P(4;3,1)$ has a natural density.

The case of the logarithmic densities has been considered in [28]. For a set of natural numbers P we write

$$\begin{aligned} \overline{\delta}(P) &= \lim_{x\to\infty} \frac{1}{\log x} \sum_{n\leq x, n\in P} \frac{1}{n}, \\ \underline{\delta}(P) &= \lim_{x\to\infty} \frac{1}{\log x} \sum_{n\leq x, n\in P} \frac{1}{n}, \end{aligned}$$

and set $\delta(P) = \underline{\delta}(P) = \overline{\delta}(P)$ if the latter two limits are equal. From the existence of the limiting distributions (under the GRH and LIH) we deduce that both sets $P(4;1,3)$ and $P(4;3,1)$ have positive logarithmic densities. One observes here the difference in the behaviour of the two kinds of densities, but in both cases Chebyshev's bias is present.

THEOREM 11 (cf. [28]) *Under the GRH and LIH we have*

$$\delta(P(4;1,3)) = 0.0040\ldots$$

and

$$\delta(P(4;3,1) = 0.9959\ldots$$

Similar results can be proved for other moduli. The fact that quadratic residues and nonresidues play a role in the distribution of primes in classes was first observed by Landau [23] a long time ago, see also [20] and [28]. In fact, results and numerical computations show that classes which are quadratic residues contain fewer primes than quadratic nonresidues (note that in Chebyshev's problem $1 (\bmod 4)$ is quadratic residue $(\bmod 4)$). Theorem 7 taken with $r = 2$ gives the explanation of this phenomenon. Indeed, the constants $c(q,a)$ cause the shift of the mean of the limiting distribution, inducing the bias towards nonresidues.

5. The race problem

The Shanks-Rényi race problem is in fact the multidimensional version of the problem of Chebyshev. This time we are interested in comparing the numbers of primes in more than two distinct arithmetic progressions. In the most classic form the race conjecture claims that for every permutation $a_1, a_2, \dots, a_{\varphi(q)}$ of the reduced set of residue classes the set of natural numbers m for which

$$\pi(m,q,a_1) > \pi(m,q,a_2) > \cdots \pi(m,q,a_{\varphi(q)}) \tag{6}$$

is infinite. We speak about the "race" because the problem has a nice interpretation as a game. Let us call each class $a(\bmod q)$ $(a,q) = 1$ a *Player*. Player $a(\bmod q)$ scores a point when in the enumeration of natural numbers in increasing order a prime $\equiv a(\bmod q)$ occurs. Then, after m steps the Player $a(\bmod q)$ has exactly $\pi(m,q,a)$ points. Inequalities in (6) can be then described by saying that a_1 is the leader of the race (after m steps), a_2 is in the second position and so on.

More generally, by a *race problem* we understand any question concerning the distribution of elements of a set $P(q; a_1, a_2, \dots, a_r)$, where $a_1, a_2, \dots, a_r$ are distinct $r \geq 3$ classes $(\bmod q)$ coprime with q.

There are no unconditional results on the general race problem. Application of Theorem 4 with $F(z,q,a)$'s with $(a,q) = 1$ as the functions F_j, $j = 1, 2, \dots, \varphi(q)$ together with (2) and (3) leads to the following result.

THEOREM 12 (cf. [14]) *Let us suppose* GRH *for L-functions* $(\bmod q)$, $q \geq 3$, *and let u denote an arbitrary non-negative real number. Then there exist constants* $b_3 = b_3(u,q) > 0$, $c_3 = c_3(u,q) > 1$, *such that for every* $T \geq 1$ *we have*

$$\#\{T \leq m \leq c_3 T : \psi(m,q,1) \geq \max_{a \not\equiv 1 (\bmod q)} \psi(m,q,a) + u\sqrt{m}\} \geq b_3 T,$$

$$\#\{T \leq m \leq c_3 T : \pi(m,q,1) \geq \max_{a \not\equiv 1 (\bmod q)} \pi(m,q,a) + u\frac{\sqrt{m}}{\log m}\} \geq b_3 T,$$

$$\#\{T \leq m \leq c_3 T : \psi(m,q,1) \leq \min_{a \not\equiv 1 (\bmod q)} \psi(m,q,a) - u\sqrt{m}\} \geq b_3 T,$$

$$\#\{T \leq m \leq c_3 T : \pi(m,q,1) \leq \min_{a \not\equiv 1 (\bmod q)} \pi(m,q,a) - u\frac{\sqrt{m}}{\log m}\} \geq b_3 T.$$

Another consequence of Theorem 4 reads as follows.

THEOREM 13 (cf. [15]) *Let us assume* GRH *for L-functions* mod 5. *Then there exist constants* $b_6 > 0$, $c_6 > 1$, $d_0 > 0$ *such that for every permutation* (a_1, a_2, a_3, a_4) *of the sequence* $(1, 2, 3, 4)$ *we have*

$$\#\{T \leq m \leq c_6 T : \quad \psi(m,q,a_1) > \psi(m,q,a_2) > \psi(m,q,a_3) > \psi(m,q,a_4),$$
$$\min_{1 \leq i < j \leq 4} |\psi(m,q,a_i) - \psi(m,q,a_j)| \geq d_0\sqrt{m}\} \geq b_6 T.$$

Roughly speaking Theorem 12 asserts that infinitely many times Player $1 (\bmod q)$ is the leader of the race $(\bmod q)$ and also infinitely many times is in the last position. Theorem 13 solves (under the GRH) the ψ-version of the race problem $(\bmod 5)$. For further results the reader is referred to [16].

Of course assuming GRH and LIH we can say more. Knowing the shape of the limiting distribution (Theorem 7) we easily see that each set $P(q; a_1, a_2, \ldots, a_r)$ has a positive logarithmic density. Hence the race problem is a corollary to these two hypotheses. Let now r be fixed and let q tend to infinity. Let $\mu(q; a_1, a_2, \ldots, a_r)$ be the measure on $\mathbf{R}^r$ whose Fourier transform is $\hat{\mathcal{F}}_{q,a_1,\ldots,a_r}(\frac{\xi}{\sqrt{\varphi(q)\log q}})$.

Theorem 14 (cf. [28]) *Assume GRH and LIH Then the measure $\mu(q; a_1, a_2, \ldots, a_r)$ converges to the Gaussian*

$$\frac{e^{-(x_1^2+\ldots+x_r^2)}}{(2\pi)^{r/2}} dx_1 \ldots dx_r$$

as $q \to \infty$, independently of the choice of $a_1, a_2, \ldots, a_r$. In particular,

$$\max_{a_1,a_2,\ldots,a_r} \left| \delta(P(q; a_1, a_2, \ldots, a_r)) - \frac{1}{r!} \right| \to 0 \quad \text{as} \quad q \to \infty.$$

This means in particular that the Chebyshev bias dissolves as q tends to infinity. Again, one can observe that the constants $c(q, a)$ in Theorem 7 are responsible for the differences in densities of the sets $P(q; a_1, a_2, \ldots, a_r)$, see [28] for more details.

6. Changes of sign

Let $V(T, q, a)$ denote the number of changes of sign of the difference $\psi(x, q, a) - \frac{1}{\varphi(q)}x$ in the interval $(0, T]$. We assume of course that $q \geq 1$ and $(a, q) = 1$.

Our estimates of $V(T, q, a)$ are based on two principal methods: the method of monotonic integral operators ([9]), and the method of nodal lines ([12]).

Let $L_{\text{loc}}(a)$, $a \in \mathbf{R}$, $a \geq 0$ denote the space of locally integrable functions defined on the interval $(0, \infty)$ and such that for t tending to 0^+ we have $f(t) = O(t^{-b})$ with $b = b(f) < a$. The operator

$$\delta_a : L_{\text{loc}}(a) \to L_{\text{loc}}(a),$$

$$\delta_a(f)(x) = x^{-a} \int_1^x f(t) t^{a-1}\, dt,$$

has the following properties.

1. Denoting by $V(T; g)$ the number of changes of sign of a real-valued function $g \in L_{loc}$ in the interval $(0, T)$ we have for arbitrary $f \in L_{loc}$:

$$V(T; f) \geq V(T; \delta_a(f)).$$

2. For a fixed complex number $\Re e(s) > -a$ the function $t \mapsto t^s$ is an eigenvector of δ_a with eigenvalue $1/(s+a)$.

If f is sufficiently regular to be written as the inverse Mellin transform

$$f(x) = \frac{1}{2\pi i}\int_{c-i\infty}^{c+i\infty} \mathbf{M}f(s)x^s\,ds,$$

then

$$\delta_a(f)(x) = \frac{1}{2\pi i}\int_{c-i\infty}^{c+i\infty} \mathbf{M}f(s)\frac{x^s}{s+a}\,ds,$$

and we see that the factor $1/(s+a)$ improves convergence making the analysis of the integral easier. On the other hand the first property of δ_a ensures that we do not lose control over $V(T,f)$ after replacing f by $\delta_a(f)$. This method proved to be very efficient in estimating the number of changes of sign from below, cf. [19] for a general result of this type. As an application to primes in progressions we have the following estimate.

Let $\theta(q,a)$ denote the least upper bound for real parts of non-trivial zeros ρ of

$$\prod_{\chi (\mathrm{mod}\ q)} L(s,\chi) \tag{7}$$

such that

$$\sum_{\chi (\mathrm{mod}\ q)} \chi(a)m(\rho,\chi) \neq 0, \tag{8}$$

where $m(\rho,\chi)$ denotes the multiplicity of a zero of $L(s,\chi)$ at $s=\rho$. Further, let $\gamma_0(q,a) = \inf|\Im m(\rho)|$, where the infimum is taken over ρ's satisfying (8) and such that $\Re e(\rho) = \theta(q,a)$. Moreover, let $\gamma_0(q,a) = \infty$ when there are no such zeros.

Theorem 15 (cf.[19])

$$\liminf_{T\to\infty} \frac{V(T,q,a)}{\log T} \geq \frac{\gamma_0(q,a)}{\pi}.$$

The second method involves nodal lines and is applicable if the Riemann Hypothesis is true. It can be best explained in general terms. Let $F(z) = \sum_{n=1}^{\infty} a_n e^{i w_n z}$ belong to the class $\mathcal{B}$. We are interested in the solutions of the equation

$$P(x,y) = \Re e F(x+iy) = 0\ , \quad (y>0).$$

The set of $z = x+iy$ satisfying this equation is the union of curves lying in the upper half-plane. The smooth curve $L : (0,1) \to H$ of solutions, which as a subset of H is maximal with respect to inclusion, is called a *nodal line*. It is easy to see that there are no closed nodal lines, and that a nodal line cannot end at a point on H. So the end point of a nodal line is at infinity or on the real axis. If $w_1 > 0$ there are infinitely many nodal lines. To see this, observe that

$$P(x,y) = e^{-w_1 y}|a_1|\{\cos(w_1 x + \phi_1) + r(x,y)\},$$

where $\phi_1 = \arg a_1$, and $r(x,y)$ tends to 0 as $y \to \infty$ uniformly in $x \in \mathbf{R}$. Hence each half-line

$$z = x_k + iy, \qquad y > 0, \qquad x_k = ((2k+1)\pi - 2\phi_1)/2w_1, \qquad k \in \mathbf{Z},$$

is a vertical asymptote for a certain nodal line. We say that these nodal lines are *infinite* and that they *begin* at $x_k + i\infty$. Suppose now that $F(z_0) = 0$ for some $z_0 \in H$. Of course at least one nodal line has to go across z_0. If it is not an infinite nodal line we say that it is *finite* and *induced* by the zero z_0.

For a nodal line L we write

$$L^- = \inf_{z \in L} \Re e(z), \qquad L^+ = \sup_{z \in L} \Re e(z),$$

and

$$\operatorname{diam}(L) = L^+ - L^-.$$

The basic fact about nodal lines is that under some mild conditions on a_n's and w_n's we always have (cf. [12], Theorem 1.3)

$$|L^-| < \infty \qquad \text{and} \qquad \operatorname{diam}(L) \ll \log(|L^-| + 2).$$

In particular, every infinite nodal line beginning at $x_k + i\infty$ ends at a point ${x_k}^*$ on the real axis and ${x_k}^* \ll \log(|x_k| + 2)$. Nodal lines are interesting because the end point of a nodal line which lies in $\mathbf{R}$ is a sign-change of the boundary function $P(x)$. In the particular case when $F(z)$ equals $F(z,q,a)$ we get the change of sign of $P(x,q,a) = E(x,q,a) + O(xe^{-x/2})$ if $x > 0$ (cf. Theorem 3). The remainder tends quickly to zero, so one can hope that for large x we detect changes of sign of $E(x,q,a)$ itself. This is really the case, and we have the following theorem.

THEOREM 16 (cf. [12]) *Let $q \geq 1$, $q \neq 2$ and let us assume the GRH for L-functions* $(\bmod q)$. *Moreover, let*

$$\prod_{\chi (\bmod q)} L(1/2, \chi) \neq 0$$

and let us define $\kappa = \kappa(q,a)$ by the formula

$$\kappa = \lim_{Y \to 0+} \lim_{T \to \infty} \frac{1}{T} \#\{z = x + iy : F_1(z,q,a) = 0, 0 < x < T, y \geq Y\}.$$

Then

$$\liminf_{T \to \infty} \frac{V(T,q,a)}{\log T} \geq \frac{\gamma_0(q,a)}{\pi} + 2\kappa.$$

We see that the term $\gamma_0(q,a)/\pi$ is the frequency of the end points of infinite nodal lines, whereas 2κ is induced by finite nodal lines induced by zeros of $F(z,q,a)$. In the case $q=1$ numerical computations with non-trivial zeros of the Riemann zeta function give a lower bound for κ.

THEOREM 17 *We have* ($\gamma_0 = 14.13\ldots$)

$$\liminf_{T\to\infty} \frac{V(T,1,1)}{\log T} \geq \frac{\gamma_0}{\pi} + 2\cdot 10^{-25}.$$

Note that a weaker result with 10^{-248} in place of 10^{-25} is proved in [12].

7. Local changes of sign

For $0 \leq \delta < 1$ let

$$E_\delta(x,q,a) = \psi(x,q,a) - \psi(\delta x,q,a) - (1-\delta)x,$$

$v(T,q,a,\delta)$ denote the number of changes of sign of $E_\delta(x,q,a)$ in $[T,eT]$, and let

$$\overline{v}(q,\delta) = \min_{(a,q)=1} \limsup_{T\to\infty} v(T,q,a,\delta).$$

Methods applied in Section 6 give $\overline{v}(1,\delta) \geq 5$ for every $0 \leq \delta < 1$. It is interesting that this estimate can be significantly improved for δ sufficiently close to 1. In the proof we apply Theorem 4 with suitably shifted F-functions $F(z+\tau_j,q,a)$ as the F_j's. Before formulating the theorem let us introduce some subsidiary notation.

Let $A(q)$ denote the least real number $\geq 2\varphi(q)$ such that every interval of the form (x,ex) with $x > A(q)$ contains $\geq \frac{1}{\varphi(q)}\frac{x}{\log x} \geq 3$ primes $p \equiv a (\operatorname{mod} q)$ for every $(a,q)=1$. Moreover, let $\Gamma_0(q)$ be the least positive imaginary part of a pole of the functions

$$\sum_{\chi(\operatorname{mod} q)} \overline{\chi(a)}\frac{L'}{L}(s,\chi),$$

$$(a,q)=1,$$

in the half plane $\sigma > 1/2$. When all these functions are regular for $\sigma > 1/2$, in particular when the GRH holds, we put $\Gamma_0(q) = \infty$. We write also

$$g(\delta) = \left(2\varphi(q)(1-\delta)\log\frac{1}{4(1-\delta)}\right)^{-1} - 4.$$

THEOREM 18 (cf. [18]) *For* $1 - \frac{1}{8A(q)} \leq \delta < 1$ *we have*

$$\overline{v}(q,\delta) \geq \min(\Gamma_0(q)/\pi, g(\delta))$$

provided

$$\prod_{\chi(\operatorname{mod} q)} L(\sigma,\chi) \neq 0, \qquad \textit{for} \quad 0 < \sigma < 1.$$

We have $A(1) = 37/e$, as can be easily seen using results of [29], and $\Gamma_0(1) \geq 545439823.215$ (cf. [24]), and hence as a corollary we obtain the following result.

THEOREM 19 (cf. [18]) *For* $1 - 1.4 \cdot 10^{-10} < \delta < 1$ *we have*

$$\overline{v}(1, \delta) \geq 1.73 \cdot 10^8.$$

8. Open problems

We end with a list of open problems. The choice is very personal and does not pretend to be either systematic or exhaustive. It partially overlaps with much larger collections of open problems from [20] and [21].

1. **Limiting distributions.** ([8], [7], [22] [25], [28], [30]). The basic open problem here is to describe $\mathcal{F}(u, q, a)$ or the measures $\mathcal{F}_{q,a_1,\ldots,a_r}$ without appealing to the hypothetical diophantine structure of the zeta zeros such as LIH. For example: prove that $\operatorname{supp}\mathcal{F}(u, q, a) = (-\infty, +\infty)$ or, what amounts to the same thing, that

$$\limsup_{x\to\infty} \frac{\log x}{\sqrt{x}}(\pi(x, q, a) - \frac{1}{\varphi(q)}\operatorname{li}x) = \infty$$

and

$$\liminf_{x\to\infty} \frac{\log x}{\sqrt{x}}(\pi(x, q, a) - \frac{1}{\varphi(q)}\operatorname{li}x) = -\infty$$

assuming the GRH alone.

Study of distribution functions of other number theoretic functions is of interest as well. One of the most interesting open problems of this type is to prove, assuming the RH, that $x^{-1/2}M(x) = x^{-1/2}\sum_{n\leq x}\mu(n)$ has a limiting distribution.

Disprove the weak Mertens conjecture showing that

$$\liminf_{x\to\infty} x^{-1/2}M(x) = -\infty \quad \text{and} \quad \limsup_{x\to\infty} x^{-1/2}M(x) = \infty.$$

(For a disproof of the classical Mertens conjecture see A. Odlyzko and H. J. J. te Riele [26].)

2. **Chebyshev and race problems.** ([14], [16], [28]). The basic open problem connected with Chebyshev's problem is to prove Theorem 8 unconditionally, or to improve significantly the lower bounds in Theorem 9.

As concerns the race problem any completely unconditional result would be of great interest.

3. **Changes of sign.** ([9], [12], [18], [19]). The basic open problem is to prove that the differences $\pi(x, q, a) - \frac{1}{\varphi(q)}\operatorname{li}x$ and $\pi(x, q, a) - \pi(x, q, b)$, $(a, q) = (b, q) = 1$, $a \not\equiv b(\bmod q)$ change sign infinitely many times as $x \to \infty$. Next, prove that

the first change of sign of this difference occurs in the interval $(2, q^C)$, where C is an absolute constant.

Find asymptotics for $V(T, q, a)$. It would be of great interest to give non-trivial upper estimates. Such estimates are known in special cases only (cf. [9], part IV).

Give asymptotics for the number of zeros of $F_1(z, q, a)$ in the rectangle $z = x + iy$, $0 < x < T$, $0 < y < y_0$ as T tends to infinity (see Theorem 16).

Prove that $\overline{v}(q, \delta) \to \infty$ as $\delta \to 1^-$.

Acknowledgment. The author would like to thank Olivier Ramaré for a stimulating discussion.

References

[1] *C. Bays, R.H. Hudson*, On the fluctuations of Littlewood for primes of the form $4n \pm 1$, Math. Comp. **32**(1978), 281–286.
[2] *S. Bochner, B. Jessen*, Distribution functions and positive-definite functions, Ann. of Math. **11**(1934), 252–257.
[3] *H. Cramér*, Studien über die Nullstellen der Riemannschen Zetafunktion, Math. Z. **4**(1919), 104–130.
[4] *C. Deninger*, Evidence for a cohomological approach to analytic number theory, Proceedings of the First European Congress of Mathematics (ECM), Paris, France, July 6-10, 1992. Volume I: Invited lectures (Part I).
[5] *P. Erdős*, Some personal reminiscences of the mathematical work of Paul Turán, Acta Arithmetica **37**(1980), 3–8.
[6] *P. Guinand*, Fourier reciprocities and the Riemann zeta-function, Proc. London Math. Soc. **2**(51)(1949), 401–414.
[7] *R. Heath-Brown*, The distribution and moments of the error term in the Dirichlet divisor problem, Acta Arithmetica, **60**(1992), 389–415.
[8] *C. Hooley*, On the Barban-Davenport-Halbertstam theorem: VII, J. London Math. Soc. **16**(1977), 1–8.
[9] *J. Kaczorowski*, On sign-changes in the remainder-term of the prime-number formula, I–IV, Acta Arithmetica **44**(1984), 365–377; **45**(1985), 65–74; **48**(1987), 347–371; **50**(1988), 15–21.
[10] *J. Kaczorowski*, The k-functions in multiplicative number theory, I; On complex explicit formulae, Acta Arithmetica **56**(1990), 195–211.
[11] *J. Kaczorowski*, The k-functions in multiplicative number theory, IV; On a method of A.E. Ingham, Acta Arithmetica **57**(1991), 231–244.
[12] *J. Kaczorowski*, The k-functions in multiplicative number theory, V; Changes of sign of some arithmetical error terms, Acta Arithmetica **59**(1991), 37–58.
[13] *J. Kaczorowski*, The boundary values of generalized Dirichlet series and a problem of Chebyshev, Astérisque, **209**(1992), 227–235.
[14] *J. Kaczorowski*, A contribution to Shanks-Rényi race problem, Quarterly Journal of Mathematics, Oxford, **44**(1993), 451–458.
[15] *J. Kaczorowski*, On Shanks-Rényi race problem mod 5, Journal of Number Theory, **50**(1995), 106–118.
[16] *J.Kaczorowski*, Results on the distribution of primes, J. reine angew. Math. **446**(1994), 89–113.

[17] *J. Kaczorowski*, On the distribution of primes (mod 4), Analysis, **15**(1995), 159–171.

[18] *J. Kaczorowski*, On the local changes of sign of the remainder term in the prime number formula, Proceedings of the Conference on Analytic and Elementary Number Theory, Vienna 1996, 128–136.

[19] *J. Kaczorowski, J. Pintz*, Oscillatory properties of arithmetic functions, I, II, Acta Math. Hung. **48**(1988), 173–185; **49**(1987), 441–453.

[20] *S. Knapowski, P. Turán*, Comparative prime-number theory, I, Acta Mathematica Hungarica **13**(1962), 299–314.

[21] *S. Knapowski, P. Turán*, Further developments in the comparative prime-number theory, I, Acta Arithmetica, **9**(1964), 23–40.

[22] *K.-L Kueh*, The moments of infinite series, J. reine angew. Math. **385**(1988), 1–9.

[23] *E. Landau*, Über einige ältere Vermutungen und Behauptungen in der Primzahltheorie, Math. Z. **1**(1918), 1–24, 213–219.

[24] *J. van de Lune, H.J.J. te Riele, D.T. Winter*, On the zeros of the Riemann zeta function in the critical strip, IV, Math. Comp. **47**(1986), 667–681.

[25] *H.L. Montgomery*, "The Zeta Function and Prime Numbers", pp. 14–24 in *Proceedings of the Queen's Number Theory Conference, 1979* (edited by P. Ribenboim), Queen's University, Kingston (Ont.), 1980.

[26] *A. Odlyzko, H.J.J. te Riele*, Disproof of the Mertens conjecture, J. reine angew. Math. **357**(1985), 138–160.

[27] *R. Phillips, P. Sarnak*, Geodesics in homology classes, Duke Math. J. **55**(1987), 287–297.

[28] *M. Rubinstein, P. Sarnak*, Chebyshev's bias, Experimental Mathematics, **3**(1994), 173–197.

[29] *J.B. Rosser, L. Schoenfeld*, Aproximate formulas for some functions of prime numbers, Illinois J. Math. **6**(1962), 64–94.

[30] *A. Wintner*, On the distribution function of the remainder term of the prime number formula, Amer. J. Math. **63**(1941), 233–248.

Progress in Mathematics, Vol. 168, © 1998 Birkhäuser Verlag Basel/Switzerland

Low Degree Polynomial Equations: Arithmetic, Geometry and Topology

János Kollár

University of Utah, Salt Lake City UT 84112
kollar@math.utah.edu

Polynomials appear in mathematics frequently, and we all know from experience that low degree polynomials are easier to deal with than high degree ones. It is, however, not clear that there is a well-defined class of "low degree" polynomials. For many questions, polynomials behave well if their degree is low enough, but the precise bound on the degree depends on the concrete problem.

My interest is in investigating polynomials through their zero sets. That is, using sets of the form

$$\{(x_1, \dots, x_n) | f(x_1, \dots, x_n) = 0\}.$$

I intentionally refrain from specifying where the coordinates x_i lie. They could be rational, real or complex numbers, but in some cases the x_i will be polynomials in a new variable t. My focus is on the polynomial f.

Consider, for instance, a polynomial

$$f := a_0 + \sum_{i=1}^{n} a_i x_i^k, \quad \text{where} \quad a_i \in \mathbb{Z} \setminus \{0\}.$$

Specifying where the coordinates lie leads us to various branches of mathematics:

Arithmetic. Choose $x_i \in \mathbb{Q}$. The solutions of these Fermat-type equations have been much studied, some cases going back to Diophantus, but we still know very little if $n > 2$.

Topology. Choose $x_i \in \mathbb{R}$ or $x_i \in \mathbb{C}$. The set of solutions is a topological manifold, and various topological properties can be related to algebraic properties of f. For instance, the dimension and the homology can be computed in terms of n, k. (Over $\mathbb{R}$ we also need to know the signs of the a_i.)

Complex manifolds. Choose $x_i \in \mathbb{C}$. The set of solutions is a complex analytic manifold. The holomorphic function theory of this complex manifold can be understood in terms of polynomials. This is especially true in the compact versions of this problem.

Finite fields. We can also look at solutions of $f = 0$ in finite fields. Centuries ago this was done by studying $f \equiv 0 \mod p$. Recently, algebraic geometry over finite fields found many connections with coding theory, combinatorics and computer science.

I like to think of any of the zero sets as a snapshot of the polynomial f. They all show something about f. Certain snapshots reveal more than others:

Do zero sets determine a polynomial? For instance, $x_1^{2k}+\cdots+x_n^{2k}+1=0$ has no solutions in $\mathbb{Q}$, nor even in $\mathbb{R}$. Thus the zero set gives essentially no information. The situation is very different over algebraically closed fields. If $f,g \in \mathbb{C}[x_1,\dots,x_n]$, then

$$\{\mathbf{x} \in \mathbb{C}^n | f(\mathbf{x})=0\} = \{\mathbf{x} \in \mathbb{C}^n | g(\mathbf{x})=0\} \quad \Leftrightarrow \quad \begin{array}{c} f \text{ and } g \text{ have the same} \\ \text{irreducible factors.} \end{array}$$

(This is an easy special case of the Nullstellensatz of [Hilbert1893].) If we want to go further, we must study solutions of $f=0$ in any commutative ring R with a unit. This approach was first adopted by Grothendieck in [EGA60-67], though in retrospect, [Weil46] and [Rilke30,vol.2.p.175] clearly pointed in this direction. We obtain that if $f,g \in \mathbb{Z}[x_1,\dots,x_n]$ are two polynomials, then

$$\begin{array}{c} \{\mathbf{x} \in R^n | f(\mathbf{x})=0\} = \{\mathbf{x} \in R^n | g(\mathbf{x})=0\} \\ \text{(for every commutative ring } R) \end{array} \quad \Leftrightarrow \quad f = \pm g.$$

Thus studying solutions in all commutative rings determines the polynomial up to a sign. This approach is very powerful, but rather technical. Therefore I will stick to studying solutions in fields for the rest of the lecture.

It turns out that there is a collection of basic questions in arithmetic, algebraic geometry and topology all of which give the same class of "low degree" polynomials. The aim of this lecture is to explain these properties and to provide a survey of the known results.

1. Introductory Remarks

We start with the observation that in some cases the degree alone does not provide a good measure of the complexity of a polynomial equation. In order to develop the correct picture, we have to understand which polynomials behave in an atypical manner.

1.1 High degree polynomials that behave like low degree ones.
There are at least three situations when the zero set of a high degree polynomial shares some of the properties of zero sets of low degree polynomials:

1.1.1 Reducible equations. If $f=gh$, then the set $(f=0)$ is the union of the sets $(g=0)$ and $(h=0)$. Thus we can restrict ourselves to the case when f is irreducible.

1.1.2 Low degree in certain variables. Let us consider an extreme case, when f has degree 1 in the variable x_n. Then f can be written as

$$f = f_1(x_1,\dots,x_{n-1}) + x_n f_2(x_1,\dots,x_{n-1}).$$

The substitution $x_n = -f_1/f_2$ shows that the set $(f = 0)$ behaves like the vector space of the first $(n-1)$ variables $\{(x_1, \ldots, x_{n-1})\}$. This is completely true if f is linear, but in general the correspondence breaks down if $f_2 = 0$. The latter equation involves one fewer variable, and therefore it is considered easier. Roughly speaking, f should be viewed as complicated as a linear equation. In general, if f has low degree in certain variables then it behaves like a low degree equation.

1.1.3 Very singular equations. Consider for instance the equation $x_1^d - x_2^{d-1} = 0$. Its degree in both variables is high. Nonetheless, the substitution

$$x_1 = t^{d-1}, \ x_2 = t^d$$

shows that solutions of $x_1^d - x_2^{d-1} = 0$ are parametrized by the values of the variable t. The same happens for any polynomial $f(x_1, x_2)$ of degree d all of whose partials up to order $d-2$ vanish at a certain point. In general, a high degree equation f behaves as a low degree equation if many of the partial derivatives of f simultaneously vanish at many points.

While all of these cases do occur, there are relatively few polynomials that behave in this way. For instance, all polynomials $f(x_1, x_2)$ of degree $\leq d$ form a vector space V_d of dimension $\binom{d+2}{2}$. The set of polynomials which are exceptional for any of the above 3 reasons is a subset of codimension $d-1$ for $d \geq 2$.

This remark shows that for most polynomials the degree is a good measurement of complexity. In order to run computer experiments, it is desirable to have a class of polynomials with very few nonzero coefficients which are nonetheless "general". A good set of examples to keep in mind is the following.

1.2 Test Examples. The equations $\sum_i c_i x_i^d = c_0$ have been much studied. Unfortunately, they are sometimes too special. It seems that the inhomogeneous version is much more indicative of the general case. Fix natural numbers $d_i : i = 1, \ldots, n$ and $c_0, \ldots, c_n$ such that $\prod_i c_i \neq 0$. Then

$$\text{(1.2.1)} \qquad \sum_{i=1}^{n} c_i x_i^{d_i} = c_0 \quad \text{has "low degree" iff} \quad \sum_{i=1}^{n} \frac{1}{d_i} \geq 1.$$

We see in (5.5) that the above condition does coorespond to the eventual definition (4.1). Moreover, I claim that the behaviour of these examples correctly predicts the broad features of the theory. You have to trust me that this purely experimental assertion is valid.

As a first example, let us see what a simple minded constant count gives about solutions of the equations (1.2.1) over $\mathbb{Q}$.

1.3 Heuristic claim. Fix natural numbers $d_i : i = 1, \ldots n$ and rational numbers $c_i : i = 0, \ldots n$. I claim that usually

$$\text{(1.3.1)} \qquad \sum_{i=1}^{n} c_i x_i^{d_i} = c_0 \quad \text{has many solutions in } \mathbb{Q} \text{ iff} \quad \sum_{i=1}^{n} \frac{1}{d_i} \geq 1.$$

Unfortunately there are large classes of equations where this is false. For instance, $\sum x_i^2 = -1$ has no solutions in $\mathbb{Q}$, not even in $\mathbb{R}$. Looking at $x_1^2 - x_2^2$ modulo 4, we see that $x_1^2 - x_2^2 = 2$ has no rational solutions. There are several approaches to correct these problems; we shall encounter two of them later. For the moment I will ignore these counterexamples, and give a proof of (1.3.1).

It is easier to look for integral solutions, so we homogenize the equation in the following (somewhat unusual) way. Set d_0 to be the least common multiple of $d_1, \dots, d_n$ and let $d_0 = d_i b_i$. Look at the equation

$$\sum_{i=1}^{n} c_i y_i^{d_i} = c_0 y_0^{d_0}. \tag{1.3.2}$$

There is a correspondence between solutions of (1.2.1) and of (1.3.2) given by

$$(x_1, \dots, x_n) \mapsto (1, x_1, \dots, x_n) \quad \text{and} \quad (y_0, y_1, \dots, y_n) \mapsto (y_1/y_0^{b_1}, \dots, y_n/y_0^{b_n}).$$

This shows that finding all rational solutions of (1.2.1) is equivalent to finding all integral solutions of (1.3.2).

Set $f = -c_0 y_0^{d_0} + \sum_{i=1}^{n} c_i y_i^{d_i}$. There is a constant C, depending on f, such that

$$|f(y_0, \dots, y_n)| \le C \cdot (\max_i |y_i|^{d_i}). \tag{1.3.3}$$

Fix a large N and let the y_i run through the set of integers in $[-N^{1/d_i}, N^{1/d_i}]$. We get

$$\text{const} \cdot N^{\sum_{i=0}^{n}(1/d_i)} \quad \text{values of } f \text{ in the interval} \quad [-C \cdot N, C \cdot N].$$

If these values are uniformly distributed, we obtain the asymptotic expression

$$\#\{\sum_i c_i y_i^{d_i} = c_0 y_0^{d_0}, |y_i| \le N^{1/d_i}\} \sim \text{const} \cdot N^{-1+\sum_{i=0}^{n}(1/d_i)} \quad \text{as } N \to \infty.$$

If $\sum_{i=1}^{n}(1/d_i) \ge 1$, then $\sum_{i=0}^{n}(1/d_i) > 1$ and the number of solutions grows as a power of N. If $\sum_{i=1}^{n}(1/d_i) < 1$ then $\sum_{i=0}^{n}(1/d_i) \le 1$ because of the special choice of d_0, thus there should be few solutions. □

For which other polynomials f does this counting method work? The main part is the estimate (1.3.3). This works if f is weighted homogeneous of degree 1 with weights $1/d_i$. That is, if we declare $\deg x_i = 1/d_i$ then $\deg f \le 1$.

There are some examples where the above simple minded counting method does work, for instance, for equations of the form

$$f(x_1, \dots, x_n) - f(y_1, \dots, y_n) = 0.$$

The above argument gives a lower bound

$$\#\{f(x_1, \dots, x_n) = f(y_1, \dots, y_n), |x_i|, |y_i| \le N\} \ge \text{const} \cdot N^{2n-d}.$$

This is interesting only if $d < n$ since the trivial solutions $x_i = y_i$ always give a lower bound const $\cdot N^n$.

In the rest of the lecture I aim to explain the various properties that lead to this class of equations, starting with the 2-variable case in section 2. This is called the theory of algebraic curves. Most of the theory was well-established in the 19th century, with the exception of the arithmetic aspects.

Section 3 is devoted to the 3-variable case, which corresponds to the theory of algebraic surfaces. The geometric aspects have been established around the turn of the century, many of the topological results are recent and most of the arithmetical questions are open.

Much less is known in higher dimensions. The open questions involve deep problems in algebraic geometry, number theory and differential topology. I am confident that these problems constitute a very interesting direction of research for a long time to come.

2. Two Variable Polynomials = Algebraic Curves

Let us consider a 2 variable polynomial $f(x,y) = \sum a_{ij}x^iy^j$ of degree d. Let C_{aff} denote its zeros, that is,

$$C_{aff} := \{(x,y) | f(x,y) = 0\}.$$

(The subscript *aff* refers to the fact that we are in affine 2-space $\mathbb{A}^2$.) This is not a set since I have not specified where the coordinates x, y are. If the coefficients a_{ij} are in a field F, then for any larger field $E \supset F$ we can look at solutions of $f = 0$ in E. The resulting set is

$$C_{aff}(E) := \{(x,y) \in E^2 | f(x,y) = 0\} \subset E^2.$$

A common case is when $a_{ij} \in \mathbb{Q}$, and for the larger field E we choose $\mathbb{Q}, \mathbb{R}$ or $\mathbb{C}$.

$C_{aff}(\mathbb{Q})$ is just a set of points, but $C_{aff}(\mathbb{R}) \subset \mathbb{R}^2$ naturally appears as a curve (that is, a 1-dimensional topological space). $C_{aff}(\mathbb{C}) \subset \mathbb{C}^2$ is a Riemann surface: a complex manifold locally like $\mathbb{C}$.

In studying the manifolds $C_{aff}(\mathbb{R})$ or $C_{aff}(\mathbb{C})$ it is frequently inconvenient that they are not compact. The usual way to deal with this problem is to introduce the projective plane $\mathbb{P}^2$ with homogeneous coordinates $(x_0 : x_1 : x_2)$. Its relationship to the old affine coordinates is $x = x_1/x_0, y = x_2/x_0$. If the coordinates x_i are in a field E, we obtain the corresponding projective plane $\mathrm{E}\mathbb{P}^2$. The most frequently used ones are $\mathbb{QP}^2, \mathbb{RP}^2$ and $\mathbb{CP}^2$.

The homogenization of f is given by

$$\bar{f}(x_0, x_1, x_2) := x_0^d f(x_1/x_0, x_2/x_0).$$

The corresponding zero set

$$C(E) := \{(x_0 : x_1 : x_2) \in \mathrm{E}\mathbb{P}^2 | \bar{f}(x_0, x_1, x_2) = 0\} \subset \mathrm{E}\mathbb{P}^2$$

turns out to be more convenient for most purposes.

Based on the real picture, algebraic geometers say that C is an *algebraic curve*. Thus we prefer to call $\mathbb{C}$ the complex line (the complex plane is of course $\mathbb{C}^2$). This leads to occasional confusion, but this is not the time to change 150 year-old terminology.

In what follows I collect certain properties of algebraic curves defined by equations of degree at most 2. In all cases I would like the properties to hold only for curves defined by equations of degree at most two (assuming the genericity conditions of (1.1)).

All of the characterizations listed here are standard results of the theory of algebraic curves and Riemann surfaces. One of the most accessible introductions to algebraic geometry is [Shafarevich94] (or any of the other editions). For algebraic curves see [Fulton69]. The analytic theory of Riemann surfaces is treated in [Siegel69; Gunning76]. For the arithmetic aspects I found [Serre73; Silverman86] especially useful.

Characterizations of "low degree " curves

I start with the algebraic geometry condition, not because it is the most obvious for curves, but because this provides the neatest definition in higher dimensions.

2.1 ALGEBRAIC GEOMETRY. *There is a one-to-one map given by rational functions $g : \mathbb{CP}^1 \to C(\mathbb{C})$.*

In this case C is called a *rational curve.*

Let $(s : t)$ be the homogeneous coordinates on $\mathbb{CP}^1$. If $f = a_0x_0 + a_1x_1 + a_2x_2$ is linear and $a_2 \neq 0$, we can choose

$$g : (s : t) \mapsto (a_2s : a_2t : -(a_0s + a_1t)) .$$

For $\deg f = 2$ assume for simplicity that $f = a_0x_0^2 + a_1x_1^2 + a_2x_2^2$. (This can always be achieved after a linear change of coordinates.) We can take

$$g : (s : t) \mapsto \Big(a_1s^2 - a_0t^2 : -2a_0st : \sqrt{-a_0/a_2}(a_1s^2 + a_0t^2)\Big) .$$

(In case you wonder where this came from, let $h : C \to L$ be the projection of C from the point $P = (\sqrt{a_2} : 0 : \sqrt{-a_0}) \in C$ to the $(x_2 = 0)$ line (stereographic projection). g is the inverse of h.)

The fact that no such g exists for higher degree equations is harder.

2.2 TOPOLOGY. *$C(\mathbb{C})$ is homeomorphic to the sphere S^2.*

The maps g from (2.1) also provide a homeomorphism; the hard part is again to see that this cannot be done for higher degree equations. The precise result is that if C is defined by a degree d equation then $C(\mathbb{C})$ is homeomorphic to a sphere with $\frac{1}{2}(d-1)(d-2)$ handles.

2.3 HARD ARITHMETIC. *$C(\mathbb{Q})$ is "large".*

For this to make sense, we should start with a curve

$$C = (\bar{f}(x_0, x_1, x_2) = 0) \subset \mathbb{P}^2,$$

where $\bar{f}$ has rational coefficients.

Unfortunately it is not easy to pin down what "large" exactly means. First of all, if $n \geq 4$ then $C(\mathbb{Q})$ is finite by [Faltings83]. Unfortunately, $C(\mathbb{Q})$ is often infinite for $n = 3$ and frequently empty for $n = 2$.

To get the right answer, we have to develop a good measure of the size of a solution. This is most conveniently done in projective coordinates.

Any point $P \in \mathbb{QP}^2$ can be represented as a triple $P = (x_0 : x_1 : x_2)$ where $x_0, x_1, x_2 \in \mathbb{Z}$ are relatively prime. This representation is unique up to sign, thus $H(P) := \max\{|x_0|, |x_1|, |x_2|\}$ is well-defined. It is called the *height* of P. One defines the counting function

$$N(C,H) := \#\{P = (x_0 : x_1 : x_2) \in \mathbb{QP}^2 | \bar{f}(x_0, x_1, x_2) = 0 \text{ and } H(P) \leq H\}.$$

Roughly speaking, we look for rational solutions of $f(x,y) = 0$ where the numerators and denominators are bounded.

This nearly gives the right answer. If $n = 2$ then $C(\mathbb{Q})$ is either empty or $N(C,H)$ grows like const $\cdot H$; if $n = 3$ then $N(C,H)$ grows slower than any power of H [Néron65].

In order to deal with the case when $C(\mathbb{Q})$ is empty, we have to count solutions in various algebraic number fields. It is not hard to generalize the notion of height to the case when the coordinates of P are in an algebraic number field $E \supset \mathbb{Q}$ (see [Silverman86,VIII.5] for a short and clear summary). We obtain a similar counting function $N_E(C,H)$. This finally gives the correct generalization:

2.3.1 Theorem. C is a rational curve iff $N_E(C,H)$ grows polynomially with H for a suitable number field E.

2.4 COMPLEX MANIFOLDS. *$C(\mathbb{C})$ has genus zero.*

Global holomorphic differential forms on a compact Riemann surface have been much studied, starting with the works of Euler, Abel and Riemann. On a Riemann surface we have only 1-forms, these are locally given as $f(z)dz$ where z is a local coordinate and $f(z)$ is holomorphic. Such forms are automatically closed, thus the integral

$$\int_\gamma f(z)dz \quad \text{over a closed loop} \quad \gamma \subset C(\mathbb{C})$$

depends only on the homology class $[\gamma] \in H_1(C(\mathbb{C}), \mathbb{Z})$. Since the fundamental studies of Riemann, these give the basic approach to finer understanding of Riemann surfaces.

By definition, the *genus* is the dimension of the vector space of global holomorphic differential forms. If there are no such forms, the above integrals give no information. Fortunately, this happens precisely when other descriptions are very simple.

2.5 EASY ARITHMETIC. *There are many solutions over function fields.*

Here we look at the behaviour of the sets $C(F)$ where $F = \mathbb{C}(t)$ is the field of rational functions in one variable. Of course $f = \sum a_{ij}(t)x^iy^j$ and the coefficients

$a_{ij}(t)$ themselves are rational functions. The field $\mathbb{C}(t)$ shares many properties of $\mathbb{Q}$, but the results are easier to state and the proofs are much simpler. (The difference between $\mathbb{Q}$ and $\mathbb{C}(t)$ becomes apparent when studying their Galois cohomology.)

The advantage of $\mathbb{C}(t)$ is that there are two ways of looking at solutions over $\mathbb{C}(t)$.

(2.5.1.1) The algebraic way. Just handle everything as quotients of polynomials in $\mathbb{C}[t]$.

(2.5.1.2) The geometric way. An equation $f(x,y)=0$ with coefficients in $\mathbb{C}(t)$ can be viewed as an equation $\tilde{f}(x,y,t)=0$ with coefficients in $\mathbb{C}$. This defines an algebraic surface $S \subset \mathbb{C}^3$ and we have a distinguished coordinate projection to the t-axis $p : S \to \mathbb{C}_t$.

A solution $(x(t), y(t))$ of $f(x,y)=0$ can be identified with a map

$$h : \mathbb{C}_t \to S \quad \text{given by} \quad t \mapsto (x(t), y(t), t).$$

h is a section of $p : S \to \mathbb{C}_t$ and every (rational) section arises as above.

The first indication that we can expect nicer results is the following theorem, which can be proved by a straightforward generalization of the counting argument (1.3). The first proof is in [Noether1871]. Later algebraic proofs, more suited to generalizations, are in [Baker22,vol.VI.p.147] and [Tsen36].

2.5.2 Theorem. If $\deg f \le 2$ then $f=0$ has a solution in $\mathbb{C}(t)$.

We may also want to know whether there are many solutions. A natural approach is to look for solutions $(x(t), y(t))$ where certain values $(x(t_k), y(t_k))$ are specified in advance. This is possible only if the points $(x(t_k), y(t_k), t_k)$ lie on the surface S, that is, if $\sum a_{ij}(t_k)x(t_k)^i y(t_k)^j = 0$. In this case we say that the pair $(x(t_k), y(t_k))$ is a solution of $f(x,y)=0$ at t_k.

As an easy exercise in the theory of algebraic surfaces we get a very strong characterization:

2.5.3 Theorem. There is a finite set $B \subset \mathbb{C}$ such that if $t_1, \dots, t_s \in \mathbb{C} \setminus B$ are arbitrary points and (x_k, y_k) any solution of f at t_k then there is a solution $(x(t), y(t))$ of $f=0$ such that $(x(t_k), y(t_k)) = (x_k, y_k)$ for $k = 1, \dots, s$.

One can reformulate the theorem to specify not just the value of $(x(t), y(t))$ at t_k but also the beginning of its Taylor expansion. With a little more care, the exceptional set B can also be eliminated (5.1).

2.5.4 Remark. More generally all of this works if $\mathbb{C}(t)$ is replaced with any finite degree extension of $\mathbb{C}(t)$. These are exactly the fields of meromorphic functions on compact Riemann surfaces.

2.6 Low degree equations. *C can be described by an equation of degree at most 2.*

This is of course our starting point, but in higher dimensions this becomes a rather nontrivial question.

It is worthwhile to note the following arithmetic implication:

2.6.1 Proposition. If $\deg f \leq 2$, then $f(x,y) = 0$ always has a solution over a degree 2 field extension.

In order to see this, pick a, b, c and consider $f(x,y) = ax + by + c = 0$. Eliminating x or y we are left with a quadratic equation in one variable.

Final remarks about curves

It should be made clear that the above properties by no means exhaust the known characterizations of curves of degree 1 and 2. Some of the others do not seem to have higher dimensional analogs. I just give a few examples:

2.7 BAD CHARACTERIZATIONS.

2.7.1 Simply connectedness.
$\pi_1(C(\mathbb{C})) = \{1\}$ iff $\deg f \leq 2$. It turns out that any smooth hypersurface $X = (f(x_0, \ldots, x_n) = 0) \subset \mathbb{CP}^n$ is simply connected for $n \geq 3$ [Lefschetz24].

2.7.2 Unique factorization in the coordinate ring.
The ring $\mathbb{C}[x,y]/f(x,y)$ is a unique factorization domain iff $\deg f \leq 2$. If $f(x_0, \ldots, x_n)$ defines a smooth hypersurface then $\mathbb{C}[x_1, \ldots, x_n]/f(x_1, \ldots, x_n)$ is a UFD for $n \geq 4$ [Grothendieck68].

2.7.3 Homogeneous spaces.
If $\deg f = 1$ then C is homogeneous under the group $SL(2)$. If $\deg f = 2$ then C is homogeneous under the group $O(\bar{f})$, the 3-variable orthogonal group of $\bar{f}$. In higher dimensions the varieties which are homogeneous under the action of a linear algebraic group give rather special examples of the class that we want.

2.7.4 Number of moduli.
Any two lines in $\mathbb{P}^2$ are equivalent under a change of coordinates, and any two smooth conics in $\mathbb{CP}^2$ are also equivalent. This fails for $\deg f \geq 3$. In all dimensions this property characterizes hypersurfaces of degree at most 2, so does not hold for most of the examples in (1.3). (We need a nondiagonal perturbation to see this.)

The above lists suggest several further possible approaches to low degree polynomials. Below I list some that do not work, even for curves.

2.8 NONCHARACTERIZATIONS.

2.8.1 Topology over $\mathbb{R}$.
One could study curves such that $C(\mathbb{R})$ is homeomorphic to S^1. If $\deg f \leq 2$ and $C(\mathbb{R})$ is not empty, this is always the case. Unfortunately, there are many other curves with this property. For instance, $(x^{2d} + y^{2d} = 1) \subset \mathbb{RP}^2$ is homeomorphic to S^1.

2.8.2 Solutions modulo p.
If f has integral coefficients, we can ask about solvability modulo p (or modulo any number).

The number of solutions in finite fields are described by the Weil conjectures (see [Freitag-Kiehl88] for a thorough treatment) and the degree of f does not affect the asymptotic behaviour much. (Though the genus can be computed if we know the exact number of solutions modulo p for many values of p.) Low degree equations have solutions in any finite field [Chevalley35], but the same holds for many other cases.

2.8.3 Solutions in p-adic fields.

An equation of degree at most two is not always solvable in p-adic fields. For equations in many variables, solvability in p-adic fields is an interesting question. The rough picture (which is not quite correct) is that if $f(x_0, \dots, x_n)$ has degree $d \le \sqrt{n}$ then f has a solution in any p-adic field and this fails for larger degree. Thus the answer does not correspond to our class. See [Greenberg69] for a discussion of these topics.

2.9 OTHER APPROACHES.

2.9.1 Holomorphic maps $h : \mathbb{C} \to C(\mathbb{C})$.

If there is a map $\mathbb{CP}^1 \to C(\mathbb{C})$, then we get plenty of holomorphic maps $\mathbb{C} \to C(\mathbb{C})$. If $\deg f \ge 4$ then there are no nonconstant holomorphic maps from $\mathbb{C}$ to $C(\mathbb{C})$. Unfortunately if $\deg f = 3$, then there are nonconstant holomorphic maps $\mathbb{C} \to C(\mathbb{C})$. Thus this property characterizes a slightly different class of curves. In higher dimensions the two classes differ substantially. See [Lang86; Vojta91] for various properties of this class.

Vojta pointed out to me that one can consider holomorphic maps $h : \mathbb{C} \to C(\mathbb{C})$ whose Nevanlinna characteristic function grows slowly, to get a characterization of rational curves in the context of the holomorphic theory. The resulting holomorphic maps are rational, so at the end this is equivalent to (2.1).

2.9.2 The Hasse principle.

One way to overcome the difficulties observed in (1.3) is to refine (1.3.1) as follows:

Assume that $f(x_1, \dots, x_n) = 0$ has a (nontrivial) solution modulo m for every m and also over $\mathbb{R}$. Does this imply that f has a solution in $\mathbb{Z}$? (Solvability modulo m for every m is equivalent to solvability in every p-adic field.)

If the answer is yes, one says that the *Hasse principle* holds for f. By the Hasse–Minkowski theorem, this is the case if f is homogeneous of degree 2.

The question for higher dimensions is very difficult. It is still not clear if the Hasse principle is connected with our class in higher dimensions or with some smaller class of varieties. See [Colliot-Thélène86,92] for surveys of this direction.

3. Algebraic Surfaces

The next step is to study zero sets of polynomials in three variables

$$S := \{(x, y, z) | f(x, y, z) = 0\} \subset \mathbb{A}^3.$$

It was noticed in the 19th century that the true measure of complexity of a system of polynomial equations is the dimension of the set of solutions over $\mathbb{C}$. Thus if we have 2 equations in 4 variables, the resulting zero set

$$(f_1(x, y, z, u) = f_2(x, y, z, u) = 0) \subset \mathbb{A}^4$$

behaves to a large extent like surfaces in 3-space. Any surface in 4-space can be made into a surface in 3-space by a generic projection. If we generically project a curve in n-space to the plane, the image has only transversal self-intersections. By contrast, if we project a surface to 3-space, the image has complicated self-intersections. According to current view, it is very hard to study a surface this way. (Earlier geometers, being ignorant of this fact, proved rather deep theorems using projections to 3-space.) Thus we are pretty much forced to look at the general case of varieties:

Algebraic varieties. Given polynomials $f_1, \dots, f_k$ in n variables, their common zero set

$$X_{\mathit{aff}} := \{(x_1, \dots, x_n) | f_1(\mathbf{x}) = \dots = f_k(\mathbf{x}) = 0\} \subset \mathbb{A}^n$$

is called an *affine* algebraic variety. Using homogeneous equations $\bar{f}_i$ we obtain *projective* varieties

$$X := \{(x_0, \dots, x_n) | \bar{f}_1(\mathbf{x}) = \dots = \bar{f}_k(\mathbf{x}) = 0\} \subset \mathbb{P}^n.$$

If the coefficients of the f_i are in a field F, we say that X is *defined over* F. X is also defined over every bigger field $E \supset F$, hence $X(E) \subset E\mathbb{P}^n$, the set of solutions in E, makes sense.

These sets can be very complicated. In order to streamline our discussions, I make two simplifying assumptions:

All varieties will be irreducible and smooth.

Over the complex numbers this means that $X(\mathbb{C})$ is a connected manifold. These assumptions are satisfied if the coefficients of the f_i are chosen at random. The general case can be reduced to this one in various ways.

The dimension of X can be defined in an abstract way. Over $\mathbb{C}$ it is one half of the topological dimension of $X(\mathbb{C})$. This gives the expected value; for instance if $X \subset \mathbb{C}\mathbb{P}^n$ is defined by a single equation then it has dimension $n-1$.

In order to decide which varieties are considered equivalent, we look at the example of the stereographic projection from (2.1)

Examples of birational maps.

(i) Let $S = (x^2 + y^2 + z^2 = 1) \subset \mathbb{R}^3$. Project S from the point $(0,0,1)$ to the (x,y)-plane P. This provides a one-to-one map

$$\pi : S \setminus (0,0,1) \xrightarrow{\cong} P \cong \mathbb{R}^2.$$

This looks good, until we notice that projectively there are problems. The plane is usually compactified as $\mathbb{R}\mathbb{P}^2$, which is not even homeomorphic to the sphere S.

(ii) $H = (x^2 - y^2 + z^2 = 1) \subset \mathbb{R}^3$ is a hyperboloid. Project H from the point $(0,0,1)$ to the (x,y)-plane P. This provides a one-to-one map

$$\pi : H \setminus \{(x,y,z) | z = 1\} \xrightarrow{\cong} P \setminus \{(x,y) | x^2 - y^2 + 1 = 0\},$$

and π and π^{-1} can not be extended to the removed sets in any reasonable way. Despite this, π is clearly very useful in understanding H. For many problems we can use π to study $H \setminus \{(x,y,z)|z=1\}$. The missing set $\{(x,y,z)|z=1\}$ is isomorphic to the plane curve $\{(x,y)|x^2-y^2=0\}$, which is a pair of lines.

(iii) For $a,b,c \in \mathbb{Q}$, $H_{abc}=(ax^2+by^2+cz^2=1) \subset \mathbb{A}^3$ is a quadric surface. As above, we would like to find a projection of H to a plane. This can be done over some field, for instance we can project from $(0,0,1/\sqrt{c})$. The formulas for π and π^{-1} involve $\sqrt{c}$, hence they are of little use if we intend to study $H(\mathbb{Q})$.

If $a,b,c<0$, then $H_{abc}(\mathbb{R})$ is empty, thus there is no map $g:\mathbb{R}^2 \to H(\mathbb{R})$.

Definition of birational maps. Let $X \subset \mathbb{A}^n$ and $Y \subset \mathbb{A}^m$ be affine varieties. Let x_i (resp. y_j) be coordinates on $\mathbb{A}^n$ (resp. $\mathbb{A}^m$).

A *rational map* $g:\mathbb{A}^n \dashrightarrow \mathbb{A}^m$ is given as

$$g:(x_1,\dots,x_n)\mapsto(g_1(\mathbf{x}),\dots,g_m(\mathbf{x})),$$

where the g_i are rational functions in the variables $x_1,\dots,x_n$. Notice that such maps need not be everywhere defined. If the coefficients of the g_i are in a field F, we say that g is defined over F.

If $g(X)\subset Y$, then g restricts to a map $g:X \dashrightarrow Y$.

We say that $g:X \dashrightarrow Y$ is *birational* if there are subvarieties $A \subsetneq X$ and $B \subsetneq Y$ such that g restricts to a one-to-one map $g:X\setminus A \to Y\setminus B$.

Informally speaking, X and Y are birational if they are isomorphic up to lower dimensional varieties.

Rational maps of projective varieties can be defined similarly. We can mimic the above definitions with projective coordinates (in which case the g_i have to be homogeneous).

A general introduction to algebraic geometry can be found in [Shafarevich94; Hartshorne77]. The analytic theory can be found in [Wells73; Griffiths-Harris78]. The books [Beauville78; BPV84] are devoted to algebraic surfaces. The topological aspects are discussed in [Donaldson-Kronheimer90; Friedman-Morgan94].

Characterizations of "low degree" surfaces

Let $S \subset \mathbb{P}^n$ be a projective surface defined by homogeneous equations $f_1=\cdots=f_k=0$. For simplicity we always assume that S is smooth and connected.

For surfaces, algebraic geometry provides the basic definition. Our task is to see to what extent the other variants (2.2–6) can be generalized to give an equivalent condition.

3.1 Algebraic geometry. *S is rational over $\mathbb{C}$.*

The precise definition of rational is the following:

3.1.1 Definition. Let S be a smooth projective surface defined over $\mathbb{C}$. We say that S is *rational* if there is a birational map $g:\mathbb{CP}^2 \dashrightarrow S(\mathbb{C})$.

If S is defined over a subfield $F\subset\mathbb{C}$, we say that S is *rational over F* if there is a birational map $g:\mathbb{P}^2 \dashrightarrow S$ defined over F.

Historically this definition appeared as a rather hard theorem. There are three classes of surfaces which are very similar to rational surfaces, but it is not obvious that they are indeed rational. These three classes are:

(3.1.2.1) cubic surfaces $S_3 \subset \mathbb{P}^3$;
(3.1.2.2) surfaces S which admit a map $f : S \to \mathbb{P}^1$ whose general fiber is $\mathbb{P}^1$;
(3.1.2.3) surfaces which are images of maps $h : \mathbb{P}^2 \dashrightarrow \mathbb{P}^n$.

Cubic surfaces were shown to be rational by [Clebsch1866]. The second case was settled in [Noether1871] and the third class was treated in [Castelnuovo1894].

3.2 TOPOLOGY. *Homeomorphism versus diffeomorphism.*

Understanding algebraic surfaces in terms of their topology turned out to be extremely difficult.

Some classical questions can be interpreted in topological terms, but this may have been first explicitly done in [Hirzebruch54]. One of the simplest problems is to give a topological characterization of the complex projective plane. This was finally done in [Yau77]:

3.2.1 Theorem. Assume that $S(\mathbb{C})$ is homeomorphic to $\mathbb{CP}^2$. Then S is also isomorphic to $\mathbb{CP}^2$.

The difficulties of this very special case discouraged attempts to move further in this direction.

A fundamental problem in general is that a birational map $g : S_1 \dashrightarrow S_2$ does not induce a homeomorphism. This question can be understood in terms of the connected sum operation as follows:

3.2.2 Proposition. If $S_1(\mathbb{C})$ and $S_2(\mathbb{C})$ are birational then there are natural numbers r, s such that

$$S_1(\mathbb{C})\#(\overline{\mathbb{CP}}^2)^r \quad \text{is diffeomorphic to} \quad S_2(\mathbb{C})\#(\overline{\mathbb{CP}}^2)^s,$$

where $\#$ denotes connected sum and $\overline{\mathbb{CP}}^2$ is $\mathbb{CP}^2$ with reversed orientation. We can assume in addition that $\min\{r, s\} \leq 1$ and even $\min\{r, s\} = 0$ with a few exceptions.

In particular we obtain:

3.2.3 Proposition. If S is rational then $S(\mathbb{C})$ is diffeomorphic to

$$\mathbb{CP}^2\#(\overline{\mathbb{CP}}^2)^r \quad \text{or to} \quad \mathbb{CP}^1 \times \mathbb{CP}^1.$$

(It is not hard to see that $(\mathbb{CP}^1 \times \mathbb{CP}^1)\#\overline{\mathbb{CP}}^2$ is diffeomorphic to $\mathbb{CP}^2\#(\overline{\mathbb{CP}}^2)^2$, that is why we have only one series in (3.2.3).)

By analogy with (2.2) one can ask if the converse is also true. It was noticed some time ago that the answer is no if we use homeomorphism instead of diffeomorphism [Dolgachev66]. As Donaldson theory started to discover the difference between diffeomorphism and homeomorphism in real dimension 4, the hope emerged that the converse of (3.2.3) holds for diffeomorphisms.

This has been one of the motivating questions of the differential topology of algebraic surfaces. After many contributions, the final step was accomplished by

[Pidstrigach95; Friedman-Qin95]. With the new methods of Seiberg-Witten theory, the proof is actually quite short [Okonek-Teleman95]:

3.2.4 Theorem. Let S be a smooth, projective algebraic surface over $\mathbb{C}$. Then

$$S \text{ is rational} \quad \Leftrightarrow \quad \begin{array}{c} S(\mathbb{C}) \text{ is diffeomorphic to} \\ \mathbb{CP}^2 \#(\overline{\mathbb{CP}}^2)^r \text{ or } \mathbb{CP}^1 \times \mathbb{CP}^1. \end{array}$$

3.3 HARD ARITHMETIC. *$S(\mathbb{Q})$ is "large".*

Let S be a surface defined over a number field F, most frequently $F = \mathbb{Q}$. As for curves, for any number field $E \supset F$ we define the counting function

$$N_E(S, H) := \#\{P \in S(E) \subset E\mathbb{P}^n | H(P) \le H\}.$$

We hope that S is rational over $\mathbb{C}$ iff $N_E(S, H)$ grows as a power of H for some E.

Unfortunately this is not quite correct, and there are two related problems.
(3.3.1.1) Look at the surface $T := (x^d + y^d = z^d + u^d) \subset \mathbb{P}^3$. One can check that $T(\mathbb{C})$ is smooth. T has high degree, but $N_{\mathbb{Q}}(T, H)$ grows quadratically with H. A closer inspection reveals that this growth is caused by (finitely many) lines on the surface (for instance $(x - z = y - u = 0)$) which contain many rational points. If we remove these lines, there are very few rational solutions left.
(3.3.1.2) The growth rate of $N_E(T, H)$ is not a birational invariant of T. Here again the problems are caused by finitely many rational curves on T.

The examples suggest that we should refine the hope as follows:

3.3.2 Conjecture. [FMT89; Batyrev-Manin90] If T is rational (over $\mathbb{C}$) then there is a number field E, $0 < \beta \in \mathbb{Q}$ and $r \in \mathbb{N}$ such that

$$N_E(T \setminus A, H) \quad \text{is asymptotic to} \quad \text{const} \cdot H^{\beta} (\log H)^r$$

for every sufficiently large subvariety $A \subsetneq T$.

It is furthermore conjectured that β and r are determined by the geometry of T in a simple way [Batyrev-Manin90]. (For higher dimensions these refinements are problematic, see (4.3).)

A weaker form of (3.3.2) is easy:

3.3.3 Theorem. If T is rational (over $\mathbb{C}$) then there is a number field E and $\epsilon > 0$ such that $N_E(T \setminus A, H) > \text{const} \cdot H^{\epsilon}$ for every subvariety $A \subsetneq T$.

The converse of (3.3.2–3) is not quite true. The conceptually correct formulation will be given in (4.3.2–3). For surfaces the following form suffices (cf. [FMT89]).

3.3.4 Conjecture. Assume that (over $\mathbb{C}$) T is not rational and not birational to $C \times \mathbb{P}^1$ where C is an elliptic curve. Then for every number field E and $0 < \epsilon$, there is a subvariety $A \subsetneq T$ such that

$$N_E(T \setminus A, H) < \text{const} \cdot H^{\epsilon}.$$

Very little is known in this direction since we have no general methods to show that nonrational surfaces have only few rational points.

3.4 COMPLEX MANIFOLDS. *Global holomorphic differential forms.*

Global holomorphic differential forms on a complex manifold have been much studied. On a surface we can have 1-forms and 2-forms. These are locally given as

$$f_1 dz_1 + f_2 dz_2, \quad \text{respectively} \quad f dz_1 \wedge dz_2,$$

where z_1, z_2 is a local coordinate system and the f_i are holomorphic. In this context, they were first considered in [Clebsch1868] and systematically studied in [Picard-Simart1897].

As in the curve case, the integrals of these forms over 1- and 2-cycles give basic invariants of a variety [Hodge41]. This approach was developed into a very powerful method of studying complex manifolds, called Hodge theory. However, if there are no global holomorphic differential forms on a surface, then Hodge theory does not say anything.

It is easy to see that if S is rational then there are no global holomorphic differential forms on $S(\mathbb{C})$. Conversely, one can hope that this property characterizes rational surfaces.

This is close to being true, and there are two ways of developing a complete answer.

(3.4.1.1) It is known that there are only finitely many families of exceptions, though the complete list is not yet known.

(3.4.1.2) The second approach, which is more promising in higher dimensions, is to study multivalued differential forms as well. On a surface a multivalued 2-form is locally written as $f(z_1, z_2)dz_1 \wedge dz_2$ where f is a multivalued analytic function. Thus we may ask about the existence of 2-valued differential forms etc. We have the following:

3.4.2 Theorem. [Castelnuovo1898] S is rational iff there are no global holomorphic 1-forms and no global holomorphic 2-valued 2-forms on $S(\mathbb{C})$.

It is technically easier to talk about global sections of symmetric or tensor powers of the cotangent bundle. In this language the above result reads:

3.4.2' Theorem. S is rational iff $H^0(S, \Omega^1_S) = 0$ and $H^0(S, (\Omega^2_S)^{\otimes 2}) = 0$.

3.5 EASY ARITHMETIC. *There are many solutions over function fields.*

Let $F = \mathbb{C}(t)$ and $S \subset F\mathbb{P}^n$ be given by the equations $f_1 = \cdots = f_k = 0$ where the f_i are homogeneous polynomials in $x_0, \dots, x_n$ with coefficients in F. Let $\bar{F}$ denote the algebraic closure of F.

The first good news is that the analog of (2.5.2) holds:

3.5.1 Theorem. [Manin66; Colliot-Thélène86] If S is rational (over $\bar{F}$) then $S(F)$ is not empty.

As for curves, we may want to prove that there are in fact many solutions. In perfect analogy with (2.5) we have:

3.5.2 Theorem. [KoMiMo92b] Assume that S is rational (over $\bar{F}$). There is a finite set $B \subset \mathbb{C}$ such that if $t_1, \dots, t_s \in \mathbb{C} \setminus B$ are arbitrary points and $(x_{0k}, \dots, x_{nk})$ is any solution of $f_1 = \cdots = f_k = 0$ at t_k, then there is a solution $(x_0(t), \ldots, x_n(t))$ of $f = 0$ such that $(x_0(t_k), \ldots, x_n(t_k)) = (x_{0k}, \ldots, x_{nk})$ for $k = 1, \ldots, s$.

It would be desirable to generalize to the case when we also specify the beginning of the Taylor expansion of $(x_0(t), \ldots, x_n(t))$ at certain points. The case when S has a conic bundle structure is quite easy (see [CTSSD87, I.3.9] for a similar hard arithmetic proof). The general case is not known.
All these results hold if $\mathbb{C}(t)$ is replaced with any finite degree extension of $\mathbb{C}(t)$.

3.6 Low degree equations.

First we may ask: is every rational surface defined by low degree equations? The answer is no, there are just too many rational surfaces. It is more reasonable to ask:

Is every rational surface T birational to a surface S which is defined by low degree equations?

By definition, any rational surface is birational to $\mathbb{CP}^2$ over $\mathbb{C}$, but this is rather useless in studying arithmetic properties of S. Thus we should be more precise and ask:

3.6.1 Question. Let T be a rational surface defined over a field F. Is T always birational over F to a surface S which is defined by low degree equations?

In this form the question is very interesting and fruitful. The answer is given in two steps.

3.6.2 Minimal models of surfaces. [Enriques1897]

The first step is to simplify the geometry of an arbitrary smooth projective surface $T(\mathbb{C})$ by birational maps. The classical name for this procedure is "adjunction". Later it was called "contraction of (-1)-curves", and the currently fashionable term is "minimal model program".

For any surface T we aim to find a birational morphism $f : T \to S$ such that S is as simple as possible. (For instance, we may want to make the Betti numbers of $S(\mathbb{C})$ small.) S is called a *minimal model* of T (in general it is not unique).

If T is defined over a field F, then we can choose S so that f and S are also defined over F. (It is remarkable that the original method of Enriques automatically works over any field, while the later variants need additional arguments.)

Next we study the geometry of the minimal models S assuming that S is rational over $\mathbb{C}$. The final result is that there are 4 classes of such surfaces.

3.6.3 Theorem. [Enriques1897; Manin66; Iskovskikh80c] Let T be a surface defined over a field $F \subset \mathbb{C}$ such that T is rational over $\mathbb{C}$. Then any minimal model of T falls in one of four classes. (For simplicity, I use affine coordinates.)

(3.6.3.1) (One low degree equation)
$S = (f(x, y, z) = 0) \subset \mathbb{A}^3$ where f satisfies one of the weighted degree conditions:

$$\deg(x,y,z) = (1,1,1) \quad \text{and} \quad \deg f \leq 3 \quad (\text{e.g. } \ x^3 + y^3 + z^3 + 1);$$
$$\deg(x,y,z) = (1,1,2) \quad \text{and} \quad \deg f \leq 4 \quad (\text{e.g. } \ x^4 + y^4 + z^2 + 1);$$
$$\deg(x,y,z) = (1,2,3) \quad \text{and} \quad \deg f \leq 6 \quad (\text{e.g. } \ x^6 + y^3 + z^2 + 1).$$

(3.6.3.2) (Two low degree equations)
$S = (f_1(x, y, z, u) = f_2(x, y, z, u) = 0) \subset \mathbb{A}^4$ where $\deg f_i = 2$.

(3.6.3.3) (Two equations with low degree in certain variables)
$S = (f_1(x, y) = f_2(x, y, z, u) = 0) \subset \mathbb{A}^4$ where $\deg f_1 = 2$ and the degree of f_2 in the (z, u) variables is 2. (The degree of f_2 in the (x, y) variables can be high.)

In these three cases a general choice of f, f_1, f_2 always gives a rational surface.

(3.6.3.4) (Miscellaneous)

These are inconvenient to pin down with equations. They are all birational to a surface $S = (f(x, y, z) = 0) \subset \mathbb{A}^3$ where $\deg f \leq 9$, but f has to be very special. It is much better to notice that all these remaining cases are birational to a homogeneous space under a linear algebraic group.

These results imply the following arithmetic assertion:

3.6.4 Theorem. Let S be a surface defined over a field $F \subset \mathbb{C}$ which is rational over $\mathbb{C}$. Then there is a field extension $E \supset F$ such that $\deg[E : F] \leq 9$ and $S(E)$ is not empty.

4. Higher Dimensional Varieties

After surfaces, the next step is the study of algebraic threefolds. The theory of threefolds is much more complicated than the theory of surfaces, but in the last 20 years a rather satisfactory approach to threefolds was developed. We know much less about higher dimensions, but all the conjectures predict that higher dimensional varieties behave exactly like threefolds, although the proofs are unknown to us.

Of course it may happen that a few examples will completely change this picture, but for the moment there is no point in discussing threefolds and higher dimensional varieties separately.

In the surface case one can always consider only irreducible and smooth surfaces. Starting with dimension three, the smoothness assumption is too strong, but this is a technical question which has very little to do with the essential points of our discussion.

For simplicity, I mostly consider smooth varieties. In a few places, where singularities do cause trouble, I mention this explicitly.

The aspects of higher dimensional algebraic geometry that are discussed here are treated in the books [CKM88; Kollár96a]. Some other works dealing with related topics are [Ueno75; Kollár et al.92]. For symplectic topology see [McDuff-Salamon94,95].

Characterizations of "low degree" varieties

Let $X \subset \mathbb{P}^n$ be a smooth projective variety defined by homogeneous equations $f_1 = \cdots = f_k = 0$.

As for surfaces, the algebraic geometry condition gives the basic concept, but here it takes some work to establish the correct definition.

4.1 ALGEBRAIC GEOMETRY. *$X(\mathbb{C})$ is rationally connected.*

Already in the surface case it is not easy to show that all low degree surfaces are rational. Therefore it did not come as a big surprise that in higher dimensions rational varieties are too special. A cubic hypersurface $X_3^n \subset \mathbb{CP}^{n+1}$ certainly has low degree. M. Noether knew that there is a map $p : \mathbb{CP}^n \dashrightarrow X_3^n$ which is generically 2:1, but nobody was able to prove that X_3^n is rational for $n \geq 3$. (And indeed, X_3^3 is not rational [Clemens-Griffiths72].) This leads to the following notion:

4.1.1 Definition. X is *unirational* (over $\mathbb{C}$) if there is a rational map $p : \mathbb{CP}^n \dashrightarrow X(\mathbb{C})$ with dense image, where $n = \dim X$.

Very low degree hypersurfaces in $\mathbb{CP}^n$ are unirational [Morin40b]. Unfortunately, it seems that the class of unirational varieties is still too restrictive.

A new concept was proposed in [KoMiMo92b]. Instead of trying to emulate global properties of $\mathbb{CP}^n$, we concentrate on rational curves. $\mathbb{CP}^n$ has lots of rational curves (lines, conics and many higher degree ones). These are images of maps $\mathbb{CP}^1 \to \mathbb{CP}^n$. The defining property of the new class should be the existence of lots of maps $\mathbb{CP}^1 \to \mathbb{CP}^n$. There are several a priori ways of making this precise. Fortunately, many of these are equivalent:

4.1.2 Theorem. [KoMiMo92b] Let X be a smooth projective variety over $\mathbb{C}$. The following are equivalent:
(4.1.2.1) There is an open subset $\emptyset \neq U \subset X(\mathbb{C})$ such that for every $x_1, x_2 \in U$ there is a morphism $f : \mathbb{CP}^1 \to X$ satisfying $x_1, x_2 \in f(\mathbb{CP}^1)$.
(4.1.2.2) For every $x_1, x_2 \in X(\mathbb{C})$ there is a morphism $f : \mathbb{CP}^1 \to X$ satisfying $x_1, x_2 \in f(\mathbb{CP}^1)$.
(4.1.2.3) For every $x_1, \dots, x_n \in X(\mathbb{C})$ there is a morphism $f : \mathbb{CP}^1 \to X$ satisfying $x_1, \dots, x_n \in f(\mathbb{CP}^1)$.
(4.1.2.4) Let $p_1, \dots, p_n \in \mathbb{CP}^1$ be distinct points. For each i let $f_i : D(p_i) \to X(\mathbb{C})$ be a holomorphic map from a small disc around p_i to $X(\mathbb{C})$. Let n_i be natural numbers. Then there is a morphism $f : \mathbb{CP}^1 \to X$ such that the Taylor series of f_i and of $f|D(p_i)$ coincide up to order n_i for every i.
(4.1.2.5) There is a morphism $f : \mathbb{CP}^1 \to X$ such that f^*T_X is ample (see [ibid] for a definition of ample).

4.1.3 Definition. A smooth projective variety X is called *rationally connected* if it satisfies the equivalent properties in (4.1.2).

Thus among n-dimensional varieties we have 3 classes, with the following easy containment relations:

$$\{\text{rational}\} \subset \{\text{unirational}\} \subset \{\text{rationally connected}\}.$$

Much effort went into understanding the precise relationship between these classes. Since 1910, several authors claimed to have produced examples of rationally connected but nonrational threefolds, but the first correct proofs appeared only around 1970. By now the situation is quite satisfactory:

4.1.4 Examples of rationally connected varieties which are not rational.
(4.1.4.1) Dimension three.

The first examples were quartic 3-folds $X_4 \subset \mathbb{CP}^4$ [Iskovskikh-Manin71] and cubic 3-folds $X_3 \subset \mathbb{CP}^4$ [Clemens-Griffiths72]. Further development by [Beauville77; Iskovskikh80b; Bardelli84] gave a quite complete picture in dimension three.

(4.1.4.2) Conic bundles.

After some very special examples [Artin-Mumford72], a general theory was developed in [Sarkisov81,82]. This shows that $X_{d,2} \subset \mathbb{CP}^n \times \mathbb{CP}^2$ is not rational for $d \gg 1$. Further examples are in [Kollár96b].

(4.1.4.3) Quadric bundles.

Only some special examples are known [CTO89; Peyre93].

(4.1.4.4) Hypersurfaces.

$X_5 \subset \mathbb{CP}^5$ is considered in [Pukhlikov87]; the method should give all $X_n \subset \mathbb{CP}^n$. These techniques also give many more examples as in (1.3), see [CPR96]. Very general hypersurfaces $X_d \subset \mathbb{CP}^{n+1}$ for $2n/3 + 2 \leq d \leq n + 1$ are treated in [Kollár95].

(4.1.4.5) Hypersurface bundles.

$X_{c,d} \subset \mathbb{CP}^m \times \mathbb{CP}^{n+1}$ where $c \geq 2m$ and $2n/3+2 \leq d \leq n+1$ are considered in [Kollár96b].

As this list suggests, most rationally connected varieties are not rational. Some of the varieties on the above list are unirational, thus rational and unirational are indeed different notions. Despite the long list of settled cases, there are many open problems. I mention two about hypersurfaces; they indicate how little is known.

4.1.5 Some unsolved cases.
(4.1.5.1) Is the general cubic n-fold $X_3^n \subset \mathbb{CP}^{n+1}$ rational for $n \geq 4$? The case of cubic 4-folds has received a lot of attention. It is known that some special ones are rational [Morin40a; Tregub93]. In particular this would show that rationality is not deformation invariant.

(4.1.5.2) Is there any rational (smooth) hypersurface of degree at least 4? There is very little evidence either way.

The biggest unsolved question in this picture is the following:

4.1.6 Conjecture. Most rationally connected varieties are not unirational.

At the moment, there is not a single example known. The simplest case to study may be general quartic threefolds $X_4 \subset \mathbb{CP}^4$.

Assume that X is unirational, that is, there is a map $p : \mathbb{CP}^n \dashrightarrow X$. The images of linear subspaces show that through a general point of $x \in X$ there are unirational subvarieties of every dimension. Even this weaker property may fail in general:

4.1.7 Question.

Let $X_d \subset \mathbb{CP}^n$ be a hypersurface of degree $d \leq n$ (thus X is rationally connected). Is it true that for every point $x \in X$ there is a rational surface $x \in S_x \subset X$?

It is easy to see that this is the case if $\binom{d+1}{2} \leq n$, and probably also for slightly larger values of d.
I do not see any obvious way to construct rational surfaces when d is close to n.

Finally I mention another problem concerning rationally connected varieties.

4.1.8 Conjecture. Let $f : X \to Z$ be a morphism between smooth projective varieties. Assume that Z and the general fiber F are rationally connected. Then X is rationally connected.

It is easy to see that the special case when $Z = \mathbb{P}^1$ implies the general one, thus (4.5.1) implies (4.1.8).

4.2 TOPOLOGY. *Diffeomorphism versus symplectomorphism.*

Guided by the results of the surface case, one can look for three types of theorems in higher dimension:

4.2.1 Basic Questions.
(4.2.1.1) Determine all algebraic varieties of a given topological type.
(4.2.1.2) Relate the topological properties of birationally equivalent varieties.
(4.2.1.3) Characterize rationally connected varieties in terms of their topology.

As in (3.2), the best example in the first direction is the following result of [Hirzebruch-Kodaira57; Yau77]

4.2.2 Theorem. If $X(\mathbb{C})$ is homeomorphic to $\mathbb{CP}^n$ then X is isomorphic to $\mathbb{CP}^n$.

There are very few such results known, and the proofs use rather lucky coincidences. One may want to have a more modest aim in mind, and try to show that the topological structure of $X(\mathbb{C})$ determines X up to finite ambiguity. I noticed the following special case some time ago (a proof is given in (5.3)):

4.2.3 Theorem. Let M be a compact differentiable manifold with $\dim H_2(M, \mathbb{Q}) = 1$. Then there are only finitely many families of algebraic varieties X such that $X(\mathbb{C})$ is diffeomorphic to M.

For M arbitrary this no longer holds. This is already shown by the example of minimal ruled surfaces, but a more convincing negative result was observed by [Friedman-Morgan88b]. This shows that diffeomorphism of algebraic 3-folds is not as strong as for surfaces:

4.2.4 Example. Let S_i be smooth projective surfaces such that $S_i(\mathbb{C})$ is simply connected. Set $X_i := S_i \times \mathbb{CP}^1$.

For differentiable manifolds of real dimension 6, homeomorphism frequently implies diffeomorphism [Wall66; Sullivan77; Zubr80]. We find that if $S_i(\mathbb{C})$ and $S_j(\mathbb{C})$ are homeomorphic, then $X_1(\mathbb{C})$ and $X_2(\mathbb{C})$ are even diffeomorphic. This gives several unpleasant examples:
(4.2.4.1) Let S_1 be a rational surface which is homeomorphic to a nonrational surface S_2 (3.2). Then X_1 is rational, hence also rationally connected and X_2 is not even rationally connected.

(4.2.4.2) One can construct infinitely many surfaces S_i such that the $S_i(\mathbb{C})$ are all homeomorphic, but the S_i are quite different as algebraic surfaces [Okonek-V.d.Ven86; Friedman-Morgan88a]. Thus the manifolds $X_i(\mathbb{C})$ are all diffeomorphic, but the varieties X_i do not fit into finitely many families.

4.2.5 The Topology of Birational Maps.
Let X_1 and X_2 be smooth projective varieties, birational to each other. In contrast with the surface case, it is not known how the manifolds $X_1(\mathbb{C})$ and $X_2(\mathbb{C})$ are related. There are certain surgery type operations, called blow-ups, that take the role of connected sum with $\overline{\mathbb{CP}}^2$. Unfortunately it is not known whether one can go from $X_1(\mathbb{C})$ to $X_2(\mathbb{C})$ by repeated application of blow-ups. This is a hard problem.

The minimal model program establishes a class of surgery type operations that can be used to go from $X_1(\mathbb{C})$ to $X_2(\mathbb{C})$. At the moment these operations are not well understood from the topological point of view. Furthermore, the intermediate stages involve singular topological spaces.

As example (4.2.4) shows, the diffeomorphism type alone does not characterize rationally connected varieties. In order to obtain a suitable analog of (3.2.4), it is necessary to study an additional structure on $X(\mathbb{C})$:

4.2.6 Symplectic manifolds.
A *symplectic* manifold is a pair (M^{2n}, ω) where M is a differentiable manifold of dimension $2n$ and ω is a 2-form $\omega \in \Gamma(M, \wedge^2 T^*)$ which is d-closed and nondegenerate. That is, $d\omega = 0$ and ω^n is nowhere zero.

Any smooth projective variety admits a symplectic structure. This can be constructed as follows. On $\mathbb{C}^{n+1}$ consider the Fubini–Study 2-form

$$\omega' := \frac{\sqrt{-1}}{2\pi}\left[\frac{\sum dz_i \wedge d\bar{z}_i}{\sum |z_i|^2} - \frac{(\sum \bar{z}_i dz_i) \wedge (\sum z_i d\bar{z}_i)}{(\sum |z_i|^2)^2}\right].$$

It is closed, nondegenerate on $\mathbb{C}^{n+1} \setminus \{0\}$ and invariant under scalar multiplication. Thus ω' descends to a symplectic 2-form ω on $\mathbb{CP}^n = (\mathbb{C}^{n+1} \setminus \{0\})/\mathbb{C}^*$.

If $X \subset \mathbb{CP}^n$ is any smooth variety, then the restriction $\omega|X$ makes $X(\mathbb{C})$ into a symplectic manifold.

The resulting symplectic manifold $(X(\mathbb{C}), \omega|X)$ depends on the embedding $X \hookrightarrow \mathbb{CP}^n$, but the dependence is rather easy to understand:

We say that two symplectic manifolds (M, ω_0) and (M, ω_1) are *symplectic deformation equivalent* if there is a continuous family of symplectic manifolds (M, ω_t) starting with (M, ω_0) and ending with (M, ω_1).

To every smooth projective variety the above construction associates a symplectic manifold $(X(\mathbb{C}), \omega|X)$ which is unique up to symplectic deformation equivalence.

This allows us to formulate the proper generalization of (3.2.4):

4.2.7 Conjecture. Let X_0 and X_1 be smooth projective varieties defined over $\mathbb{C}$ such that $(X_0(\mathbb{C}), \omega_0)$ is symplectic deformation equivalent to $(X_1(\mathbb{C}), \omega_1)$. Then X_0 is rationally connected iff X_1 is.

The evidence for this conjecture comes from three sources:

The first thing to check is that (4.2.7) holds if there is a continuous family of algebraic varieties $\{X_t, t \in [0,1]\}$. This case is settled:

4.2.8 Theorem. [KoMiMo92b, 2.4] Let $\{X_t, t \in [0,1]\}$ be a continuous family of smooth projective varieties. Then X_0 is rationally connected iff X_1 is.

Secondly, one should try to analyze the examples (4.2.4). This was studied in detail by [Ruan94] who showed that the symplectic structure of $S \times \mathbb{CP}^1$ can be used to study the differentiable structure of S in many cases.

The third piece of evidence is given by the following closely related result, whose formulation requires a definition.

4.2.9 Definition. A smooth projective variety X over $\mathbb{C}$ is called *uniruled*, if it satisfies the following equivalent conditions:

(4.2.9.1) There is an open subset $\emptyset \neq U \subset X(\mathbb{C})$ such that for every $x \in U$ there is a morphism $f : \mathbb{CP}^1 \to X$ satisfying $x \in f(\mathbb{CP}^1)$.

(4.2.9.2) For every $x \in X(\mathbb{C})$ there is a morphism $f : \mathbb{CP}^1 \to X$ satisfying $x \in f(\mathbb{CP}^1)$.

The proof of the next result is outlined in (5.4):

4.2.10 Theorem. Let X_0, X_1 be smooth projective varieties defined over $\mathbb{C}$ such that $(X_0(\mathbb{C}), \omega_0)$ is symplectic deformation equivalent to $(X_1(\mathbb{C}), \omega_1)$. Then X_0 is uniruled iff X_1 is.

(4.2.7) holds if $\dim H_2(X_0, \mathbb{Q}) = 1$, since then X is rationally connected iff it is uniruled [KoMiMo92a].

It should be noted that if X_0 is Fano (4.6.2.1), X_1 need not be Fano, as shown by the examples of rational ruled surfaces.

It would also be interesting to find some topological properties of rationally connected varieties. The only general result is the following:

4.2.11 Theorem. [Campana91b; KoMiMo92b] Let X be a rationally connected variety. Then $X(\mathbb{C})$ is simply connected.

4.3 Hard Arithmetic. *$X(\mathbb{Q})$ is "large".*

As for surfaces, the guiding principle is the following conjecture, which is a natural generalization of a problem of [Batyrev-Manin90].

4.3.1 Conjecture. If X is rationally connected (over $\mathbb{C}$) then there are $r \in \mathbb{N}$, $0 < \beta \in \mathbb{Q}$ and a number field $F' \supset F$ such that

$$N_E(X \setminus A, H) \quad \text{is asymptotic to} \quad \text{const} \cdot H^{\beta}(\log H)^r$$

for every sufficiently large subvariety $A \subsetneq X$, and for every number field $E \supset F'$.

The key point is that β is positive. Even the following weaker form is completely open:

4.3.1' Conjecture. If X is rationally connected then there is an $\epsilon > 0$ such that

$$N_E(X \setminus A, H) > H^\epsilon \quad (\text{for } H \gg 1),$$

for every subvariety $A \subsetneq X$, and for every sufficiently large number field E.

There are many special cases where (4.3.1) holds [FMT89; Batyrev-Manin90; Batyrev-Tschinkel95]. There is a more precise version of the conjecture [Batyrev-Manin90] asserting that the numbers β, r are computable from the geometry of T. This has been checked in certain cases, but a recent example of [Batyrev-Tschinkel96] shows that the conjecture for the value of r is incorrect.

A precise computation of the growth of the number of integral solutions of the equations

$$x_1^3 + x_2^3 + x_3^3 = y_1^3 + y_2^3 + y_3^3$$
$$x_1 + x_2 + x_3 = y_1 + y_2 + y_3$$

may be found in [Vaughan-Wooley95]. This corresponds to (4.3.1) for a certain singular cubic threefold. The results confirm (4.3.1), but they also seem to contradict the more refined conjecture about r. Further special cases are treated in [EMS96].

The converse of (4.3.1) again fails, but not by much:

4.3.2 Conjecture. Assume that X is not uniruled (over $\mathbb{C}$). Then for every number field E and $0 < \epsilon$, there is a subvariety $A \subsetneq X$ such that

$$N_E(X \setminus A, H) < \text{const} \cdot H^\epsilon.$$

4.3.3 The general case. The problem for a general variety X can be reduced to the above two cases as follows.

Assuming (4.1.8), there is a map $f : X \dashrightarrow Z$ such that Z is not uniruled and the fibers of f are rationally connected [KoMoMi92b].

Thus we can study the points of X in E in two steps. First we have to find the E-points of Z using (4.3.2). Then for every $P \in Z(E)$ we study the E-points in the fiber $f^{-1}(P)$, which is rationally connected.

4.4 COMPLEX MANIFOLDS. *Global holomorphic differential forms.*

As in the surface case, one can study multivalued global holomorphic differential forms on $X(\mathbb{C})$. It is easy to see that if X is rationally connected, then there are no such forms:

4.4.1 Proposition. Let X be a smooth projective variety over $\mathbb{C}$. Assume that X is rationally connected. Then

$$H^0\left(X, \left(\Omega_X^1\right)^{\otimes m}\right) = 0 \quad \text{for every } m > 0.$$

The converse is conjectured to be true, but it is known only in dimension three:

4.4.2 Theorem. [KoMiMo92b] Let X be a smooth projective threefold over $\mathbb{C}$. The following are equivalent:
(4.4.2.1) X is rationally connected;
(4.4.2.2) $H^0\big(X, (\Omega^1_X)^{\otimes m}\big) = 0$ for every $m > 0$.

In contrast with (3.5), the current proofs of (4.4.2) require the vanishing for all values of m. It is quite likely that finitely many of these values are sufficient, but there is no conjecture for the precise bound. [KoMiMo92b] contains further results in this direction.

4.5 Easy Arithmetic.

There are many solutions over function fields.

Let $F = \mathbb{C}(t)$ and $X \subset F\mathbb{P}^n$ be a subvariety. Let $\bar{F}$ denote the algebraic closure of F.

The higher dimensional analog of (3.5.1) is open:

4.5.1 Conjecture. If X is rationally connected then $X(F)$ is not empty.

This is known in many special instances (see, e.g. [Kollár96a, IV.6]), but these results give very few hints about the general case.

This of course means that we are also unable to prove that X has many points in F. Surprisingly, one can prove that if $X(F)$ is not empty, then it is very large. I formulate the result in the geometric version, which is more precise.

4.5.2 Theorem. [KoMiMo92b, 2.13] Let X be a projective variety over $\mathbb{C}$ and $f : X \to C$ a morphism onto a smooth curve. Assume that f has a section $\sigma : C \to X$. Let $c_1, \dots, c_k \in C$ be closed points such that $f^{-1}(c_i)$ are smooth and rationally connected. Pick arbitrary points $p_i \in f^{-1}(c_i)$.

Then f has a section $s = s_{p_1,\dots,p_k} : C \to X$ such that $s(c_i) = p_i$ for every i.

The following more general version is open. In analogy with the number theoretic terminology (cf. [Mazur92]), it should be called "weak approximation for rationally connected varieties over function fields".

4.5.3 Conjecture. Let X be a smooth projective variety over $\mathbb{C}$ and $f : X \to C$ a morphism onto a smooth curve whose general fibers are rationally connected. Let $c_1, \dots, c_k \in C$ be closed points and $c_i \in D(c_i) \subset C$ small discs around c_i. Pick local sections $s_i : D(c_i) \to X$ and natural numbers n_i.

Then f has a section $s : C \to X$ such that the Taylor series of $s|D(c_i)$ agrees with the Taylor series of s_i up to order n_i, for every i.

In the special case when $X = C \times Y$, this follows from (4.1.2.4).

4.6 Low degree equations.

As in the surface case, the principal question is the following:

4.6.1 Question. Let X be a rationally connected variety defined over a field F. Is X always birational over F to a variety Y which is defined by low degree equations?

In contrast with the surface case, this is interesting even for $F = \mathbb{C}$.

In analogy with (3.6), first we need:

4.6.2. Minimal model program.

This is a general method to simplify the structure of an arbitrary smooth projective variety. Already in dimension 3 it is rather complicated (cf. [Mori82,88]), and in higher dimensions remains conjectural. See [Kollár87,90] for introductions. The program can be performed over any field F with minor modifications.

For rationally connected varieties we end up with a variety Y (birational to X) satisfying one of the following conditions:
(4.6.2.1) Y is a Fano variety, that is, $-K_Y$ is ample. Unfortunately, Y may be singular. The singularities are rather mild (terminal and $\mathbb{Q}$-factorial), but they do cause certain problems.
(4.6.2.2) There is a morphism $p : Y \to Z$ such that Z and the fibers of p are rationally connected.

In the second case we hope to reduce problems about X to questions about Z and about the fibers of f. Thus we mainly concentrate on the first case. Some of the basic questions are settled:

4.6.3 Theorem. (4.6.3.1) [Nadel91; Campana91a; KoMiMo92a,c] For any n there are only finitely many families of smooth Fano varieties of dimension n.
(4.6.3.2) [Kawamata92] There are only finitely many families of singular Fano threefolds arising in (4.6.2.1).

In both cases the proof yields explicit (though huge) bounds on the number of families and also on the degrees of the defining equations of the Fano varieties.

In dimension three there is a complete list of all smooth Fano varieties, but no such list exists in the singular case. In any case, classifying Fano threefolds up to isomorphism may not be the sensible thing to do. Our original variety X is determined by Y only up to birational equivalence; thus it makes sense to classify rationally connected threefolds up to birational equivalence. [Alexev94; Corti96] contain significant steps in this direction.

4.6.4 Listing by low degree equations.

Smooth Fano threefolds were studied by G. Fano in a series of articles spanning four decades starting in 1908. A modern account of these works was given in [Iskovskikh80a,b]. The results of [Mukai89] give a better description, especially over nonclosed fields. For singular Fano threefolds there is no general theory; a series of examples can be found in [Fletcher89].

If there is a morphism $p : X \to Z$ as in (4.6.2.2), then the results of (3.6) give us defining equations as in (3.6.3). Instead of listing all cases, I just give two examples:
(4.6.4.1) $S = (f_1(u,v) = f_2(x,y,z,u,v) = 0) \subset \mathbb{A}^5$,
where $\deg f_1 = 2$ and the degree of f_2 in the (x,y,z) variables satsifies one of the conditions of (3.6.3.1) (The degree of f_2 in the (u,v) variables can be high.)
(4.6.4.2) $S = (f_1(x_1,x_2) = f_2(x_1,\dots,x_4) = f_3(x_1,\dots,x_6) = 0) \subset \mathbb{A}^6$,
where $\deg f_1 = 2$, the degree of f_2 in the (x_3,x_4) variables is 2 and the degree of f_3 in the (x_5,x_6) variables is 2. (The degrees in the other variables can be high.)

In both cases a general choice of the f_i gives a rationally connected variety.

These results imply the following arithmetic consequence:

4.6.5 Theorem. There is a constant $D(3)$ with the following property:

Let X be a rationally connected threefold defined over a field $F \subset \mathbb{C}$. Then there is a field extension $E \supset F$ such that $\deg[E : F] \leq D(3)$ and $X(E)$ is not empty.

One can write down an explicit bound for $D(3)$, though I have not done it. Conjecturally, a similar result holds in any dimension.

5. Appendix

The aim of this appendix is to outline the proofs of some statements which are new or for which I could not find a suitable reference.

5.1 Proposition. Let B be a smooth proper curve over $\mathbb{C}$ and $f : S \to B$ a proper ruled surface. Let $b_i \in B$ be different points and $D(b_i)$ a small disc around b_i. Let $s_i : D(b_i) \to S$ be holomorphic (or formal) sections and n_i natural numbers.

Then there is a section $s : B \to S$ such that $s|D(b_i)$ agrees with s_i up to order n_i for every i.

Proof. S is birationally trivial; that is, there is a birational map $\pi : \mathbb{P}^1 \times B \dashrightarrow S$. We obtain local sections

$$s_i' := \pi^{-1} \circ s_i : D(b_i) \to \mathbb{P}^1 \times B.$$

Assume that it takes k blow-ups to resolve the indeterminacies of π. Let $s' : B \to \mathbb{P}^1 \times B$ be a section such that $s'|D(b_i)$ agrees with s_i' up to order $n_i + k$ for every i. Then we can take $s := \pi \circ s'$.

Thus it is sufficient to find s'. Equivalently, we need to find a map $\bar{s} : B \to \mathbb{P}^1$ with prescribed local behavior $\bar{s}_i : D(b_i) \to \mathbb{P}^1$. By a generic coordinate change in $\mathbb{P}^1$ we can assume that $\bar{s}_i(b_i) \in \mathbb{C}$ for every i.

Choose another point b_0. One can always find regular functions on the affine curve

$B \setminus \{b_0\}$ with prescribed local behaviour at the points b_i. □

5.2 Proof of (4.1.2.4). We need to show that (4.1.2.4) is implied by (4.1.2.3). As a first step, I prove the following weaker version:

(5.2.1) Let $p_1, \ldots, p_n \in \mathbb{CP}^1$ be disctinct points. For each i let $f_i : D(p_i) \to X(\mathbb{C})$ be a holomorphic map from a small disc around p_i to $X(\mathbb{C})$. Let n_i be natural numbers. Then there is a morphism $g : \mathbb{CP}^1 \to X$ and holomorphic maps $h_i : D(p_i) \to \mathbb{CP}^1$ such that the Taylor series of f_i and of $g \circ h_i|D(p_i)$ coincide up to order n_i for every i.

To see this, let $D \subset \mathbb{C}$ be the unit disc and $f, g : D \to \mathbb{C}^n$ two holomorphic maps with coordinate functions f^j, g^j. Assume that $f(0) = g(0) = 0 \in \mathbb{C}^n$. Let $B_0\mathbb{C}^n \to \mathbb{C}^n$ be the blow-up of $0 \in \mathbb{C}^n$. f and g lift to holomorphic maps $\bar{f}, \bar{g} : D \to B_0\mathbb{C}^n$. Explicit local computation shows the following:

(5.2.2.1) If $\bar{f}$ and $\bar{g}$ agree up to order n, then so do f and g.

(5.2.2.2) If $f^1(t) = g^1(t) = t$ and $\bar{f}$ and $\bar{g}$ agree up to order $n-1$, then f and g agree up to order n.

Using (5.2.2.1) for repeated blow-ups, we first reduce (5.2.1) to the case when the f_i are immersions. Then up to a local coordinate change we may assume that $f_i^1(t) = t$ for every i. We can now prove (5.2.1) by induction on $\sum n_i$, since (4.1.2.3) gives it for $\sum n_i = 0$.

The only subtle point is the reduction step from order 1 to order 0. Let $p \in D \subset \mathbb{CP}^1$ be a disc. Given an immersion $f : D \to X$, let $x = f(p)$ and $\pi : B_x X \to X$ be the blow-up with exceptional divisor $E \subset B_x X$. Assume that we have $\bar{g} : \mathbb{CP}^1 \to B_x X$ such that $\bar{f}$ and $\bar{g}$ agree up to order 0 at p. We would like to conclude that f and $g := \bar{g} \circ \pi$ agree up to order 1 at p. (5.2.2.2) gives this, if g is an immersion. Thus we have to choose $\bar{g} : \mathbb{CP}^1 \to B_x X$ to be transversal to E. This is slightly stronger than (4.1.2.3), but can easily be arranged (see the proofs of II.3.14 and IV.3.9 in [Kollár96a]).

Once we have (5.2.1), we just need to find a map $h : \mathbb{CP}^1 \to \mathbb{CP}^1$ which approximates every h_i up to order n_i and set $f := g \circ h$.

The f we found is a multiple cover of a curve in X. As in [Kollár96a, IV.3.9] we can perturb f to obtain another solution of (4.1.2.4) where $f|\mathbb{CP}^1 \setminus \{p_1, \dots, p_n\}$ is an embedding. □

5.3 Proof of (4.2.3). Assume that $X(\mathbb{C})$ is diffeomorphic to M. We use the formula [Hirzebruch66,20.3.6*]

(5.3.1)
$$\chi(\mathcal{O}_X) = \sum_{s \geq 0} \frac{1}{2^{n+2s}(n-2s)!} c_1(X)^{n-2s} A_s(p_1, \dots, p_s)[M],$$

where the A_s are certain polynomials of the Pontrjagin classes of M and $A_0 = 1$. From Hodge theory we know that

$$|\chi(\mathcal{O}_X)| \leq \sum \dim_{\mathbb{C}} H^i(X, \mathcal{O}_X) \leq \sum \dim_{\mathbb{C}} H^i(M, \mathbb{C}),$$

and so $\chi(\mathcal{O}_X)$ is bounded in terms of M. Since $b_2(M) = 1$, we can fix an ample divisor H in $\mathrm{Pic}(X)$ and then $c_1(X) \equiv rH$ for some rational number r. (5.3.1) becomes a polynomial equation for r. As $\chi(\mathcal{O}_X)$ runs through all the possible values, we get only finitely many possible values for r. Therefore the self-intersection number (H^n) and the intersection number $(c_1(X) \cdot H^{n-1})$ are bounded depending on M only. The result now follows from Matsusaka's Big Theorem (in the form given in [Kollár-Matsusaka83]). □

The proof provides an effective bound on the number of families of algebraic structures on a given manifold M, but the bound is enormous even in the simplest cases.

5.4 Proof of (4.2.10). The proof is an application of the theory of Gromov–Witten invariants. I recall the main concepts in the needed special case. See [McDuff-Salamon94,95] for details of the general theory.

Let X be a smooth projective variety over $\mathbb{C}$. Fix a point $x \in X$, a homology class $A \in H_2(X(\mathbb{C}), \mathbb{Z})$ and very ample divisors in general position $H_i \subset X$, $i = 1, \dots, k$.

Let $y_0, \dots, y_k \in \mathbb{CP}^1$ be general points. For suitable k, there may be only finitely many maps

$$f : \mathbb{CP}^1 \to X \quad \text{such that} \quad f_*[\mathbb{CP}^1] = A, f(y_0) = x, \text{ and } f(y_i) \in H_i, \; i = 1, \dots, k.$$

We define an invariant

$$\tilde{F}_{A,X}(x, H_1, \dots, H_k; y_0, \dots, y_k) := \text{ the number of such maps.} \tag{5.4.1}$$

Gromov's theory of pseudo-holomorphic curves shows that one can make a similar definition where X is replaced by a symplectic manifold (M, ω) endowed with a general almost complex structure. The corresponding invariant is denoted by

$$\tilde{\Phi}_{A,M,\omega}(x, H_1, \dots, H_k; y_0, \dots, y_k). \tag{5.4.2}$$

It is one of the Gromov–Witten invariants of (M, ω). In fact, this is an invariant of the symplectic deformation equivalence class.

In general the algebraic number (5.4.1) and the symplectic number (5.4.2) are different. Under suitable conditions they are equal, and this means that we can get information about rational curves on X from the symplectic structure $(X(\mathbb{C}), \omega_X)$ (4.2.6). This idea was used by [Ruan93] to show that the extremal rays of Mori theory can be described using the symplectic structure. We need the following two results. (In [Ruan93] they are proved under the extra assumption that the symplectic structure is semi-positive. This is no longer necessary.)

5.4.3 Theorem. Let X be a smooth projective variety over $\mathbb{C}$ and (M, ω) the corresponding symplectic manifold.
(5.4.3.1) If $\tilde{\Phi}_{A,M,\omega}(x, H_1, \dots, H_k; y_0, \dots, y_k) \neq 0$, then there is a rational map $f : \mathbb{CP}^1 \to X$ such that $f_*[\mathbb{CP}^1] = A$, $f(y_0) = x$ and $f(y_i) \in H_i$, $i = 1, \dots, k$.
(5.4.3.2) $\tilde{F}_{A,X}(x, H_1, \dots, H_k; y_0, \dots, y_k) = \tilde{\Phi}_{A,X(\mathbb{C}),\omega_X}(x, H_1, \dots, H_k; y_0, \dots, y_k)$ if the following conditions are satisfied:

(5.4.3.2.1) If $g : \mathbb{CP}^1 \to X$ is any map such that $g_*[\mathbb{CP}^1] = A$ and $g(y_0) = x$, then $H^1(\mathbb{CP}^1, g^*T_X) = 0$.

(5.4.3.2.2) If $C_1, \dots, C_m \subset X$ are rational curves such that $\sum [C_i] = A$ and $x \in C_1$, then $m = 1$.

We can now prove (4.2.10).

Let (M, ω) be the common symplectic structure of X_0 and of X_1. Fix a very general point $x \in X_0$. Fix a very ample divisor $H \subset X_0$ and let $x \in C \subset X$ be a rational curve such that $(C \cdot H)$ is minimal (C exists since X_0 is uniruled). Set $A := [C]$. By [KoMiMo92c,1.1], the condition (5.4.3.2.1) holds and (5.4.3.2.2) follows from the minimality of $(C \cdot H)$. Let k be the dimension of the space of maps $g : \mathbb{CP}^1 \to X$ such that $g_*[\mathbb{CP}^1] = A$ and $g(y_0) = x$. Let $H_1, \dots, H_k \subset X_0$ be general divisors linearly equivalent to H. By construction, $\tilde{F}_{A,X}(x, H_1, \dots, H_k; y_0, \dots, y_k)$ is defined and is nonzero. Thus $\tilde{\Phi}_{A,M,\omega}(x, H_1, \dots, H_k; y_0, \dots, y_k) \neq 0$.

By (5.4.3.1) this implies that there is a rational curve through any very general point of X_1, and thus X_1 is also uniruled. □

Finally we prove that condition (1.2) correctly identifies the class of rationally connected varieties among diagonal hypersurfaces.

5.5 Proposition. Let X be any smooth compactification of the affine hypersurface

$$(\sum_{i=1}^{n} c_i x_i^{d_i} + c_0 = 0) \subset \mathbb{C}^n.$$

(5.5.1) X is rationally connected iff $\sum 1/d_i \geq 1$.
(5.5.2) The Kodaira dimension of X is nonnegative iff $\sum 1/d_i < 1$.

Proof. Consider first the case $n = 2$, assuming $d_1 \leq d_2$. View X as a d_1-sheeted cover of the line ramified along $c_2 x_2^{d_2} + c_0 = 0$. The Hurwitz formula gives that

$$2g(X) = (d_1 - 1)(d_2 - 2) + (\text{ramification at infinity}).$$

This implies (5.5) for $n = 2$.

If $n \geq 3$ then as in (1.3), we view these as hypersurfaces in weighted projective spaces.

Let $d = lcm(d_i)$, $d = d_i a_i$ and set $a_0 = 1$, $d_0 = d$. A (nonsmooth) compactification is given by the projective weighted hypersurface

$$Y := \sum_{i=0}^{n} c_i x_i^{d_i} \subset \mathbb{P}(a_0, \dots, a_n).$$

As long as $\prod c_i \neq 0$, these hypersurfaces are isomorphic (over $\mathbb{C}$), thus Y can be viewed as a general member of the linear system $|x_0^{d_0}, \dots, x_n^{d_n}|$. This implies that Y has only quotient singularities and Picard number 1 for $n \geq 4$.

Assume that $d < \sum a_i$. $K_Y = \mathcal{O}(d - \sum a_i)$, thus Y is $\mathbb{Q}$-Fano. Therefore Y is uniruled by [Miyaoka-Mori86]. Let $p : \bar{Y} \to Y$ be a desingularization and $\bar{f} : \bar{Y}^0 \to Z$ the MRC fibration [KoMiMo92b]. The fibers of p are all rationally connected (cf. [Kollár96a, VI.1.6.2]), thus $\bar{f}$ descends to $f : Y^0 \to Z$. If $n \geq 4$, then as in [Kollár96a,IV.4.14], we obtain that Z is a point, hence Y is rationally connected. If $n = 3$ then we use that $h^1(X, \mathcal{O}_X) = h^1(Y, \mathcal{O}_Y) = 0$. A smooth uniruled surface S with $h^1(S, \mathcal{O}_S) = 0$ is rational, hence X is rational.

Next assume that $d \geq \sum a_i$. Let $e = d - \sum a_i$ and introduce e new coordinates $x_{n+1}, \dots, x_{n+e}$ of weight $a_{n+1} = \dots = a_{n+e} = 1$. Consider the hypersurface

$$Z := \sum_{i=0}^{n} c_i x_i^{d_i} + \sum_{j=1}^{e} c_{n+j} x_{n+j}^{d} \subset \mathbb{P}(a_0, \dots, a_{n+e}).$$

Here every a_i divides $d = \sum a_i$, thus $\mathbb{P}(a_0, \dots, a_{n+e})$ has only index one canonical singularities. Therefore the same holds for Z. But $\omega_Z \cong \mathcal{O}_Z$, and this implies that $\kappa(Z) = 0$.

General fibers of the projection map

$$Z \dashrightarrow \mathbb{P}^e \quad \text{given by} \quad (x_0, \dots, x_{n+e}) \mapsto (x_0, x_{n+1}, \dots, x_{n+e})$$

are isomorphic to Y. This shows that $\kappa(Y) \geq 0$.

Since a variety can not be rationally connected and have nonnegative Kodaira dimension at the same time, this proves (5.5). □

5.5.3 Remark. It is not true that X is of general type if $d > \sum a_i$. For instance, $x_1^2 + x_2^3 + \sum c_i x_i^{d_i} + c_0 = 0$ has an elliptic fiberspace structure (projection to the $(x_3, \ldots, x_n)$-subspace) for every value of the d_i.

Acknowledgements. I would like to thank S. Gersten, G. Mess, M. Reid, K. Ribet, Y. Ruan and P. Vojta for useful conversations and e-mails. J.-L. Colliot-Thélène sent very detailed comments which helped me to understand the arithmetical questions much better. Partial financial support was provided by the NSF under grant number DMS-9102866. These notes were typeset by $\mathcal{A}_{\mathcal{M}}\mathcal{S}$-TEX, the TEX macro system of the American Mathematical Society.

References

[Alexeev94] V. A. Alexeev, *General elephants on Q-Fano 3-folds*, Comp. Math. **91** (1994), 91–116.

[Artin-Mumford72] M. Artin - D. Mumford, *Some elementary examples of uniruled varieties which are not rational*, Proc. London. Math. Soc. **25** (1972), 75–95.

[Baker22] H. Baker, Principles of geometry, Vols. I–VI, Cambridge Univ. Press., 1922–1933.

[Bardelli84] F. Bardelli, *Polarized mixed Hodge structures*, Annali di Math. pura e appl. **137** (1984), 287–369.

[BPV84] W. Barth - C. Peters - A. Van de Ven, Compact Complex Surfaces, Springer, 1984.

[Batyrev-Manin90] V. V. Batyrev - Y. I. Manin, *Sur les nombres des points rationnels de hauteur bornée des variétés algébriques*, Math. Ann. **286** (1990), 27–43.

[Batyrev-Tschinkel95] V. V. Batyrev - Y. Tschinkel, *Rational points on toric varieties*, Number Theory, CMS Conf. Proc. **15** (1995), 39–48.

[Batyrev-Tschinkel96] V. V. Batyrev - Y. Tschinkel, *Rational points of some Fano cubic bundles*, C.R. Acad. Sci. Paris **323** (1996), 41–46.

[Beauville77] A. Beauville, *Variétés de Prym et jacobiennes intermédiaires*, Ann. Sci. E. N. S. **10** (1977), 309–391.

[Beauville78] A. Beauville, Surfaces algébriques complexes, Astérisque, vol.54, 1978.

[Campana91a] F. Campana, *Une version géométrique généralisée du théorème du produit de Nadel*, C. R. Acad. Sci. Paris **312** (1991), 853–856.

[Campana91b] F. Campana, *On twistor spaces of the class C*, J. Diff. Geom. **33** (1991), 541–549.

[Castelnuovo1894] G. Castelnuovo, *Sulla razionalità delle involuzioni piane*, Math. Ann. **44** (1894), 125–155.

[Castelnuovo1898] G. Castelnuovo, *Sulle superficie di genere zero*, Mem. Soc. Ital. Sci. **10** (1898), 103–126.

[Chevalley35] C. Chevalley, *Démonstration d'une hypothèse de E. Artin*, Abh. Math. Sem. Hansischen Univ. **11** (1935), 73.

[Clebsch1866] A. Clebsch, *Die Geometrie auf den Flächen dritter Ordnung*, J. f.r.u.a. Math. **65** (1866), 359–380.

[Clebsch1868] A. Clebsch, *Sur les surfaces algébriques*, C.R. Acad. Sci. Paris **67** (1868), 1238–1239.

[Clemens-Griffiths72] H. Clemens - P. Griffiths, *The intermediate Jacobian of the cubic threefold*, Ann. Math. **95** (1972), 281–356.

[CKM88] H. Clemens - J. Kollár - S. Mori, Higher Dimensional Complex Geometry, Astérisque 166, 1988.

[Colliot-Thélène86] J.-L. Colliot-Thélène, *Arithmétique des variétés rationnelles et problèmes birationnels*, Proc. Int. Congr. Math., 1986, pp. 641–653.

[Colliot-Thélène92] J.-L. Colliot-Thélène, *L'arithmétique des variétés rationnelles*, Ann. Fac. Sci. Toulouse **1** (1992), 295–336.

[CTO89] J.-L. Colliot-Thélène - M. Ojanguren, *Variétés unirationnelles non rationnelles : au-delà de l'exemple d'Artin et Mumford*, Inv. Math. **97** (1989), 141–158.

[CTSSD87] J.-L. Colliot-Thélène - J.-J. Sansuc - P. Swinnerton–Dyer, *Intersections of two quadrics and Châtelet surfaces I.*, J. f.r.u.a. Math. **373** (1987), 37–107; ... *II.*, J. f.r.u.a. Math. **374** (1987), 72–168.

[Corti96] A. Corti, *Del Pezzo surfaces over Dedekind schemes*, Ann. Math. **144** (1996), 641–683.

[CPR96] A. Corti - A. Pukhlikov - M. Reid, *(in preparation)* (1996).

[Dolgachev66] I. Dolgachev, *On Severi's conjecture concerning simply connected algebraic surfaces*, Soviet Math. Dokl. **7** (1966), 1169–1171.

[Donaldson-Kronheimer90] S. Donaldson - P. Kronheimer, The geometry of four-manifolds, Clarendon, 1990.

[EGA60-67] A. Grothendieck – J. Dieudonné, Eléments de Géométrie Algébrique, vol. 4, 8, 11, 17, 20, 24, 28, 32, Publ. Math. IHES, 1960–67.

[Enriques1897] F. Enriques, *Sulle irrazionalità da cui può farsi dipendere la risoluzione di un'equazione algebrica ...*, Math. Ann. **49** (1897), 1–23.

[EMS96] A. Eskin - S. Mozes - N. Shaf, *Unipotent flows and counting lattice points on homogeneous varieties*, Ann. Math. **143** (1996), 253–299.

[Faltings83] G. Faltings, *Endlichkeitssätze für abelsche Varietäten über Zahlkörpern*, Inv. Math. (1983), 349–366.

[Fiorot-Jeannin92] J.C. Fiorot - P. Jeannin, Rational curves and surfaces, Applications to CAD, Wiley, 1992.

[Fletcher89] A. Fletcher, *Working with weighted complete intersections*, MPI Preprint (1989).

[FMT89] J. Franke - Y. I. Manin - Y. Tschinkel, *Rational points of bounded height on Fano varieties*, Inv. Math. **95** (1989), 421–436.

[Freitag-Kiehl88] E. Freitag - R. Kiehl, Etale cohomology and the Weil conjecture, Springer, 1988.

[Friedman-Morgan88a] R. Friedman - J. Morgan, *On the diffeomorphism types of certain algebraic surfaces I-II*, J. Diff. Geom. **27** (1988), 297–68 and 371–398.

[Friedman-Morgan88b] R. Friedman - J. Morgan, *Algebraic surfaces and 4-manifolds*, Bull. A.M.S. **18** (1988), 1–9.

[Friedman-Morgan94] R. Friedman - J. Morgan, Smooth four-manifolds and complex surfaces, Springer, 1994.

[Friedman-Qin95] R. Friedman - Z. Qin, *On complex surfaces diffeomorphic to rational surfaces*, Inv. Math. **120** (1995), 81–117.

[Fulton69] W. Fulton, Algebraic curves, Benjamin, 1969.

[Greenberg69] M. Greenberg, Lectures on forms in many variables, Benjamin, 1969.

[Griffiths-Harris78] P. Griffiths - J. Harris, Principles of Algebraic Geometry, John Wiley and Sons, Inc., 1978.

[Grothendieck68] A. Grothendieck, Cohomologie Locale des Faisceaux Cohérents et Théorèmes de Lefschetz Locaux et Globaux - SGA 2, North Holland, 1968.

[Gunning76] R. Gunning, Riemann surfaces and generalized theta functions, Springer, 1976.

[Hardy-Wright79] G. Hardy - E. Wright, An introduction to the theory of numbers, 5th ed., Clarendon, Oxford, 1979.

[Hartshorne77] R. Hartshorne, Algebraic Geometry, Springer, 1977.

[Hilbert1893] D. Hilbert, *Ueber die vollen Invariantensysteme*, Math. Ann. **42** (1893), 313–373.

[Hirzebruch54] F. Hirzebruch, *Some problems on differential and complex manifolds*, Ann. Math. **60** (1954), 213–236.

[Hirzebruch66] F. Hirzebruch, Topological methods in algebraic geometry, Springer, 1966.

[Hirzebruch-Kodaira57] F. Hirzebruch - K. Kodaira, *On the complex projective spaces*, J. Math. Pure. Appl. **36** (1957), 201–216.

[Hodge41] W. Hodge, The theory and applications of harmonic integrals, Cambridge Univ. Press, 1941.

[Iskovskikh80a] V. A. Iskovskikh, *Anticanonical models of three-dimensional algebraic varieties*, J. Soviet Math **13** (1980), 745–814.

[Iskovskikh80b] V. A. Iskovskikh, *Birational automorphisms of three-dimensional algebraic varieties*, J. Soviet Math **13** (1980), 815–868.

[Iskovskikh80c] V. A. Iskovskikh, *Minimal models of rational surfaces over arbitrary fields*, Math. USSR Izv. **14** (1980), 17–39.

[Iskovskikh-Manin71] V. A. Iskovskikh - Ju. I. Manin, *Three-dimensional quartics and counterexamples to the Lüroth problem*, Math. USSR Sbornik **15** (1971), 141–166.

[Kawamata92] Y. Kawamata, *Boundedness of Q-Fano threefolds*, Proc. Int. Conf. Algebra, Contemp. Math. vol 131, 1992, pp. 439–445.

[Kollár87] J. Kollár, *The structure of algebraic threefolds - an introduction to Mori's program*, Bull. AMS **17** (1987), 211–273.

[Kollár90] J. Kollár, *Minimal Models of Algebraic Threefolds: Mori's Program*, Astérisque **177–78** (1990), 303–26.

[Kollár91] J. Kollár, *Flips, Flops, Minimal Models, etc.*, Surv. in Diff. Geom. **1** (1991), 113–199.

[Kollár95] J. Kollár, *Nonrational hypersurfaces*, Jour. AMS **8** (1995), 241–249.

[Kollár96a] J. Kollár, Rational Curves on Algebraic Varieties, Springer Verlag, Ergebnisse der Math. vol. 32, 1996.

[Kollár96b] J. Kollár, *Nonrational covers of* $CP^m \times CP^n$ (1996), (to appear).

[Kollár et al.92] J. Kollár (with 14 coauthors), Flips and Abundance for Algebraic Threefolds, Astérisque, vol 211, 1992.

[Kollár-Matsusaka83] J. Kollár - T. Matsusaka, *Riemann-Roch type inequalities*, Amer. J. Math. **105** (1983), 229–252.

[KoMiMo92a] J. Kollár - Y. Miyaoka - S. Mori, *Rational Curves on Fano Varieties*, Proc. Alg. Geom. Conf. Trento, Springer Lecture Notes 1515, 1992, pp. 100–105.

[KoMiMo92b] J. Kollár - Y. Miyaoka - S. Mori, *Rationally Connected Varieties*, J. Alg. Geom. **1** (1992), 429–448.

[KoMiMo92c] J. Kollár - Y. Miyaoka - S. Mori, *Rational Connectedness and Boundedness of Fano Manifolds*, J. Diff. Geom. **36** (1992), 765–769.

[Lang86] S. Lang, *Hyperbolic and diophantine analysis*, Bull. AMS **14** (1986), 159–205.

[Lefschetz24] S. Lefschetz, L'Analysis Situs et la géometrie algébrique, Gauthier-Villars, 1924.

[Manin66] Yu. I. Manin, *Rational surfaces over perfect fields*, Publ. Math. IHES **30** (1966), 55–114.

[Manin72] Yu. I. Manin, Cubic forms (in Russian), Nauka, 1972; English translation, North-Holland, 1974; second enlarged edition, 1986.

[Manin93] Y. I. Manin, *Notes on the arithmetic of Fano threefolds*, Comp. Math. **85** (1993), 37–56.

[Manin-Tschinkel93] Y. I. Manin - Y. Tschinkel, *Points of bounded height on del Pezzo surfaces*, Comp. Math. **85** (1993), 315–332.

[Mazur92] B. Mazur, *The topology of rational points*, Exper. Math. **1** (1992), 35–46.

[McDuff-Salamon94] D. McDuff - D. Salamon, J-holomorphic curves and quantum cohomology, Univ. Lect. Notes, AMS, 1994.

[McDuff-Salamon95] D. McDuff - D. Salamon, Introduction to symplectic topology, Clarendon, 1995.

[Mori82] S. Mori, *Threefolds whose Canonical Bundles are not Numerically Effective*, Ann. of Math. **116** (1982), 133–176.

[Mori88] S. Mori, *Flip theorem and the existence of minimal models for 3-folds*, Journal AMS **1** (1988), 117–253.

[Morin40a] U. Morin, *Sulla razionalità dell' ipersuperficie cubica ...*, Rend. Sem. Math. Univ. Padova (1940), 108–112.

[Morin40b] U. Morin, *Sull' unirazionalità dell' ipersuperficie algebrica di qualunque ordine e dimensione sufficientemente alta*, Atti dell II Congresso Unione Math. Ital., 1940, pp. 298–302.

[Mukai89] S. Mukai, *Biregular classification of Fano threefolds*, Proc. Natl. Acad. Sci. **86** (1989), 3000–3002.

[Nadel91] A. M. Nadel, *The boundedness of degree of Fano varieties with Picard number one*, Jour. AMS **4** (1991), 681–692.

[Néron65] A. Néron, *Quasi-fonctions et hauteurs sur les variétés abéliennes*, Ann. Math. **82** (1965), 249–331.

[Noether1871] M. Noether, *Über Flächen, welche Schaaren rationaler Curven besitzen*, Math. Ann. **3** (1871), 161–227.

[Okonek-Teleman95] C. Okonek - A. Teleman, *Les invariants de Seiberg–Witten et la conjecture de Van de Ven*, C. R. Acad. Sci. **321** (1995), 457–461.

[Okonek-V.d.Ven86] C. Okonek - A. Van de Ven, *Stable bundles and differentiable structures on certain elliptic surfaces*, Inv. Math. **86** (1986), 357–370.

[Peyre93] E. Peyre, *Unramified cohomology and rationality problems*, Math. Ann. **296** (1993), 247–268.

[Picard-Simart1897] É. Picard - G. Simart, Théorie des fonctions algébriques, Gauthiers-Villars, 1897.

[Pidstrigach95] V. Pidstrigach, *Patching formulas for spin polynomials and a proof of the Van de Ven conjecture*, Izvestiya Russ. A.S. **45** (1995), 529–544.

[Pukhlikov87] A. V. Pukhlikov, *Birational isomorphisms of four dimensional quintics*, Inv. Math. **87** (1987), 303–329.

[Rilke30] R. M. Rilke, Gesammelte Werke, Insel–Verlag, Leipzig, 1930.

[Ruan93] Y. Ruan, *Symplectic topology and extremal rays*, Geometry and Functional Analysis **3** (1993), 395–430.

[Ruan94] Y. Ruan, *Symplectic topology on algebraic 3-folds*, J. Diff. Geom. **39** (1994), 215–227.

[Sarkisov81] V. G. Sarkisov, *Birational automorphisms of conic bundles*, Math. USSR Izv. **17** (1981), 177–202.

[Sarkisov82] V. G. Sarkisov, *On the structure of conic bundles*, Math. USSR Izv. **20** (1982), 355–390.

[Segre43] B. Segre, *A note on arithmetical properties of cubic surfaces*, J. London Math. Soc. **18** (1943), 24–31.

[Segre50] B. Segre, *Questions arithmétiques sur les variétés algèbriques*, Algèbre et Théorie des Nombres, CNRS, 1950, pp. 83–91.

[Segre51] B. Segre, *The rational solutions of homogeneous cubic equations in four variables*, Notae Univ. Rosario **2** (1951), 1–68.

[Serre73] J.-P. Serre, A course in arithmetic, Springer Verlag, 1973.

[Severi50] F. Severi, *La géométrie algébrique italienne*, Colloque de géométrie algébrique, Liège 1949, Masson, Paris, 1950, pp. 9–55.

[Shafarevich94] R. I. Shafarevich, Basic Algebraic Geometry I-II, Springer, 1994.

[Siegel69] C. L. Siegel, Topics in Complex Function Theory, I–III, Wiley, 1969.

[Silverman86] J. Silverman, The arithmetic of elliptic curves, Springer, 1986.

[Sullivan77] D. Sullivan, *Infinitesimal computations in topology*, Publ. Math. IHES **47** (1977), 269–332.

[Tregub93] S. Tregub, *Two remarks on four dimensional cubics*, Russ. Math. Surv. **48:2** (l993), 206–208.

[Tsen36] C. Tsen, *Quasi-algebraisch-abgeschlossene Funktionenkörper*, J. Chin. Math. **1** (1936), 81–92.

[Ueno75] K. Ueno, Classification Theory of Algebraic Varieties and Compact Complex Spaces, Springer Lecture Notes vol. 439, 1975.

[Vaughan-Wooley95] R. Vaughan - T. Wooley, *On a certain nonary cubic forms and related equations*, Duke Math. J. **80** (1995), 669–735.

[Vojta91] P. Vojta, *Arithmetic and hyperbolic geometry*, Proc. Int. Congr. Math. Kyoto 1990, Springer Verlag, pp. 757–765.

[Wall66] C.T.C. Wall, *Classification problems in differential topology V.*, Inv. Math. **1** (1966), 355–374.

[Weil46] A. Weil, Foundations of algebraic geometry, AMS, 1946.

[Wells73] R. Wells, Differential analysis on complex manifolds, Prentice-Hall, 1973.

[Yau77] S. T. Yau, *Calabi's conjecture and some new results in algebraic geometry*, Proc. Nat. Acad. USA . **74** (1977), 1789–1799.

[Zubr80] A. Zubr, *Classification of simply connected six-dimensional manifolds*, Dokl. A.N. CCCP **225** (1980), 1312–1315.

Progress in Mathematics, Vol. 168, © 1998 Birkhäuser Verlag Basel/Switzerland

Sufficient Conditions of the Uniform Integrability of Exponential Martingales

D.O. KRAMKOV[1] AND A.N. SHIRYAEV[2]

[1] Tokyo-Mitsubishi International plc, 6 Broadgate
London EC2M 2AA, GB
[2] Steklov Mathematical Institute, ul. Gubkina 8
Moscow GSP-1, 117966, Russia

1. Let $W = (W_t, \mathcal{F}_t, \mathbf{P})$ be the standard Wiener process on a filtered probability space $(\Omega, \mathcal{F}, \mathcal{F}_t)_{t\geq 0}, \mathbf{P})$ and let

$$Z_t = \exp\left(W_t - \frac{1}{2}t\right), \quad t \geq 0. \tag{1}$$

The process $Z = (Z_t, \mathcal{F}_t, \mathbf{P})_{t\geq 0}$ is a positive (exponential) martingale and by Doob's convergence theorem ([1], Chapter 1, §1e), with $\mathbf{P}$-probability one there exists a limit $Z_\infty = \lim_{t\to\infty} Z_t$; in addition, $Z_\infty = 0$ ($\mathbf{P}$-a.s.) and therefore $\mathbf{E}Z_\infty = 0$.

It is of interest in many issues of the stochastic calculus to describe the Markov times $\tau = \tau(\omega)$ such that $\mathbf{E}Z_\tau = 1$, which is equivalent to the condition that the family of random variables $\{Z_{t\wedge\tau}, t \geq 0\}$ be uniformly integrable with respect to $\mathbf{P}$.

Obviously, $\mathbf{E}Z_\tau = 1$ for each bounded Markov time τ, that is, for a time such that $\tau(\omega) \leq N$ for some constant N and all $\omega \in \Omega$.

Is was shown in [2] that if there exists $\varepsilon > 0$ that $\mathbf{E}\exp((1+\varepsilon)\tau) < \infty$, then $\mathbf{E}Z_\tau = 1$.

This condition was weakened in [3]; namely, it was shown there that if $\mathbf{E}\exp((1/2+\varepsilon)\tau) < \infty$ for some $\varepsilon > 0$, then $\mathbf{E}Z_\tau = 1$.

Novikov [4] showed that one can set $\varepsilon = 0$ in the above condition, that is,

$$\mathbf{E}\exp\left(\frac{1}{2}\tau\right) < +\infty \Rightarrow \mathbf{E}Z_\tau = 1. \tag{2}$$

Some time later Kazamaki [6] proved that

$$\sup_{t\geq 0} \mathbf{E}\exp\left(\frac{1}{2}W_{\tau\wedge t}\right) < +\infty \Rightarrow \mathbf{E}Z_\tau = 1. \tag{3}$$

Moreover, the 'Kazamaki condition' turned out to be weaker than the 'Novikov condition' since

$$\sup_{t\geq 0} \mathbf{E}\exp\left(\frac{1}{2}W_{\tau\wedge t}\right) \leq \left[\mathbf{E}\exp\left(\frac{\tau}{2}\right)\right]^{1/2}.$$

In [5], Novikov weakened the assumption $\mathbf{E}\exp(\frac{1}{2}\tau) < +\infty$ by showing that if $\phi = \phi(t)$ is a *lower* function for the Wiener process (that is, $\mathbf{P}(W_t < \phi(t), t \to \infty) = 0$), then

$$\mathbf{E}\exp\left(\frac{1}{2}\tau - \phi(\tau)\right) < +\infty \Rightarrow \mathbf{E}Z_\tau = 1.$$

(See, for example, [7] for the well-known 'Kolmogorov test' for lower and upper functions.)

2. The aim of this paper is to present a fairly elementary proof of the above results and some of their generalizations under the assumption that we restrict ourselves to the consideration of Markov moments $\tau = \tau(\omega)$ that are measurable with respect to the flow of σ-algebras $(\mathcal{F}_t^W)_{t\geq 0}$ generated by the Wiener process and to (lower) functions ϕ in the class Φ_* of non-negative measurable functions $\phi = \phi(t)$ such that

$$\limsup_{t\to\infty}(W_t - \phi(t)) = +\infty \quad (\mathbf{P}\text{-a.s.}). \tag{4}$$

Now let $\mathcal{M}_N = \{\sigma = \sigma(\omega) : \ \sigma(\omega) \leq N, \omega \in \Omega\}$ be the set of bounded (by a constant N) Markov times (with respect to $(\mathcal{F}_t^W)_{t\geq 0}$).

Theorem 1. *Let τ be a Markov time (with respect to $(\mathcal{F}_t^W)_{t\geq 0}$) and let $\phi \in \Phi_*$. Then each of the conditions*

$$\lim_{N\to\infty} \sup_{\sigma\in\mathcal{M}_N} \mathbf{E}\exp\left\{\frac{1}{2}(\tau\wedge\sigma) - \phi(\tau\wedge\sigma)\right\} < +\infty \tag{5}$$

and

$$\lim_{N\to\infty} \sup_{\sigma\in\mathcal{M}_N} \mathbf{E}\exp\left\{\frac{1}{2}W_{\tau\wedge\sigma} - \phi(\tau\wedge\sigma)\right\} < +\infty \tag{6}$$

is sufficient for the equality $\mathbf{E}Z_\tau = 1$.

3. Before proceeding to the proof itself we give several explanations for our assumption that the Markov moments in question must be $(\mathcal{F}_t^W)_{t\geq 0}$-measurable. The reasons for this are as follows.

Traditionally, one proves the equality $\mathbf{E}Z_\tau = 1$ by verifying directly that the family $\{Z_{t\wedge\tau}, t \geq 0\}$ is uniformly integrable with respect to $\mathbf{P}$. Our proof of this equality is based on another idea.

Since we assume that all the Markov moments in question are $(\mathcal{F}_t^W)_{t\geq 0}$-measurable, it suffices to consider the probability distribution μ_0 of the Wiener process in the space of continuous functions $x = (x_t)_{t\geq 0}$ rather than a measure on $(\Omega, \mathcal{F})$. The key point here is the fact that, in this space, there also exist measures μ_a that are the distributions of Wiener processes with drift at, $t \geq 0$. If $a \neq 0$, then the measures μ_a and μ_0 are singular; however, they are locally equivalent in the following sense: their restrictions μ_0^t and μ_a^t to the σ-algebras $\sigma(x : \ x_s, s \leq t)$ are equivalent for each $t \geq 0$, with Radon-Nikodým derivative

$$\frac{d\mu_a^t}{\mu_0^t}(x) = e^{ax_t - \frac{a^2}{2}t}$$

cf. (1)).

To avoid new notation we shall assume that the original probability space $(\Omega, \mathcal{F}, (\mathcal{F}_t)_{t\geq 0}, \mathbf{P})$ is the coordinate space of continuous functions $\omega = (\omega_t)_{t\geq 0}$, $\mathbf{P}$ is the Wiener measure, $W_t(\omega) = \omega_t$, and $\mathcal{F} = \vee \mathcal{F}_t$. We also set $\widetilde{W_t}(\omega) = W_t(\omega) - t$ and we denote by $\tilde{\mathbf{P}}$ the probability distribution of the process $\widetilde{W} = (\widetilde{W_t})_{t\geq 0}$.

Also, let $\mathbf{P}_t = \mathbf{P}|\mathcal{F}_t$ and $\tilde{\mathbf{P}}_t = \tilde{\mathbf{P}}|\mathcal{F}_t$, $t \geq 0$. As pointed out above, $\tilde{\mathbf{P}} \perp \mathbf{P}$, but $\tilde{\mathbf{P}} \overset{\text{loc}}{\sim} \mathbf{P}$ in the sense of the equivalence $\tilde{\mathbf{P}}_t \sim \mathbf{P}_t$, $t \geq 0$; in addition,

$$\frac{d\tilde{P}_t}{d\mathbf{P}_t}(\omega) = e^{W_t(\omega) - \frac{1}{2}t} \quad (= Z_t(\omega)) \tag{7}$$

Remark 1. Rather than assuming that the initial probability space has 'coordinate' structure we could assume that $(\Omega, \mathcal{F}, (\mathcal{F}_t)_{t\geq 0}, \mathbf{P})$ is an arbitrary space, but $\mathcal{F} = \vee \mathcal{F}_t$ and there exists a measure $\tilde{\mathbf{P}}$ in this space such that $\tilde{\mathbf{P}} \overset{\text{loc}}{\sim} \mathbf{P}$ and (7) holds. Under this assumption we can consider arbitrary Markov moments (that is, they must be $(\mathcal{F}_t)_{t\geq 0}$-measurable, but not necessarily $(\mathcal{F}_t^W)_{t\geq 0}$-measurable).

4. By our assumption that $\tilde{\mathbf{P}} \overset{\text{loc}}{\sim} \mathbf{P}$, for each $t \geq 0$ and $A \in \mathcal{F}_t$ we have

$$\tilde{\mathbf{E}} I_A = \mathbf{E} I_A \cdot Z_t,$$

where $\tilde{\mathbf{E}}$ is the expectation with respect to $\tilde{\mathbf{P}}$, and

$$\tilde{\mathbf{E}} I_A I_{\tau<\infty} = \mathbf{E} I_A I_{\tau<\infty} \cdot Z_\tau \tag{8}$$

for each τ and all $A \in \mathcal{F}_\tau$. (See, for example, [1], Chapter 3, Theorem 3.4 for greater detail.)

By (8) we obtain

$$\tilde{\mathbf{E}}\left(f I_{(\tau<\infty)}\right) = \mathbf{E}\left(f I_{(\tau<\infty)} Z_\tau\right) \tag{9}$$

and

$$\mathbf{E}\left(f I_{(\tau<\infty)}\right) = \tilde{\mathbf{E}}\left(f I_{(\tau<\infty)} Z_\tau^{-1}\right) \tag{10}$$

for each non-negative $\mathcal{F}_\tau$-measurable function $f = f(\omega)$.

In particular, by (9) we obtain

$$\tilde{\mathbf{P}}(\tau < \infty) = \mathbf{E}\left(I_{(\tau<\infty)}\right) = \mathbf{E} Z_\tau \tag{11}$$

because $Z_\infty = 0$ ($\{\tau = \infty\}$; $\mathbf{P}$-a.s.).

By (11) we obtain

$$\mathbf{E} Z_\tau = 1 \Leftrightarrow \tilde{\mathbf{P}}(\tau < \infty) = 1. \tag{12}$$

By (10) with $f(\omega) = \exp\left(\frac{1}{2}\tau - \phi(\tau)\right)$, bearing in mind that τ is bounded ($\tau \in \mathcal{M}_N$ for some $N < \infty$), we obtain

$$\begin{aligned} \mathbf{E} e^{\frac{1}{2}\tau - \phi(\tau)} &= \tilde{\mathbf{E}} e^{\frac{1}{2}\tau - \phi(\tau)} e^{-W_\tau + \frac{1}{2}\tau} = \\ \tilde{\mathbf{E}} e^{\tau - W_\tau - \phi(\tau)} &= \tilde{\mathbf{E}} e^{-\tilde{W}_\tau - \phi(\tau)} \end{aligned} \tag{13}$$

and

$$\mathbf{E}e^{\frac{1}{2}(W_\tau-\phi(\tau)} = \tilde{\mathbf{E}}e^{\frac{1}{2}(W_\tau-\phi(\tau))}e^{-W_\tau+\frac{1}{2}\tau} = \tilde{\mathbf{E}}e^{-\frac{1}{2}(W_\tau-\tau)-\frac{1}{2}\phi(\tau)} = \tilde{\mathbf{E}}e^{-\frac{1}{2}\tilde{W}_\tau-\frac{1}{2}\phi(\tau)}. \tag{14}$$

5. To prove the theorem we require the following result.

Lemma 1. *Let $\phi \in \Phi_*$ and let $A \in \mathcal{F} = \vee\mathcal{F}_t$. Then the following results are equivalent:*

1. $\mathbf{P}(A) = 0$;
2. $$\sup_{N\geq 0,\sigma\in\mathcal{M}_N} \mathbf{E}e^{a(W_\sigma-\phi(\sigma))}I_A < \infty$$

where a is a positive number.

Proof. Clearly, condition (1) yields (2).

Assume now that (2) holds, but $\mathbf{P}(A) = \varepsilon > 0$ for some $\varepsilon > 0$. From definition (4) of the class of lower functions $\phi \in \Phi_*$ we see that for each $L > 0$ there exists a finite Markov moment $\gamma = \gamma(L, \varepsilon)$ such that $\mathbf{P}(W_\gamma - \phi(\gamma) \geq L) \geq 1 - \varepsilon/4$.

Now let $N(L, \varepsilon)$ be a number such that $\mathbf{P}(\gamma \geq N(L, \varepsilon)) \leq \varepsilon/4$. Then $\mathbf{P}((W_\gamma - \phi(\gamma) \geq L) \cap A \cap (\gamma \leq N(L, \varepsilon))) \geq \varepsilon/2$ and therefore

$$\mathbf{P}\left((W_{\gamma\wedge N(L,\varepsilon)} - \phi(\gamma \wedge N(L,\varepsilon)) \geq L)\bigcap A\right) \geq \varepsilon/2.$$

Hence

$$\sup_{N\geq 0,\sigma\in\mathcal{M}_N} \mathbf{E}e^{a(W_\sigma-\phi(\sigma))}I_A \geq e^L\frac{\varepsilon}{2} \to \infty, \quad L \to \infty,$$

which contradicts assumption (2).

6. *Proof of Theorem 1* As is clear from (12), we must prove that the equality $\tilde{\mathbf{P}}(\tau = \infty) = 0$ is a consequence of each of the conditions (5) and (6).

By (13),

$$\begin{aligned}\infty &> \lim_{N\to\infty}\sup_{\sigma\in\mathcal{M}_N} \mathbf{E}e^{\frac{1}{2}(\tau\wedge\sigma)-\phi(\tau\wedge\sigma)} \\ &= \lim_{N\to\infty}\sup_{\sigma\in\mathcal{M}_N} \tilde{\mathbf{E}}e^{-\tilde{W}_{\tau\wedge\sigma}-\phi(\tau\wedge\sigma)} \\ &\geq \lim_{N\to\infty}\sup_{\sigma\in\mathcal{M}_N} \tilde{\mathbf{E}}e^{-\tilde{W}_\sigma-\phi(\sigma)}I(\tau=\infty)\end{aligned} \tag{15}$$

Since $\widehat{W} = -\widetilde{W}$ is the Wiener process with respect to the measure $\tilde{\mathbf{P}}$, it follows from the lemma that $\tilde{\mathbf{P}}(\tau = \infty) = 0$, and therefore $\mathbf{E}Z_\tau = 1$.

Further, by (14) we obtain

$$\begin{aligned}\infty &> \lim_{N\to\infty}\sup_{\sigma\in\mathcal{M}_N} \mathbf{E}e^{\frac{1}{2}(W_{\tau\wedge\sigma}-\phi(\tau\wedge\sigma))} \\ &= \lim_{N\to\infty}\sup_{\sigma\in\mathcal{M}_N} \tilde{\mathbf{E}}e^{-\frac{1}{2}\tilde{W}_{\tau\wedge\sigma}-\frac{1}{2}\phi(\tau\wedge\sigma)} \\ &\geq \lim_{N\to\infty}\sup_{\sigma\in\mathcal{M}_N} \tilde{\mathbf{E}}e^{-\frac{1}{2}\tilde{W}_\sigma-\frac{1}{2}\phi(\sigma)}I(\tau=\infty)\end{aligned}$$

and again, we see in the same way as in (15) that $\tilde{\mathbf{P}}(\tau = \infty) = 0$ and therefore $\mathbf{E}Z_\tau = 1$. This completes the proof.

7. Corollary 1. *Let $\phi \in \Phi_*$ and assume that $\frac{1}{2}t - \phi(t)$ is a non-decreasing function. Then*

$$\mathbf{E}\exp\left(\frac{1}{2}\tau - \phi(\tau)\right) < \infty \Rightarrow \mathbf{E}Z_\tau = 1.$$

Corollary 2. *Assume that $\phi_t \equiv a > 0$. Then $\phi = (\phi_t)_{t\geq 0} \in \Phi_*$ and*

$$\mathbf{E}\sqrt{\tau_a} = \infty$$

for $\tau_a = \inf\{t:\ W_t = a\}$. Hence $\mathbf{E}e^{\frac{1}{2}\tau_a} = \infty$ and one cannot use (2) to find out whether $\mathbf{E}Z_\tau = 1$. However, the 'Kazamaki condition' is well known to hold in this case since $W_{\tau_a\wedge t} \leq a$; consequently, $\mathbf{E}Z_{\tau_a} = 1$ by (3).

Corollary 3. *Let $\phi = (\phi_t)_{t\geq 0}$ be a positive continuous function such that $\phi(t) \leq \phi_\varepsilon(t)$ for large t, where*

$$\phi_\varepsilon(t) = (1-\varepsilon)\sqrt{2t\log\log t}$$

for some ε, $0 < \varepsilon < 1$. Then the function $\phi_\varepsilon = \phi_\varepsilon(t)$ is in the class Φ_* and if

$$\tau_\varepsilon = \inf\{t:\ W_t = \phi_\varepsilon(t)\},$$

then $W_{\tau_\varepsilon\wedge\sigma} \leq \phi_\varepsilon(\tau\wedge\sigma)$ for each $\sigma \in \mathcal{M}_N$, $N \geq 0$, therefore condition (6) holds and $\mathbf{E}Z_{\tau_\varepsilon} = 1$. Thus, for $\phi(t)$ under consideration we obtain by the law of the iterated logarithm that $\mathbf{E}Z_\tau = 1$, where τ is the first hitting time.

8. The following example is very interesting in the sense that it shows that if ϕ in (5) is not a lower function, then the equality $\mathbf{E}Z_\tau = 1$ may fail for some Markov moments τ.

Actually, for each continuous function $\phi = \phi(t)$ we have with probability one that either $\limsup_{t\to\infty}(W_t - \phi(t)) = +\infty$, or $\limsup_{t\to\infty}(W_t - \phi(t)) = -\infty$.

Hence it is easy to see that if $\tau_N = \inf\{t:\ W_t - t - \phi(t) \geq N\}$ is a Markov moment, then the probability $\tilde{\mathbf{P}}(\tau_N = \infty) = \tilde{\mathbf{P}}(\widetilde{W}_t - \phi(t) < N,\ t \geq 0)$ is positive for N sufficiently large and therefore $\mathbf{E}Z_{\tau_N} < 1$, while both conditions (5) and (6) are satisfied.

9. We now present several properties of $m(a) = \mathbf{E}e^{W_\tau + a\tau}$ regarded as a function of $a \in \mathbf{R}$ for fixed time τ.

First, it is clear that $m(a) \leq 1$ for each $a \leq -1/2$. We claim that $m(a) \geq 1$ for each $a > 0$.

We set

$$X_t = e^{\frac{1}{p}W_t + \frac{a}{p}t}, \quad Y_t = e^{-\frac{1}{p}W_t - \frac{q}{2p^2}t}$$

where $\frac{1}{p} + \frac{1}{q} = 1$ and $1 < q < 1 + 2a$. Then

$$X_tY_t = e^{(\frac{a}{p} - \frac{q}{2p^2})t} \geq 1$$

because $\frac{a}{p} - \frac{q}{2p^2} \geq 0$.

The process $(Y_t^q)_{t\geq 0}$ is a supermartingale with $\|Y_\tau\|_q \leq 1$. Hence

$$\left(\mathbf{E}e^{W_\tau + a\tau}\right)^{1/p} = \|X_\tau\|_p \geq \|X_\tau\|_p \|Y_\tau\|_q \geq \mathbf{E}X_\tau Y_\tau \geq 1, \tag{16}$$

by the Hölder inequality, which just shows that $m(a) \geq 1$ for $a > 0$.

Further, we note that

$$\mathbf{E}X_\tau Y_\tau = \exp\left(\frac{a}{p} - \frac{q}{2p^2}\right)\tau\,.$$

If $a \geq \frac{3}{2}$, then taking $p = q = 2$ we deduce from (16) that

$$\left(\mathbf{E}e^{(W_\tau + a\tau)}\right)^{1/p} \geq \mathbf{E}e^{\frac{1}{2}\tau}\,.$$

Consequently, if $m(-\frac{1}{2}) < 1$, then $m(a) = \infty$ for all $a \geq \frac{3}{2}$, because if $m(-\frac{1}{2}) < 1$, then $\mathbf{E}e^{\frac{1}{2}\tau} = \infty$.

10. In conclusion, we point out that the method used in the proof of the above theory 'works' also for an arbitrary continuous local martingale $M = (M_t, \mathcal{F}_t, \mathbf{P})_{t\geq 0}$ with $\langle M\rangle_\infty = \infty$ in place of the Wiener process.

Let

$$Z_t = e^{M_t - \frac{1}{2}\langle M\rangle_t}$$

and assume that there exists a measure $\tilde{\mathbf{P}}$ such that its restrictions $\tilde{\mathbf{P}}_t = \tilde{\mathbf{P}}|\mathcal{F}_t$ are equivalent to $\mathbf{P} = \mathbf{P}|\mathcal{F}_t$ and

$$\frac{d\tilde{\mathbf{P}}_t}{d\mathbf{P}_t} = Z_t.$$

Then assuming that $\mathcal{F} = \vee\mathcal{F}_t$ we obtain the following result: in order that $\mathbf{E}Z_\tau = 1$, where τ is some Markov moment with respect to $(\mathcal{F}_t)_{t\geq 0}$, either of the two conditions

$$\lim_{N\to\infty} \sup_{\sigma\in\mathcal{M}_N} \mathbf{E}\exp\left\{\frac{1}{2}\langle M\rangle_{\tau\wedge\sigma} - \phi(\langle M\rangle_{\tau\wedge\sigma})\right\} < \infty,$$

$$\lim_{N\to\infty} \sup_{\sigma\in\mathcal{M}_N} \mathbf{E}\exp\left\{\frac{1}{2}\langle M\rangle_{\tau\wedge\sigma} - \phi(\langle M\rangle_{\tau\wedge\sigma})\right\} < \infty,$$

is sufficient, where $\phi = \phi(t)$ is some function in the class Φ_*.

Remark 2. It is worth noting that if $\gamma_t = \inf\{u : \langle M\rangle_u \geq t\}$, then $W = (W_t)_{t\geq 0}$ with $W_t = M_{\gamma_t}$ is a Wiener process. Hence if $\phi \in \Phi_*$, then $\limsup_{t\to\infty}(M_t - \phi(\langle M\rangle_t)) = \infty$ ($\mathbf{P}$-almost surely).

References

[1] J. Jacod and A.N. Shiryaev, Limit Theorems for Stochastic Processes, Springer-Verlag, Berlin–New York, 1987.

[2] I.I. Gikhman and A.V. Skorokhod, Stochastic differential equations and their applications, Naukova Dumka, Kiev, 1982 (in Russian).

[3] R.Sh. Liptzer and A.N. Shiryaev, On the absolute continuity of measures corresponding to processes of diffusion type with respect to the Wiener measures, Izv. Akad. Nauk SSSR Ser. Mat., v. 36 (1972), p. 847–889.

[4] A.A. Novikov, On discontinuous martingales, Theory Probab. Appl., v. 20 (1975), p. 13–28.

[5] A.A. Novikov, Condition for uniform integrability of continuous nonnegative martingales, Theory Probab. Appl., v. 24 (1979).

[6] N. Kazamaki, On a problem of Grisanov, Tôhoku Math. J., v. 29 (1977), p. 597–600.

[7] N.V. Krylov, Introduction to the theory of Diffusion Processes (Transl. Math. Monogr. v. 142), Amer. Math. Soc., Princeton, NJ.

Progress in Mathematics, Vol. 168, © 1998 Birkhäuser Verlag Basel/Switzerland

On the Casson Invariant

CHRISTINE LESCOP

CNRS, Institut Fourier (UMR 5582), B.P.74
38402 Saint-Martin-d'Hères cedex, France
e-mail: lescop@fourier.ujf-grenoble.fr

ABSTRACT. The Casson invariant is a topological invariant of closed oriented 3-manifolds. It is an integer that counts the $SU(2)$-representations of the fundamental group of these manifolds in a sense introduced by Casson in 1985. Its first properties allowed Casson to solve famous problems in 3-dimensional topology.

The Casson invariant can also be independently defined in a combinatorial way as a function of Alexander polynomials of framed links presenting the 3-manifolds. It has numerous interesting properties: It behaves nicely under most topological mutations such as orientation reversal, connected sum, surgery, regluing along surfaces... This makes it easy to compute and to use. The Casson invariant contains the Rohlin invariant of $\mathbf{Z}/2\mathbf{Z}$-homology spheres, that is the signature of any smooth spin 4-manifold bounded by such a sphere. It is also explicitly related to quantum invariants and is the first finite type invariant in the sense of Ohtsuki. (This Ohtsuki notion of finite type invariants for 3-manifolds is analogous to the Vassiliev notion for knots.)

This talk will be a general presentation of the Casson invariant where its newest properties and developments will be emphasized.

1. Introduction

The Casson invariant is the simplest topological invariant of 3-manifolds among the invariants introduced after 1984. It interacts with many domains of mathematics. Casson defined it using the differential topology of $SU(2)$-representation spaces. It can be extended in the spirit of this definition using symplectic geometry [W, CLM, C]. It can also be defined using infinite-dimensional analysis and gauge theory [Ta]: it is the Euler characteristic of the Floer homology [BD]. Its natural connections with the structure of the mapping class group have been investigated in a series of articles of S. Morita [Mo2] who interpreted it as a certain secondary invariant associated with the first characteristic class of surface bundles. Its relationship with quantum invariants [Mu] links it to quantum physics; and the Dedekind sums which often show up in the study of its topological properties show that it even interacts with arithmetic.

Here, with the aim of being as elementary as possible, we will present a combinatorial definition of the Casson invariant, and describe most of its topological properties.

2. Surgery presentations of 3-manifolds

Here, all the manifolds are compact and oriented (unless otherwise mentioned). Boundaries are oriented with the "outward normal first" convention. We usually work with topological manifolds, but since any topological manifold of dimension less or equal than 3 has exactly one C^∞-structure (see [Ku]), we will sometimes make incursions in the smooth category. Manifolds are always considered up to oriented homeomorphism and embeddings and homeomorphisms are considered up to ambient isotopy.

Given a 3-manifold M, a knot K of M (that is an embedding of the circle S^1 in the interior of M), and a parallel μ of K (that is a closed curve on the boundary of a tubular neighborhood $T(K)$ of K which runs parallel to K), we can define the *manifold* $\chi(M;(K,\mu))$ *obtained from* M *by surgery on* K *with respect to* μ by the following construction: remove the interior of $T(K)$ from M and replace it by another solid torus $D^2 \times S^1$ glued along the boundary $\partial T(K)$ of $T(K)$ by a homeomorphism from $\partial T(K)$ to $\partial D^2 \times S^1$ which maps μ to $\partial D^2 \times \{1\}$.

$$\chi(M;(K,\mu)) = \overline{M \setminus T(K)} \cup_{\partial T(K) \sim \partial D^2 \times S^1} D^2 \times S^1$$

Note that $\chi(M;(K,\mu))$ is well-defined. Indeed, it is obtained by first gluing a thickened disk to $\overline{M \setminus T(K)}$ along an annulus around μ in $\partial T(K)$ and by next filling in the resulting sphere S^2 in the boundary by a standard 3-ball B^3.

(K,μ) is called a *framed knot*. A collection of disjoint framed knots is called a framed link. Surgery on framed links is the natural generalization of surgery on framed knots. The surgery is performed on each component of the link. The first reason for taking this operation into consideration is the following theorem which is proved in a very elegant way in [Ro].

THEOREM 2.1 (LICKORISH [LI], WALLACE [W] 1960) *Any closed (i.e. compact, connected, without boundary) 3-manifold can be obtained from the standard 3-sphere* S^3 *by surgery on a framed link.*

We now define linking numbers to help parametrize surgeries. Let J and K be two disjoint knots in a 3-manifold M which are rationally null-homologous (i.e. null in $H_1(M;\mathbf{Q})$). Then there is a surface Σ embedded in $\overline{M \setminus T(K)}$, whose boundary lies in $\partial T(K)$ and is homologous to $d[K]$ in $T(K)$ for some nonzero $d \in \mathbf{Z}$; the *linking number* of J and K, $lk(J,K)$, is unambiguously defined as the algebraic intersection number of J and Σ divided out by d. It is symmetric.

For $R = \mathbf{Z}, \mathbf{Q}$ or $\mathbf{Z}/2\mathbf{Z}$, a 3-manifold with the same R-homology $H_*(.;R)$ as S^3 is called an R-sphere. When M is a $\mathbf{Q}$-sphere, the isotopy class in $\partial T(K)$ of the *characteristic curve* μ of the surgery is specified by the linking number of μ and K in M and the framed knot (K,μ) is also denoted by $(K, lk(K,\mu))$. In particular, a framed link in S^3 is a link of S^3, each component of which is equipped with an integer.

The second reason for studying surgeries is the Kirby calculus which relates two surgery presentations of the same 3-manifold. The following is the Fenn and Rourke version of the Kirby calculus:

Figure 1: A surgery presentation of the Poincaré sphere (see [R])

THEOREM 2.2 (FENN-ROURKE [FR], KIRBY [K] 1978) *Any two framed links of S^3 presenting the same 3-manifold can be obtained from each other by a finite number of FR-moves, w.r.t. the following description of FR-moves.*

Let $\mathbf{L}$ be a framed link in S^3 such that a component $\mathbf{U}$ of $\mathbf{L}$ is a trivial knot U equipped with a parallel μ_U satisfying $lk(U, \mu_U) = \varepsilon = \pm 1$. Consider a cylinder $I \times D^2$ embedded in $\overline{S^3 \setminus T(U)}$ so that $I \times S^1$ is embedded in $\partial T(U)$. Let τ be the homeomorphism of $\overline{S^3 \setminus T(U)}$ which is the identity outside the cylinder, and which twists the cylinder around its axis so that μ_U is mapped to the meridian of U. Clearly, $\tau(\mathbf{L} \setminus \mathbf{U})$ presents the same 3-manifold as $\mathbf{L}$ (where we think of framed links as links equipped with curves to give a meaning to $\tau(\mathbf{L} \setminus \mathbf{U})$). We define an FR-move as the operation described above, which transforms $\mathbf{L}$ into $\tau(\mathbf{L} \setminus \mathbf{U})$, or its inverse.

According to the above theorem, in order to define an invariant of closed 3-manifolds, it suffices to find a function of surgery presentations invariant under FR-moves. Due to a lack of good candidates, this process had not been used before 1988. Since then, with the invasion of quantum invariants, there are a lot of 3-manifolds invariants which have been proved to be invariant using this simple principle ([RT], [W],...) but for most of them a topological interpretation is still to be found. Here, I propose to introduce such an invariant function, and then give the topological interpretation of the yielded invariant of 3-manifolds. This will be a generalization of the Casson invariant.

3. A combinatorial definition of the Casson invariant

In order to introduce our invariant function $\mathbf{F}$, we need some notation. Let $\mathbf{L} = (\mathbf{K}_i)_{i \in N}$ be a framed link in a $\mathbf{Q}$-sphere M, $\mathbf{K}_i = (K_i, \mu_i) = (K_i, lk(\mu_i, K_i))$. $N = \{1, \dots, n\}$ is the set of indices of the components of $\mathbf{L}$. For a subset I of N, $\mathbf{L}_I = (\mathbf{K}_i)_{i \in I}$. $E(\mathbf{L}) = [\ell_{ij} = lk(\mu_i, K_j)]_{i,j=1,\dots,n}$ denotes the symmetric *linking matrix* of $\mathbf{L}$. $b^-(\mathbf{L})$ (resp. $b^+(\mathbf{L})$) is the number of negative (resp. positive) eigenvalues of $E(\mathbf{L})$. signature$(E(\mathbf{L})) = b^+(\mathbf{L}) - b^-(\mathbf{L})$. For a $\mathbf{Z}$-module A, $|A|$ denotes the order of A, that is its cardinality if A is finite and 0 otherwise. Note that

$$|H_1(\chi(M; \mathbf{L}))| = (-1)^{b^-(\mathbf{L})} \det(E(\mathbf{L}))|H_1(M)|$$

(Unless otherwise mentioned, the homology coefficients are the integers.)

Now, we can set:

$$\mathbf{F}_M(\mathbf{L}) = (-1)^{b^-(\mathbf{L})} \sum_{I \subset N, I \neq \emptyset} \det(E(\mathbf{L}_{N \setminus I}))\alpha(\mathbf{L}_I) + |H_1(\chi(M; \mathbf{L}))| \frac{\text{signature}(E(\mathbf{L}))}{8}$$

with

$$\alpha(\mathbf{L}_I) = |H_1(M)| \left(\tilde{\zeta}(L_I) + \frac{(-1)^{\sharp I}}{24} L_8(\mathbf{L}_I) \right)$$

where $L_8(\mathbf{L})$ and $\tilde{\zeta}(L)$ are described below.

$L_8(\mathbf{L})$ is the following homogeneous polynomial in the coefficients of the linking matrix. Let G be a graph whose vertices are indexed by N; for an edge e of G whose ends are indexed by i and j, we set $lk(\mathbf{L}; e) = \ell_{ij}$. Next we define $lk(\mathbf{L}; G)$ as the product running over all edges e of G of the $lk(\mathbf{L}; e)$. Now, $L_8(\mathbf{L})$ is the sum of the $lk(\mathbf{L}; G)$ running over all graphs G whose vertices are the elements of N, and whose underlying spaces have the form of a figure eight made of two oriented distinguished circles (North and South) with one common vertex.

The coefficient $\tilde{\zeta}$ can be defined from the several-variable Alexander polynomial Δ (as defined and normalized in [Ha] and [BL2] or in Section 6 below) for several component links, and from the classical Alexander polynomial Δ of knots which is the order of the H_1 of the infinite cyclic covering of $M \setminus K$, viewed as a natural $\mathbf{Z}[t, t^{-1}]$-module, normalized in such a way that $\Delta(1) > 0$ and Δ is symmetric.

$$\tilde{\zeta}(L) = \begin{cases} (-1)^{n-1} \frac{\partial^n \Delta}{\partial t_1 \ldots \partial t_n}(L)(1, \ldots, 1) & \text{if } n > 1 \\ \frac{O_M(K_1)}{2|H_1(M)|} \Delta''(K_1)(1) + \frac{(-1)}{24} \left(1 + \frac{1}{O_M(K_1)^2}\right) & \text{if } n = 1 \end{cases}$$

where $O_M(K_1) = |H_1(M)| / |\mathrm{Torsion}(H_1(M \setminus K_1))|$ is the order of the class of K_1 in $|H_1(M)|$. It is one if M is a $\mathbf{Z}$-sphere.

We can now state the theorem:

THEOREM 3.1 ([L3]) *There exists a rational topological invariant λ of closed 3-manifolds such that: for any framed link $\mathbf{L}$ in S^3,*

$$\lambda(\chi(S^3; \mathbf{L})) = \mathbf{F}_{S^3}(\mathbf{L}).$$

The so-defined λ-invariant satisfies the more general surgery formula:

PROPERTY 1 *For any framed link $\mathbf{H}$ in a $\mathbf{Q}$-sphere M,*

$$\lambda(\chi(M; \mathbf{H})) = \frac{|H_1(\chi(M; \mathbf{H}))|}{|H_1(M)|} \lambda(M) + \mathbf{F}_M(\mathbf{H}).$$

The principle behind the proof of the theorem is very simple. According to the Fenn and Rourke version of the Kirby theorem, it suffices to show the invariance of $\mathbf{F}$ under a FR-move. Now, the function $\mathbf{F}$ is a function of homological invariants of the exterior of the framed link whose variation under a homeomorphism of this exterior can be followed (with some combinatorial efforts). See [L3].

The proof of the general surgery formula rests on the same remark. Take a surgery presentation $\mathbf{L}$ of the $\mathbf{Q}$-sphere M. By transversality, we may assume that $\mathbf{H}$ is disjoint from the link $\hat{L}$ made of the cores of the new solid tori glued by the surgery. Then the surgery presentation $\mathbf{H}$ can be seen in S^3. (Again we think of it as a link equipped with characteristic curves.) Now, the equality to be shown is:

$$\mathbf{F}_{S^3}((\mathbf{H} \subset S^3) \cup \mathbf{L}) = \frac{|H_1(\chi(M; \mathbf{H}))|}{|H_1(M)|} \mathbf{F}_{S^3}(\mathbf{L}) + \mathbf{F}_M(\mathbf{H})$$

where both sides are functions of homological invariants of

$$S^3 \setminus (H \cup L) = M \setminus (H \cup \hat{L}),$$

equipped with the surgery curves.

Because of the form of the surgery formula, it is easy to compare the λ-invariant with the Rohlin invariant for $\mathbf{Z}/2\mathbf{Z}$-spheres. Before stating the comparison property, let us give a definition of this invariant discovered in 1952. A spin structure on a smooth manifold of dimension greater than or equal to 3 is a homotopy class of trivializations of its tangent bundle over its 2-skeleton. (See [Mi2] for other definitions.) The Rohlin invariant σ of a $\mathbf{Z}/2\mathbf{Z}$-sphere M is the signature mod 16 of (the intersection form on the H_2 of) a smooth spin (i.e. equipped with a spin structure) 4-manifold bounded by M.

PROPERTY 2 *For any $\mathbf{Z}/2\mathbf{Z}$-sphere M,*

$$\sigma(M) = 8|H_1(M)|\lambda(M) \bmod 16$$

To a surgery presentation ($\mathbf{L} \subset S^3$) of a 3-manifold M, we may associate the following natural 4-manifold $W_{\mathbf{L}}$ bounded by M: $W_{\mathbf{L}}$ is constructed from the standard 4-dimensional ball B^4 by gluing 2-handles $D^2 \times D^2$ to each component of the tubular neighborhood of L, $T(L) \subset S^3 = \partial B^4$, with respect to the trivialization given by the characteristic curves (which allows to identify a component of $T(L)$ to $(D^2 \times S^1 \subset D^2 \times D^2)$). $W_{\mathbf{L}}$ is next smoothed in a standard way. The linking matrix $E(\mathbf{L})$ is the matrix of the intersection form on $H_2(W_{\mathbf{L}})$ w.r.t. the basis of $H_2(W_{\mathbf{L}})$ associated to its handle decomposition above. A necessary and sufficient condition for $W_{\mathbf{L}}$ to be spin is that the diagonal of $E(\mathbf{L})$ is even (see [GM1, p.43]), and this can always be realized by FR moves (see [Ka]). In this case, it is easy to check that when $\det(E(\mathbf{L}))$ is odd (that is when M is a $\mathbf{Z}/2\mathbf{Z}$-sphere), $8|\det(E(\mathbf{L}))|\mathbf{F}_{S^3}(\mathbf{L}) - \text{signature}(E(\mathbf{L}))$ belongs to $16\mathbf{Z}$. This proves the congruence with the Rohlin invariant stated above. A few classical easy arguments show that this also gives a proof of the original Rohlin theorem asserting that the signature of a closed smooth spin 4-manifold is divisible by 16 (this Rohlin theorem yields the well-definedness of the Rohlin invariant as a direct corollary) (see [L3, Sec. 6.3]).

The following properties of the λ-invariant can also be checked very easily:

PROPERTY 3 *For any closed 3-manifold M, the λ-invariant of the manifold $-M$ obtained from M by orientation reversal satisfies:*

$$\lambda(-M) = (-1)^{\beta_1(M)+1}\lambda(M)$$

where $\beta_1(M)$ is the first Betti number of M.

PROPERTY 4 *For any two closed 3-manifolds M_1 and M_2, the λ-invariant of their connected sum $M_1 \sharp M_2 \overset{def}{=} \overline{M_1 \setminus B^3} \cup_{S^2} \overline{M_2 \setminus B^3}$ satisfies*

$$\lambda(M_1 \sharp M_2) = |H_1(M_2)|\lambda(M_1) + |H_1(M_1)|\lambda(M_2)$$

Now, the main property of λ is that it can be expressed in terms of previously known invariants:

PROPERTY 5 *Let M be a closed 3-manifold.*

- *If $\beta_1(M) \geq 4$, then*

$$\lambda(M) = 0.$$

- *If $\beta_1(M) = 3$, let (a, b, c) be a basis of $H^1(M)$ and let $\cup$ denote the cup product, then*

$$\lambda(M) = |\mathrm{Torsion}(H_1(M))|(a \cup b \cup c)([M])^2.$$

- *If $\beta_1(M) = 2$, let $([F], [G])$ be a basis of $H_2(M)$, represent it by two closed surfaces F and G embedded in general position in M, call γ their oriented intersection, call γ' the parallel of γ w.r.t. the trivialization of the normal bundle of γ induced by F and G, then*

$$\lambda(M) = -|\mathrm{Torsion}(H_1(M))| lk(\gamma, \gamma').$$

- *If $\beta_1(M) = 1$, let $\Delta(M)$ be the Alexander polynomial of M, that is (again) the order of the H_1 of the infinite cyclic covering of M, viewed as a natural $\mathbf{Z}[t, t^{-1}]$-module, normalized in such a way that $\Delta(M)(1) > 0$ and $\Delta(M)(t) = \Delta(M)(t^{-1})$, then*

$$\lambda(M) = \frac{\Delta''(M)(1)}{2} - \frac{|\mathrm{Torsion}(H_1(M))|}{12}.$$

- *If $\beta_1(M) = 0$ (i.e. if M is a $\mathbf{Q}$-sphere), then $\lambda(M)$ is the Casson-Walker invariant of M. More precisely, if M is a $\mathbf{Z}$-sphere, $\lambda(M)$ is the Casson invariant of M as normalized in [AM, GM2], and in general, if λ_W denotes the normalization of the Walker invariant used in [W], then*

$$\lambda(M) = \frac{|H_1(M)|}{2} \lambda_W(M).$$

It is now time to describe the Casson invariant of $\mathbf{Z}$-spheres as introduced by Casson in 1985.

4. The Casson invariant after Casson

Let M be a $\mathbf{Z}$-sphere. A. Casson defined $\lambda(M)$ as an algebraic number of conjugacy classes of irreducible $SU(2)$-representations of $\pi_1(M)$ as follows (details can be found in [AM] or [GM2]).

Like any closed 3-manifold, M can be decomposed into two handlebodies A and B glued along a genus g surface $\Sigma = \partial A = -\partial B$. (A handlebody is a regular neighborhood of a wedge of circles in a 3-manifold.) Such a decomposition $M = A \cup_\Sigma B$ is called a Heegaard splitting of M.

For a topological space X, call $R(X)$ the space of $SU(2)$-representations of the discrete group $\pi_1(X)$ equipped with the compact open topology. The subspace of $R(X)$ consisting of irreducible representations is an open set in $R(X)$ denoted by $\tilde{R}(X)$. When $\pi_1(X)$ is a free group of rank g, for example, when $X{=}A$ or B, $R(X)$ has a natural smooth structure which makes it diffeomorphic to $SU(2)^g \cong (S^3)^g$. $\tilde{R}(\Sigma)$ also has a natural smooth structure: call Σ_* the surface obtained from Σ by removing an open disk, choose a basepoint of Σ_* on $\partial\Sigma_*$ and call $(\partial : R(\Sigma_*) \to S^3)$ the evaluation of a representation of $R(\Sigma_*)$ at $\partial\Sigma_*$. The restriction of ∂ to $\tilde{R}(\Sigma_*)$ is a submersion. Thus, $\tilde{R}(\Sigma) = \tilde{R}(\Sigma_*)\cap\partial^{-1}(1)$ becomes a natural smooth $(6g-3)$-submanifold of $\tilde{R}(\Sigma_*)$. Let $X = A, B, \Sigma$ or Σ_*. The free smooth action of $SO(3) = SU(2)/\{-1,1\}$ by right conjugation on $\tilde{R}(X)$ identifies $\tilde{R}(X)$ with the total space of a principal $SO(3)$-bundle whose base is a smooth open manifold denoted by $\hat{R}(X)$. $\hat{R}(X)$ is the space of conjugacy classes of irreducible $SU(2)$-representations of $\pi_1(X)$.

The inclusions of Σ_* into A and B identify $R(A)$ and $R(B)$ with submanifolds of $R(\Sigma_*)$, and the Van Kampen theorem identifies $R(M)$ with $R(A) \cap R(B)$.

Since M is a **Z**-sphere, the only reducible $SU(2)$-representation of $\pi_1(M)$ is the trivial one, ρ_0, and it can be shown that $R(A)$ and $R(B)$ intersect transversally at ρ_0. Thus, $\tilde{R}(A) \cap \tilde{R}(B)$ and hence $\hat{R}(A) \cap \hat{R}(B)$ are compact. Therefore, an isotopy with compact support perturbing the inclusion of $\hat{R}(A)$ into $\hat{R}(\Sigma)$ can make $\hat{R}(A)$ transverse to $\hat{R}(B)$ inside $\hat{R}(\Sigma)$. Now, since $\hat{R}(A)$ and $\hat{R}(B)$ are of complementary dimension in $\hat{R}(\Sigma)$, their intersection is a finite number of points which can be given signs (+1) or (-1) once $\hat{R}(A)$, $\hat{R}(B)$ and $\hat{R}(\Sigma)$ are oriented. The sum of these signs is denoted by $< \hat{R}(A), \hat{R}(B) >_{\hat{R}(\Sigma)}$. It is, up to sign, twice the Casson invariant.

In order to suppress the sign indetermination we must specify orientations. $SU(2)$, $R(A)$, $R(B)$ and $R(\Sigma_*)$ are oriented arbitrarily. $SO(3)$ is oriented by the double covering $SU(2) \to SO(3)$. $\tilde{R}(\Sigma)$ is oriented as the fiber of ∂ with the convention ($base \oplus fiber$). Once $R(X)$ is oriented, $\hat{R}(X)$ is oriented as the base of an $SO(3)$-bundle with the convention ($base \oplus fiber$). It can be shown that the (classical) algebraic intersection number $< R(A), R(B) >_{R(\Sigma_*)}$ is ± 1.

Now, with all the notation above, we can state Casson's original definition of λ:

$$\lambda(M) = \frac{(-1)^g}{2} \frac{< \hat{R}(A), \hat{R}(B) >_{\hat{R}(\Sigma)}}{< R(A), R(B) >_{R(\Sigma_*)}}.$$

Casson proved the invariance of λ using the Reidemeister-Singer theorem which asserts that two Heegaard splittings of the same manifold become isomorphic after a finite number of stabilizations (that are connected sums with the genus one Heegaard splitting of S^3), and following the transformation of the above definition under a stabilization.

Casson's theorem was:

THEOREM 4.1 (CASSON, 1985) *There exists an integral topological invariant λ of* **Z***-spheres such that*

1. *If the trivial representation is the only representation of* $\pi_1(M)$ *into* $SU(2)$, *then* $\lambda(M) = 0$.

2. $\lambda(-M) = -\lambda(M)$.

3. $\lambda(M_1 \sharp M_2) = \lambda(M_1) + \lambda(M_2)$.

4. *For any knot* K *in a homology sphere* M, *for any* $\varepsilon = \pm 1$,

$$\lambda(\chi(M;(K,\varepsilon))) = \lambda(M) + \frac{\varepsilon}{2}\Delta(K)''(1).$$

5. $\sigma(M) = 8\lambda(M) \bmod 16$.

The immediate corollaries of this theorem, *'The Rohlin invariant of a homotopy sphere is null'* and *'The Rohlin invariant of an amphicheral* **Z***-sphere is null'* answered two long-unsolved questions in low-dimensional topology and allowed Casson to show the existence of a topological 4-manifold which cannot be triangulated (see [AM]) as a simplicial complex.

Note that the first assertion of the theorem is a direct corollary of Casson's definition of λ. The second and third assertions can also be proved very easily from this definition. Since any **Z**-sphere can be obtained from S^3 by a sequence of surgeries on knots framed by ± 1 (see [GM2]), and because an analogous surgery formula was known for the Rohlin μ-invariant $\mu = \frac{\sigma}{8}$, the fifth assertion is a direct consequence of the surgery formula. Thus, the only difficulty in the proof of Casson's theorem (in addition to inventing this definition ...) is to prove the surgery formula from the definition above.

To prove the surgery formula Casson proved the following lemma. A *boundary link* is a link whose components bound disjoint surfaces in the ambient manifold; T denotes the trefoil knot in S^3 pictured in Figure 1.

LEMMA 4.2 *Let* ν *be a rational invariant of* **Z***-spheres such that for any 2-component boundary link* **L** *in a* **Z***-sphere* M *whose components are framed by* ± 1 *the following holds*

$$\sum_{I \subset \{1,2\}} (-1)^{\sharp I} \nu(\chi(M; \mathbf{L}_I)) = 0.$$

Then

$$\nu(\chi(M;(K,\varepsilon))) = \nu(M) + \frac{\varepsilon}{2}\Delta(K)''(1)(\nu(\chi(S^3;(T,1))) - \nu(S^3)).$$

He then showed that $\lambda(\chi(S^3;(T,1))) = 1$ and proved that λ satisfied the hypothesis of the lemma from his definition. In fact, it is possible [GM2] to compute directly from Casson's definition the Casson invariant of Seifert fibered **Z**-spheres with 3 exceptional fibers, and the variation of the Casson invariant under surgery along a knot bounding an unknotted genus one Seifert surface. Both computations give $\lambda(\chi(S^3;(T,1)))$.

Remark 4.3 In [Lin], X. S. Lin proved that the signature of a knot can also be obtained by counting some $SU(2)$-representations of the π_1 of its exterior 'à la Casson'. Like Casson's comparison of his representation number with the Rohlin invariant, Lin's proof that his representation number coincides with the signature is not direct. In both cases, it would be interesting to have a more direct identification.

Note also that Lemma 4.2 provides a nice characterization of the Casson invariant. In the same spirit, it can be shown [L5]:

Property 6 *Any two* $\mathbf{Z}$*-spheres which have the same Casson invariant can be obtained one from the other by a sequence of surgeries on knots with trivial Alexander polynomial framed by* ± 1.

In 1988, K. Walker [W] used the stratified symplectic structure of the representation spaces [Go] to give a complete generalization of Casson's work to $\mathbf{Q}$-spheres. In this case, reducible representations cannot be ignored, and basic differential topology does not suffice anymore to provide a powerful generalization. (A weaker generalization of the Casson invariant to $\mathbf{Q}$-spheres had been proposed by S. Boyer and A. Nicas [BN].). Furthermore, K. Walker gave a very nice proof based on Kirby calculus that his one-component surgery formula gives a consistent definition of his invariant λ_W.

Next, S. Cappell, R. Lee and E. Miller [CLM] generalized Walker's definition to other Lie groups like $SU(n)$, but they have not yet found interesting properties for their invariants. C. Curtis [C] studied the $SO(3)$, $U(2)$, $Spin(4)$ and $SO(4)$-invariants more precisely and proved that they are functions of the Walker $SU(2)$-invariant.

Of course, the combinatorial extension of the Casson invariant described in Section 3 is also a development of Casson's work. Indeed, without Walker's generalization of the Casson theorem above, and Boyer-Lines's work [BL1] the author would not have been able to find the general surgery formula of Property 1. In their work, independent from Walker's, S. Boyer and D. Lines gave a combinatorial definition of the restriction of the Walker invariant λ_W to homology lens spaces, they proved a two-component surgery formula for the Casson invariant, they exhibited the first part $\mathbf{F}_1$ of the surgery function $\mathbf{F}$, the combination of the coefficients $\tilde{\zeta}$, and they proved that $(\lambda(\chi(S^3;.)) - \mathbf{F}_1)$ is invariant under link homotopy. It should also be mentioned that the surgery formula for algebraically split links, that are links whose components do not algebraically link each other, is due to Hoste [Ho].

5. Further topological properties of the Casson invariant

Since the Alexander-Conway polynomial is a well-understood knot invariant, it is easy to apply the surgery formula satisfied by λ in order to:

- compute the λ-invariant of any manifold presented by surgery [L2, L1],
- study the behaviour of λ under other topological mutations as in [D, Ki, Wo] or in the properties 7, 8 and 9 described below,

- compare λ with other invariants as H. Murakami did in proving that the Walker invariant is equal to an appropriately defined function of the Reshetikhin and Turaev invariants [Mu].

REMARK 5.1 In [O1], T. Ohtsuki generalized Murakami's work and renormalized the Reshetikhin and Turaev invariants into an invariant series of $\mathbf{Q}$-spheres whose first coefficients are $|H_1(.)|$ and λ. It would be interesting to know whether the other coefficients of this series are related to Casson-type invariants. To study his series, Ohtsuki [O2] defined the notion of finite type invariant for $\mathbf{Z}$-spheres. This notion is analogous to the notion of Vassiliev invariants of knots. Say that a rational invariant ν of $\mathbf{Z}$-spheres is of AS-type (resp. of B-type) less or equal than n if for any $(n+1)$-component algebraically split (resp. boundary) link L in a $\mathbf{Z}$-sphere M whose components are framed by ± 1:

$$\sum_{I \subset \{1,\dots,n+1\}} (-1)^{\sharp I} \nu(\chi(M; \mathbf{L}_I)) = 0.$$

Note that Casson's lemma (4.2) proves that the B-type 1 invariants are exactly the degree 1 polynomials in λ while the Hoste surgery formula [Ho] shows that λ is of AS-type 3. In fact, it is proved that the AS-type is always a multiple of 3, and it is conjectured (proved ?) that the two mentioned notions of finite type invariants coincide and that the AS-type is three times the B-type. It is not hard to see that for any integer n, a degree n polynomial in λ is an invariant of B-type n and of AS-type $3n$. Thus, the polynomials in λ are nice prototypes for finite type invariants. But T. T. Q. Le proved [Le] that they are not the only ones. It would be interesting to place the $SU(n)$-invariants of Cappell, Lee and Miller among these finite type invariants.

It is worth mentioning the existence of some variants of the surgery formula (Property 1), that have not yet been mentioned to avoid introducing too much notation. Note that in the surgery definition, we do not need the characteristic curve μ of the surgery to be parallel to the knot K. Any non-separating simple closed curve of $\partial T(K)$ can play the role of the characteristic curve, and the surgery defined by such a curve is (sometimes) called a rational surgery. The surgery formula extends naturally to rational surgeries. For surgeries starting from $\mathbf{Z}$-spheres the surgery function $\mathbf{F}$ can be expressed only in terms of linking numbers and one-variable Alexander-Conway polynomials. (See [L3].)

There are also some formulae for the Casson invariant of cyclic p-fold branched coverings. For a link L in a $\mathbf{Z}$-sphere M, let $R_p(M; L)$ be the cyclic p-fold covering of M branched along L, obtained from the covering of the exterior of L associated with the 'linking number with L modulo p' by filling it in by solid tori whose meridians are sent to old meridians of L.

PROPERTY 7 (HOSTE [HO]) *Let K be a knot in a $\mathbf{Z}$-sphere M. Let $D_\varepsilon K$ be the untwisted double of K with an ε-clasp. Then*

$$\lambda(R_p(M; D_\varepsilon K)) = p\lambda(M) + \varepsilon p \Delta''(K)(1).$$

The following Mullins property relates the Walker invariant of 2-fold branched coverings to the Jones polynomial V and the oriented signature σ of links:

PROPERTY 8 (MULLINS [MUL]) *Let L be a link in S^3 such that $R_2(S^3;L)$ is a $\mathbf{Q}$-sphere. Then*

$$\lambda_W(R_2(S^3;L)) = \frac{\sigma(L)}{4} - \frac{V'(L)(-1)}{6V(L)(-1)}.$$

To prove this formula, Mullins studied the variation of $\lambda_W(R_2(S^3;L))$ under a crossing change of L. Owing to the fact that the 2-fold branched covering of the ball of the crossing change is a solid torus, such a crossing change induces a surgery on $R_2(S^3;L)$.

For other cyclic p-fold branched coverings, a crossing change induces a handlebody replacement. This leads us to the following natural question. What can we say about $\lambda(A \cup_\Sigma B)$ for a $\mathbf{Q}$-sphere obtained by gluing two pieces A and B along a genus g surface Σ?

Our partial answer is the following property of λ [L4]. A $\mathbf{Q}$-*handlebody* is a 3-manifold with the same rational homology as a standard handlebody. For a 3-manifold A with boundary, the kernel $\mathcal{L}_A$ of the map from $H_1(\partial A;\mathbf{Q})$ to $H_1(A;\mathbf{Q})$ induced by the inclusion is called the *Lagrangian* of A.

PROPERTY 9 *Let A, A', B and B' be four $\mathbf{Q}$-handlebodies such that ∂A, $\partial A'$, $-\partial B$ and $-\partial B'$ are identified via orientation-preserving homeomorphisms with a genus g surface Σ. Assume that $\mathcal{L}_A = \mathcal{L}_{A'}$ and $\mathcal{L}_B = \mathcal{L}_{B'}$ and that $\mathcal{L}_A \cap \mathcal{L}_B = \{0\}$ inside $H_1(\Sigma;\mathbf{Q})$. Then*

$$\lambda_W(A \cup_\Sigma B) - \lambda_W(A' \cup_\Sigma B) - \lambda_W(A \cup_\Sigma B') + \lambda_W(A' \cup_\Sigma B') = R(A,A',B,B')$$

where $R(A,A',B,B')$, described below in general, is zero if $g \leq 2$.

Before describing $R(A,A',B,B')$ in general, note that, for $g=0$ and $A' = B' = B^3$, this property is nothing but the additivity of λ_W under connected sum. The genus one formula, when A' and B' are solid tori, is the splicing formula, shown by several authors [BN, FM] for the Casson invariant, and generalized by Fujita to the Walker invariant [F]. In this case, starting with $A \cup_\Sigma B$, there is a unique way of filling in A with a solid torus B' having the right Lagrangian. $A' \cup B$ and $A' \cup B'$ are similarly well-determined, and the Walker invariant of the lens space $A' \cup B'$ is a known Dedekind sum.

Now, we describe $R(A,A',B,B')$ under the hypotheses of Property 9. The isomorphism $\partial_{AA'}$ from $H_2(A \cup_\Sigma -A';\mathbf{Q})$ to $\mathcal{L}_A$ which maps the homology class of a surface S of $A \cup_\Sigma -A'$ (transverse to ∂A) to the class of $\partial(S \cap A)$ carries the algebraic intersection defined on $\bigwedge^3 H_2(A \cup_\Sigma -A';\mathbf{Q})$ to a form $\mathcal{I}_{AA'}$ defined on $\bigwedge^3 \mathcal{L}_A$. Define $\mathcal{I}_{BB'}$ similarly. Let $(\alpha_1,\dots,\alpha_g)$ and $(\beta_1,\dots,\beta_g)$ be two bases for $\mathcal{L}_A$ and $\mathcal{L}_B$, respectively, that are dual for the intersection form $<,>_\Sigma$ on Σ ($<\alpha_i,\beta_j>_\Sigma = \delta_{ij}$). Then

$$R(A,A',B,B') = -4 \sum_{\{i,j,k\}\subset\{1,\dots,g\}} \mathcal{I}_{AA'}(\alpha_i \wedge \alpha_j \wedge \alpha_k)\mathcal{I}_{BB'}(\beta_i \wedge \beta_j \wedge \beta_k).$$

REMARK 5.2 Let $(\Sigma, \mathcal{L}_A)$ be a closed, connected surface equipped with a rational Lagrangian (as above). In [S], D. Sullivan proved that any integral form on $\bigwedge^3(H_1(\Sigma;\mathbf{Z}) \cap \mathcal{L}_A)$ may be realized as a $\mathcal{I}_{AA'}$ for two standard handlebodies A and A' with boundary Σ and Lagrangian $\mathcal{L}_A$.

A splitting $A \cup_\Sigma B$ of a $\mathbf{Q}$-sphere induces the following function λ_{AB} on the Torelli group of Σ. The Torelli group is the group of the (isotopy classes of) homeomorphisms of Σ which induce the identity on $H_1(\Sigma)$. For a homeomorphism f of the Torelli group, $A \cup_f B$ denotes the manifold obtained by replacing the (underlying) identification $j_B : \Sigma \longrightarrow -\partial B$ by $j_B \circ f$,

$$\lambda_{AB}(f) = \frac{1}{2}(\lambda_W(A \cup_f B) - \lambda_W(A \cup_\Sigma B)).$$

As a direct corollary of Property 9, we see that $\lambda_{AB}(g \circ f) - \lambda_{AB}(g) - \lambda_{AB}(f)$ is a function of the evaluations of the Johnson homomorphism at f and g (see [J, Second definition, p.170] for a definition of the Johnson homomorphism which is a homomorphism from the Torelli group to $\bigwedge^3 H^1(\Sigma)$). With completely different methods (based mainly on Johnson's study of the Torelli group), S. Morita proved this corollary for Heegaard splittings of $\mathbf{Z}$-spheres [Mo, Theorem 4.3], but he did not think that it extended to general embeddings [Mo, Remark 4.7].

The above corollary also proves, that when $A \cup B$ is a $\mathbf{Z}$-sphere, the function μ_{AB} induced by the Rohlin μ-invariant $\mu = \frac{\sigma}{8}$ defines a homomorphism from the Torelli group to $\mathbf{Z}/2\mathbf{Z}$. These homomorphisms were first studied by J. Birman and R. Craggs [BC]. They are the so-called Birman-Craggs homomorphisms.

It is worth mentioning that the best natural generalization of Property 9 that may be expected for the generalized Casson invariant of Section 3 is true [L4]. This generalized Casson invariant also admits a homogeneous definition via Kontsevich integrals [LMMO]. Both of these properties together with the homogeneous surgery formula enhance the naturality of the generalization of λ proposed in Section 3.

To prove Property 9, we first find a sequence of simple surgeries on links transforming A into A' and staying among the $\mathbf{Q}$-handlebodies with Lagrangian $\mathcal{L}_A$. Then we apply the surgery formula of [L3, BL1] to these surgeries and we analyse how the involved formulae depend on B when B varies among the $\mathbf{Q}$-handlebodies with boundary $-\partial A$ and with fixed Lagrangian.

This analysis led us [L4] to construct a tautological generalization of Alexander polynomials to 3-manifolds with boundary which may be useful to prove other properties of the Casson invariant. We conclude this article with a brief presentation of this function called the Alexander function which will allow us to define the normalized several variable Alexander polynomial.

6. More about Alexander polynomials: the Alexander function

All the assertions of this section are proved in [L4, Section 3]. Here, A denotes a connected 3-manifold with non-empty boundary and with non-negative *genus* $g = g(A) = 1 - \chi(A)$. Λ_A denotes the group ring:

$$\Lambda_A = \mathbf{Z}\left[\frac{H_1(A)}{\mathrm{Torsion}(H_1(A))}\right]$$

Recall that $\Lambda_A = \bigoplus_{x\in\frac{H_1(A)}{\text{Torsion}}} \mathbf{Z}\exp(x)$ as a $\mathbf{Z}$-module, that its multiplication sends $(\exp(x), \exp(y))$ to $\exp(x+y)$, and that the units of Λ_A are its elements of the form $\pm\exp(x \in H_1(A)/\text{Torsion})$.

The maximal free abelian covering of A is denoted by $\tilde{A}$ and the covering map from $\tilde{A}$ to A by p_A. We fix a basepoint $\star$ in A. The Λ_A-module $H_1(\tilde{A}, p_A^{-1}(\star); \mathbf{Z})$ is denoted by $\mathcal{H}_A$.

DEFINITION 6.1 The *Alexander function* $\mathcal{A}_A$ of A is the Λ_A-morphism

$$\mathcal{A}_A : \bigwedge^g \mathcal{H}_A \longrightarrow \Lambda_A$$

which is defined up to a (global) multiplication by a unit of Λ_A as follows. Take a presentation of $\mathcal{H}_A$ over Λ_A with $(r+g)$ generators $\gamma_1, \ldots, \gamma_{r+g}$ and r relators $\rho_1, \ldots, \rho_r$ (which are Λ_A-linear combinations of the γ_i). Let $\hat{u} = u_1 \wedge \ldots \wedge u_g$ be an element of $\bigwedge^g \mathcal{H}_A$. Then $\mathcal{A}_A(\hat{u})$ is defined by the equality:

$$\mathcal{A}_A(\hat{u})\hat{\gamma} = \hat{\rho} \wedge \hat{u}$$

where $\hat{\rho} = \rho_1 \wedge \ldots \wedge \rho_r$, $\hat{\gamma} = \gamma_1 \wedge \ldots \wedge \gamma_{r+g}$, the u_i are represented as combinations of the γ_j, and the exterior products are to be taken in $\bigwedge^{r+g}\left(\bigoplus_{i=1}^{r+g} \Lambda_A \gamma_i\right)$.

Of course, $\mathcal{A}_A(\hat{u})$ is just the order of the Λ_A-module $\mathcal{H}_A/(\oplus \Lambda_A u_i)$. But, hopefully, some of the properties of $\mathcal{A}_A$ mentioned below will convince the reader that it may be interesting to work with a fixed normalization of $\mathcal{A}_A$.

Fix a preferred lift $\star_0$ of $\star$ in $\tilde{A}$. Let ∂ denote the boundary map from $\mathcal{H}_A$ to $H_0(p_A^{-1}(\star)) = \Lambda_A[\star_0] = \Lambda_A$. Once a normalization of $\mathcal{A}_A$ is fixed, $\mathcal{A}_A$ satisfies the easy property: for any $v = (v_1, \ldots, v_g) \in \mathcal{H}_A^g$, and for any $u \in \mathcal{H}_A$,

$$\sum_{i=1}^{g} \partial(v_i)\mathcal{A}_A(\hat{v}(\frac{u}{v_i})) = \mathcal{A}_A(\hat{v})\partial(u)$$

where $\hat{v} = v_1 \wedge \ldots \wedge v_g$ and $\hat{v}(\frac{u}{v_i}) = v_1 \wedge \ldots \wedge v_{i-1} \wedge u \wedge v_{i+1} \wedge \ldots \wedge v_g$.

This property shows that the next property of the Alexander function gives a consistent definition of the Reidemeister torsion τ (which yields the Alexander polynomial). If A is a link exterior, then for any element u of $\mathcal{H}_A$,

$$\mathcal{A}_A(u) = \partial(u)\tau(A).$$

If A is a many-component link exterior, then $\tau(A)$ belongs to Λ_A. It is defined up to multiplication by a unit of Λ_A.

In fact, a well-chosen multiplication by an element of the form $\exp(\frac{1}{2}x)$ makes the Reidemeister torsion satisfy $\tau(A) = \pm\overline{\tau(A)}$ where the conjugation sends $\exp(x)$ to $\exp(-x)$ [Mi]. Thus, the Reidemeister torsion is an element defined up to sign in $\mathbf{Z}[\frac{1}{2}H_1(A)/\text{Torsion}] \subset \mathbf{Z}[H_1(A;\mathbf{Q})]$. The choice of an orientation $\mathcal{O}$ of the vector space $H_1(A;\mathbf{R}) \oplus H_2(A;\mathbf{R})$ suppresses the sign indeterminacy and allows one to define $\tau(A,\mathcal{O}) \in \mathbf{Z}[H_1(A;\mathbf{Q})]$ unambiguously. (See [T, L3].) If

A is the exterior of an n-component link L in a $\mathbf{Q}$-sphere M, such an orientation $\mathcal{O}_L$ is unambiguously defined by a basis of $H_1(A;\mathbf{R}) \oplus H_2(A;\mathbf{R})$ of the form $(m_1,\dots,m_n,\partial T(K_1),\dots,\partial T(K_{n-1}))$ where m_i and $\partial T(K_i)$ denote the oriented meridian and the boundary of the tubular neighborhood of the ith component of L, respectively.

Note that for a general A, a basis $\mathcal{M} = \{m_1,\dots,m_n\}$ of $H_1(A;\mathbf{Q})$ induces the natural ring inclusion $\psi_{\mathcal{M}}$ from $\mathbf{Z}[H_1(A;\mathbf{Q})]$ into the ring $\mathbf{Q}[[x_1,\dots,x_n]]$ of formal series in the x_i: $\psi_{\mathcal{M}}(\exp(m_i)) = \exp(x_i)$.

In particular, if A is the exterior of a many-component link L in a $\mathbf{Q}$-sphere M, then we use the natural basis $\mathcal{M}$ of the meridians of L to define the Alexander series

$$\mathcal{D}(L) = \psi_{\mathcal{M}}(\tau(M\setminus \overset{\circ}{T}(L),\mathcal{O}_L))$$

which is equivalent to the several variable Alexander polynomial

$$\Delta(L)\,(t_1=\exp(x_1),\dots,t_n=\exp(x_n)) = (-1)^{n-1}\frac{\mathcal{D}(L)}{|H_1(M)|}.$$

In general, a morphism $\psi_{\mathcal{M}}$ allows us to define the *order* of an element of Λ_A as the order of its image under $\psi_{\mathcal{M}}$. It does not depend on $\mathcal{M}$. Similarly, we will speak of the *low degree parts* of the elements of Λ_A. Indeed, the information required to compute the coefficients $\tilde{\zeta}$ of the surgery formula (Property 1) is contained in the low-degree parts of Alexander functions images. Thus, it is worth noting that the degree 0 part of $\mathcal{A}_A(\hat{u})$ is $\varepsilon(\mathcal{A}_A(\hat{u})) = |H_1(A)/(\oplus \mathbf{Z}p_{A*}(u_i))|$, and that the order of $\mathcal{A}_A(\hat{u})$ is greater or equal than the dimension of $H_1(A;\mathbf{Q})/(\oplus \mathbf{Q}p_{A*}(u_i))$. The Alexander function also satisfies the following interesting property which relates the low degree parts of some of its images to algebraic intersections.

Proposition 6.2 *For any (A,ℓ,m), where A is a $\mathbf{Q}$-handlebody whose boundary is equipped with two systems of curves $\ell = (\ell_1,\dots,\ell_g)$ and $m = (m_1,\dots,m_g)$ as in Figure 2, such that the homology classes of the ℓ_i generate $\mathcal{L}_A$,*

$$\mathcal{A}_A(\hat{m}(\frac{\ell_j}{m_k})) = \varepsilon(\mathcal{A}_A(\hat{m}))\sum_{i=1}^{g}\mathcal{I}_{A\Sigma_\ell}(\ell_i\wedge\ell_j\wedge\ell_k)(\exp(m_i)-1)+O(2)$$

where $O(2)$ makes up for an element of Λ_A of order greater or equal than 2, and Σ_ℓ is the standard handlebody with boundary ∂A where the ℓ_i bound disks.

Recall that $\mathcal{I}$ is defined in Section 5. The statement is unambiguous despite the fact that ℓ_i denotes the curve ℓ_i, its homology class, and the class of a lifting of the curve ℓ_i (joined to the basepoint) in $\mathcal{H}_A$, depending on the context.

It is also worth observing the natural good behaviour of Alexander functions under the two operations: (1) adding a 2-handle to A, (2) performing a connected sum along the boundary of two 3-manifolds A and B. A lot of properties of Alexander polynomials can be derived from this natural behaviour. More generally, if A is a submanifold of the interior of a 3-manifold B, in order to compute $\mathcal{A}_B$, it is enough to know $\overline{B\setminus A}$, $\mathcal{A}_A$ and the inclusion from ∂A into $\overline{B\setminus A}$.

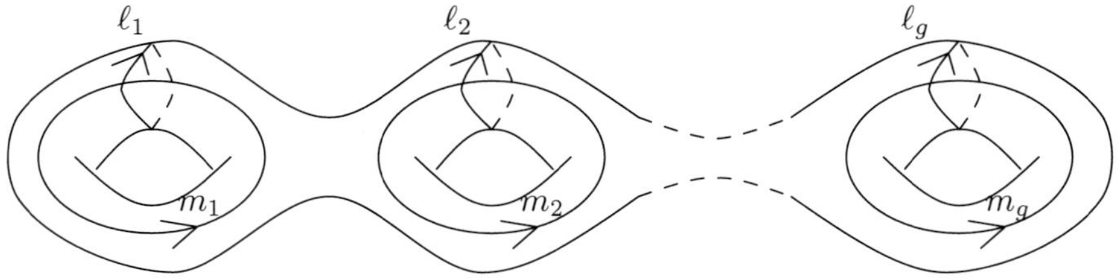

Figure 2: Two systems of curves on ∂A

Let us use these remarks to be more specific about the sign determination of the Alexander series.

Let $L = (K_i)_{i\in\{1,\dots,g\}}$ be a link in a **Q**-sphere M with $g \geq 2$. Consider a regular neighborhood of a graph made of the K_i and paths joining them to the basepoint. This is a handlebody which is a connected sum along boundaries of the $T(K_i)$. Removing the interior of this handlebody from M yields a **Q**-handlebody A whose boundary is equipped with the meridians m_i and some longitudes ℓ_i of the K_i which sit there as in Figure 2. We let δ_i denote the boundary of the genus one subsurface of ∂A with connected boundary containing m_i and ℓ_i.

Then, up to units of the form $\exp(x \in \frac{1}{2}H_1(A)/\text{Torsion})$, for any $j,k \in \{1,\dots,g\}$,

$$\mathcal{D}(L) = sign(\varepsilon(\mathcal{A}_A(\hat{m})))\psi_L\left(\frac{\mathcal{A}_A(\hat{\delta}(\frac{m_j}{\delta_k}))}{\partial(m_j)}\right).$$

Now, the definition of the coefficient $\tilde{\zeta}$ is complete and we know enough about the surgery formula. Thus, we can apply it together with the helpful formalism introduced above, and we are hopefully ready to discover more properties for the Casson invariant.

References

[AM] S. Akbulut and J. McCarthy, *Casson's invariant for oriented homology 3-spheres, an exposition.* Mathematical Notes **36**, Princeton University Press, Princeton 1990.

[BC] J. Birman and R. Craggs, *The μ-invariant of 3-manifolds and certain structural properties of the group of homeomorphisms of a closed, oriented 2-manifold.* Trans. Amer. Math. Soc. **237**, (1978), 283–309.

[BL1] S. Boyer and D. Lines, *Surgery formulae for Casson's invariant and extensions to homology lens spaces,* J. Reine Angew. Math. **405**, (1990), 181–220.

[BL2] S. Boyer and D. Lines, *Conway potential functions for links in **Q**-homology 3-spheres,* Proc. of the Edinburgh Math. Soc. **35**, (1992), 53–69.

[BN] S. Boyer and A. Nicas. *Varieties of group representations and Casson's invariant for rational homology 3-spheres.* Trans. Amer. Math. Soc. **322**, (1990), 507–522.

[BD] P.J. Braam and S.K. Donaldson, *Floer's work on instanton homology, knots and surgery.* The Floer memorial volume (H. Hofer, C. H. Taubes, A. Weinstein, E. Zehnder (ed.)), Prog. Math. **133**, Birkhäuser Boston, Basel, Berlin (1995), 195–256.

[CLM] S. E. Cappell, R. Lee, E. Y. Miller, *A symplectic geometry approach to generalized Casson's invariants of 3-manifolds,* Bull. Am. Math. Soc., New Ser. **22**, (1990), 269–275.

[CF] R. H. Crowell and R. H. Fox, *Introduction to knot theory.* Ginn, 1963.

[C] C. L. Curtis, *Generalized Casson invariants for* $SO(3)$, $U(2)$, Spin(4), *and* $SO(4)$, Trans. Am. Math. Soc. **343**, No.1, (1994), 49–86.

[D] A. Davidow, *Casson's invariant and twisted double knots,* Topology Appl. **58**, No.2, (1994), 93–101.

[FR] R. Fenn and C. Rourke, *On Kirby's calculus of links,* Topology **18**, (1979), 1–15.

[F] G. Fujita, *A splicing formula for Casson-Walker's invariant,* Math. Ann. **296**, (1993), 327–338.

[FM] S. Fukuhara and N. Maruyama, *A Sum Formula for Casson's* λ*-Invariant.* Tokyo J. Math. **11**, (1988), 281–287.

[FMS] S. Fukuhara, Y. Matsumoto, K. Sakamoto, *Casson's invariant of Seifert homology 3-spheres,* Math. Ann. **287**, No.2 (1990), 275–285.

[Go] W. Goldman, *The symplectic nature of fundamental groups of surfaces,* Adv. Math., **54**, (1984), 200–225.

[GM1] L. Guillou and A. Marin, *A la recherche de la topologie perdue,* Progress in Math. **62**, Birkhäuser Boston, Basel, Berlin 1986.

[GM2] L. Guillou and A. Marin, *Notes sur l'invariant de Casson des sphères d'homologie de dimension 3,* L'Enseignement Math. **38**, (1992), 233–290.

[Ha] R. Hartley, *The Conway potential function for links,* Comment. Math. Helvetici **58**, (1983), 365–378.

[Ho] J. Hoste, *A formula for Casson's invariant,* Trans. Am. Math. Soc. **297**, (1986), 547–562.

[J] D. Johnson, *A survey of the Torelli group,* Contemporary Mathematics **20**, (1983), 165–179.

[Ka] S. Kaplan, *Constructing framed 4-manifolds with given almost framed boundaries.* Trans. A. M. S. **254**, (1979), 237–263.

[K] R. Kirby, *A calculus for framed links in* S^3. Inventiones Math. **45**, (1978), 35–56.

[Ki] P. A. Kirk, *Mutations of homology spheres and Casson's invariant.* Math. Proc. Camb. Philos. Soc. **105**, (1989), 313–318.

[Ku] N. H. Kuiper, *A short history of triangulations and related matters.* Proc. bicenten. Congr. Wiskd. Genoot., Part I, Amsterdam 1978, Math. Cent. Tracts **100**, (1979), 61–79.

[Le] T. T. Q. Le, *An invariant of integral homology 3-spheres which is universal for all finite type invariants,* Amer. Math. Soc. Transl. **179**, No. 2 (1997), 75–100.

[LMMO] T. T. Q. Le, H. Murakami, J. Murakami, T. Ohtsuki *A three-manifold invariant derived from the universal Vassiliev-Kontsevich invariant,* Proc. Japan Acad., **71**, Ser. A, (1995), 125-127.

[L1] C. Lescop, *Invariant de Casson-Walker des sphères d'homologie rationnelle fibrées de Seifert,* C. R. Acad. Sci. Paris **310**, (1990), 727–730.

[L2] C. Lescop, *Un algorithme pour calculer l'invariant de Walker,* Bulletin de la Société Mathématique de France **118**, (1990), 363–376.

[L3] C. Lescop, *Global surgery formula for the Casson-Walker invariant,* Annals of Mathematics Studies **140**, Princeton University Press, Princeton 1996.

[L4'] C. Lescop, *Une formule de somme pour l'invariant de Casson-Walker,* C. R. Acad. Sci. Paris **320**, Série I, (1995), 843–846.

[L4] C. Lescop, *A sum formula for the Casson-Walker invariant,* to appear in Inventiones Math.

[L5] C. Lescop, *On the kernel of the Casson invariant,* Topology **37**, (1998), 25–38.
[Li] W. B. R. Lickorish, *A representation of orientable combinatorial 3-manifolds,* Ann. Math. **76**, (1962), 531–540.
[Lin] X. S. Lin, *A knot invariant via representation spaces,* J. Differ. Geom. **35**, (1992), 337–357.
[M] A. Marin, *Un nouvel invariant pour les sphères d'homologie de dimension 3 (d'après Casson),* Semin. Bourbaki, 40ème Année, Vol. 1987/88, Exp. No.693, Astérisque **161-162**, (1988), 151–164.
[Mi] J. Milnor, *A duality theorem for Reidemeister torsion,* Ann. Math. **76**, (1962), 137–147.
[Mi2] J. Milnor, *Spin structures on manifolds,* L'Enseignement Math. **9**, (1963), 198–203.
[Mo] S. Morita, *On the structure of the Torelli group and the Casson invariant,* Topology **30**, (1991), 603–621.
[Mo2] S. Morita, *Characteristic classes of surface bundles and the Casson invariant,* Sugaku Expositions **7**, (1994), 59–79.
[Mul] D. Mullins, *The generalized Casson invariant for 2-fold branched covers of* S^3 *and the Jones polynomial,* Topology **32**, No.2, (1993), 419–438.
[Mu] H. Murakami, *Quantum* $SO(3)$*-invariants dominate the* $SU(2)$*-invariant of Casson and Walker,* Math. Proc. Camb. Philos. Soc. **117**, No.2, (1995), 237–249.
[NW] W. Neumann and J. Wahl, *Casson invariant of links of singularities,* Comment. Math. Helvetici **65**, (1990), 58–78.
[O1] T. Ohtsuki, *A polynomial invariant of rational homology 3-spheres,* Invent. math. **123**, (1996), 241–257.
[O2] T. Ohtsuki, *Finite type invariants of integral homology 3-spheres.* Journal of knot theory and its ramifications **5**, (1996) 101–115.
[RT] N. Reshetikhin and V. Turaev, *Invariants of 3-manifolds via link polynomials and quantum groups,* Invent. math. **103**, (1991), 547–597.
[R] D. Rolfsen, *Knots and links,* Publish or Perish, Berkeley 1976.
[Ro] C. Rourke, *A new proof that* Ω_3 *is zero,* Jour. London Math. Soc. **31**, (1985), 373–376.
[S] D. Sullivan, *On the intersection ring of compact three manifolds,* Topology **14**, (1975), 275–277.
[Ta] C. H. Taubes, *Casson's invariant and gauge theory,* J. Differ. Geom. **31**, (1990), 547–599.
[T] V. G. Turaev, *Reidemeister torsion in knot theory,* Russian Math. Surveys, 41:1, (1986), 119–182.
[Wa] A. W. Wallace, *Modifications and cobounding manifolds. I, II,* Can. J. Math, **12**, (1960), 503–528.
[W] K. Walker, *An Extension of Casson's Invariant,* Annals of Mathematics Studies, **126**, Princeton University Press, Princeton 1992.
[Wo] M. R. Woodard, *The Casson invariant of surgered, sewn link exteriors,* Topology Appl. **46**, No.1, (1992), 1–12.

Progress in Mathematics, Vol. 168, © 1998 Birkhäuser Verlag Basel/Switzerland

EXTRA-ordinary Differential Equations: Attempts to an Analysis of Differential-algebraic Systems

Roswitha März

Humboldt-Universität Berlin, Math.-Naturw. Fakultät II
Institut für Mathematik
D-10099 Berlin, Germany

1. Introduction

For about 15 years differential-algebraic equations (DAE's) have been an intensively discussed field of applied mathematics. DAE's arise in models that couple dynamical parts with constraints and invariants. The most popular fields of application are the simulation of electric circuits, chemical reactions and vehicle system dynamics, but also optimal control problems. There is also a close relationship with singular perturbation problems.

Formally, most DAE's are of the form

$$A(x,t)x' + g(x,t) = 0 \tag{1.1}$$

with an everywhere singular leading coefficient matrix $A(x,t)$ of constant rank. In 1971, *C. W. Gear* proposed to integrate DAE's numerically like regular differential equations by means of the so-called backward differentiation formulae (BDF). Up to now, this has been practice with some great successes, e.g. in the electrical industry, where large systems have to be treated. It was only 10 years later, after some inexplicable and unexpected failures, that a more detailed mathematical investigation of DAE's was undertaken. This was initiated in a lecture by C.W. Gear in Oberwolfach in 1981 ([2]), and by L.R. Petzold's paper ([3]) with the provocative title

DAE's are not ODE's,

which, however, only meant that DAE's do not behave numerically like (regular) ordinary differential equations (ODE's). Of course, equations of the form (1.1) *are* ODE's, too, but not regular ones.

Let us consider two very simple (perhaps too simple) examples.

Example 1: On $x_2 > 0$, consider the system

$$\left.\begin{array}{rcl} x_1' & = & \alpha x_1 \\ x_2' & = & x_3/x_2 \\ x_1^2 & + & x_2^2 = 1 \end{array}\right\} \tag{1.2}$$

Obviously, the flow is restricted to the (obvious) constraint manifold given by the derivative-free third equation. Differentiating this equation leads to a second, "hidden" constraint given by

$$\alpha x_1^2 + x_3 = 0 \, .$$

Clearly, initial values should meet both constraints.

Example 2: Consider the special linear system in constant coefficient Kronecker normal form

$$\left. \begin{array}{lcl} x_1' \quad +Wx_1 & = & q_1(t) \\ x_3' \quad +x_2 & = & q_2(t) \\ \qquad x_3 & = & q_3(t) \end{array} \right\} . \tag{1.3}$$

Again we have two constraints: the obvious one $x_3 = q_3(t)$ and the hidden one $x_2 = q_2(t) - q_3'(t)$. For solvability, say for $x_1, x_3 \in C^1$, $x_2 \in C$, the righthand-side q has to be continuous as usual, but q_3 has to be in addition C^1.

Consequently, a linear map L representing the related IVP's on a compact interval may be assumed to be injective. However, in its natural setting $(q \in C)$ L is not a Fredholm map ([4]). Since L^{-1} is unbounded, the IVP is ill-posed in the sense of Tikhonov. It is this which makes the numerical treatment so difficult.

In both examples the DAE's have index 2. Roughly speaking, the index μ is the maximal number of steps of the nested constraint manifolds, and $\mu - 1$ is the number of the inherent differentiations. The higher the index, the more complex is the problem.

From the point of view of applications, DAE's with low index (1-3) and with describing functions of low smoothness are of special interest. To an increasing extent the models are generated automatically by complex algorithms, where the dimensions are often very large and the equations are not given explicitly. Looking at the long equations that frequently cover several pages, if they are explicitly available, seems to be of little profit. It is strenuous but also challenging to reveal those pieces of information and those model structures that are valuable for mathematical characterizations. A reliable numerical treatment of the models generated is increasingly combined with more exact information on the structure, based on a thorough DAE-analysis in the given coordinates, which represent physical quantities like voltage etc. in most cases. Models of practical interest contain functions which are far from being analytic or meromorphic. Sometimes, difficulties already occur with the second derivative. There is little possibility of getting the hidden constraints explicitly.

From this point of view the two examples above are already too trivial and misleading.

2. Briefly on Reduction Methods and Transformation into Formally Integrable Systems

If one is tempted by nice simple examples, one will think of applying the formal theory of differential systems ([5]). By means of prolongations and projections into lower-dimensional jet spaces the original system is transposed into a formally integrable system. Then, for sufficiently regular systems

$$f(x', x, t) = 0 \,, \tag{2.1}$$

a formal index is defined to be the number of prolongation and projection steps needed to transform (2.1) into a formally integrable system. Finally, the formal index is found out to be finite. The resulting formally integrable system is an overdetermined system constituting a regular ODE as well as constraints.

The approach described above is closely related to the notion of the differentiation index (e.g. [6]). This notion is based on the so-called derivative array or compound function obtained from (2.1) by k-times formal differentiations

$$\left.\begin{array}{l} f(x^1, x, t) = 0 \\ f'_{x'}(x^1, x, t)x^2 + f'_x(x^1, x, t)x^1 + f'_t(x^1, x, t) = 0 \\ \cdots \\ f'_{x'}(x^1, x, t)x^{k+1} + \cdots = 0 \end{array}\right\} \tag{2.2}$$

System (2.2) is treated as a nonlinear equation in separate variables $x^i \in \mathbb{R}^m$, $i = 1, \ldots, k+1$. The differentiation index μ is the smallest number such that (2.2) with $k = \mu$ can be solved for

$$x^1 = \mathcal{S}(x, t) \,.$$

The resulting regular ODE

$$x' = \mathcal{S}(x, t) \tag{2.3}$$

is called the ODE underlying (2.1). Unfortunately, the ODE (2.3) is not equivalent to (2.1). The system (2.2) contains crucial information, which was not taken into account here: namely the equations for the constraints. E. Griepentrog ([7]) has thoroughly worked out how to deal with the constraints. Finally, the vector field of (2.3) is restricted to a constraint manifold given by a certain equation $r(x, t) = 0$. Hence, one does not have the underlying ODE in $\mathbb{R}^m$, but instead

$$x' = \mathcal{S}(x, t) \,, \quad r(x, t) = 0 \,. \tag{2.4}$$

(2.1) and (2.2) are equivalent under correspondingly strong regularity conditions.

Reduction methods (e.g. [6],[7]) successively realize the transition process to (2.4). Those reduction steps are similar to the transformation steps to a formally integrable system, where usually not all equations are differentiated, but only the derivative-free ones that have been filtered out by projection. In this connection, consider the simple linear DAE with constant coefficients

$$Ax' + Bx = q \,. \tag{2.5}$$

Prolongation gives

$$\left.\begin{array}{lcl} Ax^1 + Bx & = & q \\ Ax^2 + Bx^1 & = & q' \end{array}\right\} \,,$$

and after the projection step this yields the system

$$\left.\begin{array}{rcl} Ax^1 + Bx & = & q \\ (I - AA^+)Bx^1 & = & (I - AA^+)q' \end{array}\right\} .$$

On the other hand, in a reduction step, the derivative-free part of (2.5) is filtered out first, i.e.,

$$(I - AA^+)Bx = (I - AA^+)q$$

and the result is then added to (2.5) in differentiated form. In both cases the DAE (2.5) is transformed into the system

$$\left.\begin{array}{rcl} Ax' + AA^+Bx & = & AA^+q \\ (I - AA^+)Bx' & = & (I - AA^+)q' \\ (I - AA^+)Bx & = & (I - AA^+)q \end{array}\right\} .$$

If we suppose that the matrix pencil $\{A, B\}$ is regular and has index 1, the matrix $A + (I - AA^+)B$ becomes nonsingular. Hence we arrive at a special form of (2.4), namely

$$\left.\begin{array}{rcl} x' & = & (A + (I - AA^+)B)^{-1}(-AA^+Bx + AA^+q(t) + (I - AA^+)q'(t)) \\ & & (I - AA^+)(Bx - q(t)) = 0 \end{array}\right\} .$$

On the basis of reduction steps P.J. Rabier and W.C. Rheinboldt (e.g. [8]) have studied DAE's and found, among other things, nice solvability statements.

All these approaches require f to be highly regular and that, recursively, new strong conditions of rank constancy have to be agreed upon again and again. Moreover, the vector field in (2.4) is relevant for the DAE on the corresponding constraint manifold only. Outside this invariant manifold this vector field is absolutely irrelevant, which becomes obvious e.g. in the asymptotic stability behaviour, which has nothing in common with that of (2.1).

Even if one can realize this reduction procedure (which is very doubtful for serious applications), one will be confronted with the overdetermined system (2.4) in the given coordinates. Transition to the coordinates of the constraint manifold might be successful in practice in trivial cases only.

For quite a long time numerical analysts have been confronted with the problem of integrating regular ODE's numerically in such a way that invariants are taken into account and maintained. C.W. Gear ([9]) has shown that this problem actually requires the numerical solution of a DAE with index 2.

Altogether, the reduction to (2.4) involves unacceptably high demands on regularity. Example (1.3) makes very clear that C^1 does not represent the appropriate class for the solution and that a larger class has to be considered. For semi-explicit systems

$$\left.\begin{array}{rcl} x_1' + g_1(x_1, x_2, t) & = & 0 \\ g_2(x_1, x_2, t) & = & 0 \end{array}\right\} , \tag{2.6}$$

with index 1, i.e., with an everywhere nonsingular partial Jacobian $g'_{22}(x_1, x_2, t)$, the above fact becomes obvious. In this case (2.4) is of the form

$$\left.\begin{array}{l} x'_1 + g_1(x_1, x_2, t) = 0 \\ x'_2 + g'_{22}(x_1, x_2, t)^{-1}\{-g'_{21}(x_1, x_2, t)g_1(x_1, x_2, t) + g'_t(x_1, x_2, t)\} = 0 \\ g_2(x_1, x_2, t) = 0 \end{array}\right\} . \qquad (2.7)$$

A consistent initial value (x_0, t_0) has to fulfil the condition $g_2(x_{0,1}, x_{0,2}, t_0) = 0$. On the other hand, in a neighbourhood of (x_0, t_0), the second equation of (2.6) can be solved directly by means of the implicit function theorem to give

$$x_2 = h(x_1, t) .$$

Now it becomes clearer what the DAE is really made of, namely

$$\left.\begin{array}{rcl} x'_1 + g_1(x_1, h(x_1, t), t) & = & 0 \\ x_2 & = & h(x_1, t) \end{array}\right\} . \qquad (2.8)$$

No doubt, (2.8) is simpler than (2.7). We want to go on from that analytical local decoupling in the following. It seems to be possible to develop such an analysis for DAEs with low index.

Several investigations (e.g. [10], [11]) have been available for the simpler case of DAE's with index 1 since the middle of the eighties. Further, let us mention Takens ([12]), who in 1976 investigated small gradient systems (index 1)

$$x'_1 = f(x_1, x_2), \quad g(x_1, x_2) = 0$$

in connection with the approximating singularly perturbed system

$$x'_1 = f(x_1, x_2), \quad \varepsilon x'_2 = g(x_1, x_2) .$$

In particular, singularities of the vector field were characterized in the case of $g : \mathbb{R} \times \mathbb{R} \to \mathbb{R}$.

We are aiming at a DAE analysis which should characterize the DAE (2.1) in terms of the original model, that is, in actual information on f. What we need, e.g. for creating appropriate numerical methods, are criteria that *guarantee* the regularity of the inherent ODE, its stability etc., and which may be checked in practice. From this point of view, assertions like "if (2.1) represents a regular ODE on a smooth manifold, then ..." are nice, but *l'art pour l'art.* Now we show that a direct analysis based on local decoupling like (2.8) provides nice nontrivial extensions of the standard ODE theory for lower index DAEs, which are also important for applications. Hence, *l'art pour la practique.*

3. Linear Continuous Coefficient DAE's

In this section we try to decouple the DAE

$$A(t)x'(t) + B(t)x(t) = q(t), \quad t \in J , \qquad (3.1)$$

into its characteristic parts by means of certain projections. Not surprisingly, certain subspaces play an important role in this approach.

Since we have in mind applications to nonlinear DAEs (2.1), we set great store by having continuous coefficients A, B, but not smoother ones in general. Then we may expect to handle linearizations of (2.1) along solutions, i.e., equations (3.1) with $A(t) = f'_{x'}(x'(t), x(t), t)$, $B(t) = f'_x(x'(t), x(t), t)$. More precisely, for a given interval $J \subseteq \mathbb{R}$, the coefficients A, B are assumed to belong to $C(J, L(\mathbb{R}^m))$, but the leading nullspace

$$N(t) := \ker A(t), \quad t \in J\,, \tag{3.2}$$

is supposed to depend smoothly on t, i.e., to be spanned by a C^1-basis, or, equivalently, there is a projector function $Q \in C^1(J, L(\mathbb{R}^m))$ such that $Q(t)^2 = Q(t)$, $imQ(t) = N(t)$, $t \in J$. In addition to Q we introduce $P := I - Q$. Taking into account the trivial relations

$$A(t)Q(t) \equiv 0, \quad A(t)P(t) \equiv A(t)$$

and further

$$A(t)x'(t) = A(t)\{(Px)'(t) - P'(t)x(t)\} \tag{3.3}$$

we know that (3.1) involves the derivative of $(Px)(t) = P(t)x(t)$, but that of the nullspace component $(Qx)(t)$ is not involved at all. Naturally, we should look for solutions of (3.1) from the function space C^1_N,

$$C^1_N(J, \mathbb{R}^m) := \{w \in C(J, \mathbb{R}^m) : Pw \in C^1(J, \mathbb{R}^m)\}\,.$$

Note that C^1_N does not depend on the choice of the projector function since $P = P\bar{P}$, $\bar{P} = \bar{P}P$ are true for any two projectors P and $\bar{P}$ along N. Moreover, due to

$$A\{(Px)' - P'x\} = A\{(\bar{P}x)' - \bar{P}'x\}$$

we may agree to use the expression Ax' as an abbreviation of $A\{(Px)' - P'x\}$ with any C^1 projector P. Next, considering the homogeneous system

$$A(t)x'(t) + B(t)x(t) = 0 \tag{3.4}$$

we find the subspace

$$S(t) := \{z \in \mathbb{R}^m : B(t)z \in imA(t)\}$$

to be relevant because it contains all solutions. However, $S(t)$ is filled by solutions of (3.4) only in case of index-1 DAEs.

Definition: The DAE (3.1) is index-1 tractable on J if and only if

$$N(t) \cap S(t) = \{0\}, \quad t \in J\,. \tag{3.5}$$

The index-1 condition (3.5) is well-known to be equivalent to the full-rank of the matrix (cf. Appendix A)

$$G_1(t) := A(t) + B(t)Q(t), \quad t \in J\,. \tag{3.6}$$

Hence, with $G_1^{-1}A = G_1^{-1}G_1P = P$, $G_1^{-1}B = G_1^{-1}BQ + G_1^{-1}BP = Q + G_1^{-1}BP$, equation (3.1) scaled by G_1^{-1} reads

$$Px' + Qx + G_1^{-1}BPx = G_1^{-1}q\,.$$

Multiplying by P and Q, respectively, we arrive at the decoupled version

$$\left.\begin{array}{rcl} Px' + PG_1^{-1}BPx &=& PG_1^{-1}q \\ Qx + QG_1^{-1}BPx &=& QG_1^{-1}q \end{array}\right\},$$

and, more precisely, at

$$\left.\begin{array}{r} (Px)' - P'(I - QG_1^{-1}B)Px + PG_1^{-1}BPx = PG_1^{-1}q + P'QG_1^{-1}q \\ Qx = -QG_1^{-1}BPx + QG_1^{-1}q \end{array}\right\}. \tag{3.7}$$

It should be mentioned that

$$Q_{\text{can}} := QG_1^{-1}B, \quad P_{\text{can}} := I - Q_{\text{can}}$$

again represent projectors. Q_{can} projects onto N along S and is said to be the canonical projector for the index 1 case. Note that Q_{can} is continuous.

System (3.7) shows how to state an initial condition, namely

$$P(t_0)x(t_0) = P(t_0)x^0, \quad x^0 \in \mathbb{R}^m \tag{3.8}$$

i.e., the initial condition should fix the free integration constants of the inherent in (3.7) regular ODE for the component $u := Px$,

$$u' - P'P_{\text{can}}u + PG_1^{-1}Bu = PG_1^{-1}q + P'QG_1^{-1}q\,. \tag{3.9}$$

The subspace $imP(t)$ is easily checked to be invariant for the regular ODE (3.9), that is, $u(t_0) \in imP(t_0)$ implies $Q(t)u(t) \equiv 0$. Now the solutions of the IVP (3.1), (3.8) are represented by

$$\begin{array}{rcl} x &=& Px + Qx = u + Qx \\ &=& P_{\text{can}}u + QG_1^{-1}q\,, \end{array}$$

where $u \in C^1$ solves the inherent regular ODE (3.9), but also the initial condition $u(t_0) = P(t_0)x^0$. Obviously, the consistent initial value is

$$x_0 := x(t_0) = P_{\text{can}}(t_0)x^0 + Q(t_0)G_1(t_0)q(t_0)\,.$$

We have $P(t_0)x_0 = P(t_0)x^0$, but not $x_0 = x^0$, in general.

For solutions of the homogeneous system (3.4) we find the expression

$$x = P_{\text{can}}u = P_{\text{can}}\mathcal{U}P(t_0)x^0 =: Xx^0\,,$$

with the fundamental solution matrix $\mathcal{U}$ of (3.9), $\mathcal{U}(t_0) = I$. For $x^0 \in S(t_0)$, the following holds

$$x(t_0) = P_{\text{can}}(t_0)P(t_0)x^0 = P_{\text{can}}(t_0)x^0 = x^0\,,$$

that is, $S(t_0)$ is the set of consistent initial values for the homogeneous system. Let us summarize what we know in the following theorem.

THEOREM 3.1 *Let the DAE* (3.1) *have index* 1.

(i) *For each* $t_0 \in J$, $x^0 \in \mathbb{R}^m$, $q \in C(J, \mathbb{R}^m)$, *the IVP* (3.1)(3.8) *is uniquely solvable on the given interval* J.

(ii) *Exactly one solution of the homogeneous equation* (3.4) *passes through each* $x_0 \in S(t_0)$, *at* t_0.

The matrix function used above

$$X(t) := P_{\mathrm{can}}(t)\mathcal{U}(t)P(t_0) \tag{3.10}$$

is said to be the fundamental solution matrix of the DAE. It is uniquely determined by the IVP

$$AX' + BX = 0, \quad P(t_0)(X(t_0) - I) = 0 .$$

The problem with that fundamental solution matrix lies in its singularity. It is easily verified that $\ker X(t) \equiv N(t_0)$ holds. Thus, instead of an inverse we are confronted with generalized inverses, say X^- defined by the relations $X^-XX^- = X^-$, $XX^-X = X$, $XX^- = P_{\mathrm{can}}$, $X^-X = P_{\mathrm{can}}(t_0)$. In particular, Green functions are developed in this way (e.g. [13]).

If the index-1 condition (3.5) does not hold, the situation is much more complicated. Interrupting (3.5) at isolated points may cause bifurcations etc. Up to now, there has been no comprehensive systematic analysis of such singularities. On the other hand, if (3.5) fails on the whole interval J, we may be confronted with a higher index DAE. The best understood higher index DAEs are those having index 2.

To give a precise definition we introduce additional matrix functions and subspaces, namely

$$A_1 := G_1(I - PP'Q) ,$$

which has the same rank as G_1,

$$N_1(t) := \ker A_1(t) ,$$

$$S_1(t) := \{z \in \mathbb{R}^m : B(t)P(t)z \in im A_1(t)\}, \quad t \in J .$$

The subspace $N_1(t)$ has the same dimension as the intersection $N(t) \cap S(t)$.

Definition: The DAE (3.1) is index-2 tractable on J if and only if the intersection $N(t) \cap S(t)$ has constant dimension on J and further

$$N_1(t) \cap S_1(t) = \{0\}, \quad t \in J . \tag{3.11}$$

Relation (3.11) allows the use of the further projector $Q_1(t)$ onto $N_1(t)$ along $S_1(t)$, and $P_1(t) := I - Q_1(t)$.

By basic linear algebra (Appendix A), the matrix function

$$G_2 := A_1 + BPQ_1 = A + (B - AP')Q + BPQ_1$$

remains nonsingular. Moreover, the relations

$$Q_1 = Q_1 G_2^{-1} BP, \quad Q_1 Q = 0$$

become true. As a consequence, the decomposition $x = PP_1x + PQ_1x + Qx$ makes sense. Taking into account the further identities

$$G_2^{-1}A = P_1P, \quad G_2^{-1}B = G_2^{-1}BPP_1 + Q_1 + Q + P_1PP'Q$$

we decouple the DAE (3.1) into its essential parts in a similar way to the index-1 case. Then, the following system results (cf. [14]):

$$\left.\begin{array}{rcl} PP_1x' + PP_1P'Qx + PP_1G_2^{-1}BPP_1x & = & PP_1G_2^{-1}q \\ -QQ_1PQ_1x' + QQ_1Q'x + Qx + QP_1G_2^{-1}BPP_1x & = & QP_1G_2^{-1}q \\ PQ_1x & = & PQ_1G_2^{-1}q \end{array}\right\} . \tag{3.12}$$

A priori, $Q_1(t)$ depends continuously on t because $A_1(t)$ has constant rank.

Suppose additionally that Q_1 belongs to C_N^1 so that PQ_1 and $PP_1 = P - PQ_1$ are from C^1. Then, with the denotations $u := PP_1x$, $v := PQ_1x$, $w := Qx$, system (3.12) transforms into

$$\left.\begin{array}{rcl} u' - (PP_1)'(u+v) + PP_1G_2^{-1}Bu & = & PP_1G_2^{-1}q \\ -QQ_1v' + QQ_1(PQ_1)'(u+v) + w + QP_1G_2^{-1}Bu & = & QP_1G_2^{-1}q \\ v & = & PQ_1G_2^{-1}q \end{array}\right\} . \tag{3.13}$$

Now it is evident that initial conditions should be directed to the component $u = PP_1x$, say in the form

$$(PP_1)(t_0)x(t_0) = (PP_1)(t_0)x^0, \quad x^0 \in \mathbb{R}^m. \tag{3.14}$$

By similar arguments as for the index-1 case the next assertion may be proved.

THEOREM 3.2 *Given an index-2 DAE* (3.1) *with* $Q_1 \in C_N^1$.

(i) For each $t_0 \in J$, $x^0 \in \mathbb{R}^m$, $q \in C(J, \mathbb{R}^m)$, $PQ_1G_2^{-1}q \in C^1(J, \mathbb{R}^m)$ *the IVP* (3.1), (3.14) *is uniquely solvable on the given interval.*

(ii) At t_0 *exactly one solution of the homogeneous DAE passes through each* $x_0 \in S^{[1]}(t_0)$, *where*

$$\begin{array}{rcl} S^{[1]}(t) & := & im\Pi_{\text{can}}(t), \\ \Pi_{\text{can}} & := & (I - QQ_1(PQ_1)' - QP_1G_2^{-1}B)PP_1. \end{array}$$

It should be mentioned that $S^{[1]}(t) \subset S(t)$ is a proper subspace. Π_{can} is a projector function, $\Pi_{\text{can}}(t)$ projects along $N(t) \oplus N_1(t)$. Both the subspace $S^{[1]}(t)$ as well as the projector function Π_{can} may be shown to be independent of the choice of the projectors P, Q we started with.

The fundamental solution matrix X given by

$$AX' + BX = 0, \; (PP_1)(t_0)(X(t_0) - I) = 0$$

has the representation

$$X = \Pi_{\text{can}} \mathcal{U} P(t_0) P_1(t_0), \; \ker X(t) = N(t_0) \oplus N_1(t_0),$$

where $\mathcal{U}$ denotes the fundamental matrix of the inherent regular ODE

$$u' - (PP_1)'u + PP_1 G_2^{-1} Bu = PP_1 G_2^{-1} q + (PP_1)' PQ_1 G_2^{-1} q \, .$$

Apart from the greater technical expense, Theorem 3.2 sounds as simple as Theorem 3.1, and index-2 DAEs behave quite similar to index-1 DAEs from this point of view. On the other hand, there is an essential difference relative to the linear map $\mathfrak{A}$ given by

$$\mathfrak{A}x := Ax' + Bx, \quad x \in C_N^1(J, \mathbb{R}^m) \, . \tag{3.15}$$

Theorem 3.3

(i) If the index-1 condition holds true, then $\mathfrak{A}$ is surjective.

(ii) If the index-2 condition holds true and $Q_1 \in C_N^1$, then

$$im \; \mathfrak{A} = \{q \in C(J, \mathbb{R}^m) : PQ_1 G_2^{-1} q \in C^1(J, \mathbb{R}^m)\}$$

is a proper but nonclosed subset of $C(J, \mathbb{R}^m)$.

With respect to a compact interval J, $\mathfrak{A}$ is a Fredholm map in the index-1 case, but $\mathfrak{A}$ is no longer Fredholm for index-2 problems. Due to the inherent differentiation of the component $PQ_1 G_2^{-1} q$, the index-2 IVP (3.1), (3.14) is essentially ill-posed in the sense of Tikhonov while the index-1 IVP remains well-posed. Fortunately, for index-2 problems the ill-posedness is somewhat harmless, and we are able to manage the numerical problems well in many cases.

In view of the asymptotical stability and further questions, transformations of the unknown functions $x(t) = F(t)\bar{x}(t)$ and scalings of the DAE (3.1) by $E(t)$ are of particular interest. The coefficients of the resulting DAE for $\bar{x}(\cdot)$ are

$$\bar{A} := EAF, \quad \bar{B} := EBF + EAF' \, . \tag{3.16}$$

Theorem 3.4 *Let $F \in C_N^1(J, L(\mathbb{R}^m))$, $E \in C(J, L(\mathbb{R}^m))$ be nonsingular.*

(i) Then $\bar{A}$ has a smooth nullspace $\bar{N}$.

(ii) $\bar{N}(t) \cap \bar{S}(t) = F(t)^{-1}(N(t) \cap S(t))$, i.e., the index-1 property is invariant.

(iii) $Q_1 \in C_N^1$ implies $\bar{Q}_1 \in C_{\bar{N}}^1$.

(iv) The index-2 property is invariant.

It should be mentioned that index-μ-tractability generalizes the notion of global index μ introduced in [22] in terms of a possible reduction of the DAE to Kronecker normal form by making a linear smooth transformation of the variable and scaling the DAE.

A linear DAE (3.1) is said to be in Kronecker normal form if it has the special coefficients

$$A(t) = \begin{pmatrix} I & 0 \\ 0 & J \end{pmatrix}, \quad B(t) = \begin{pmatrix} W(t) & 0 \\ 0 & I \end{pmatrix},$$

and J represents a constant nilpotent block, say with index μ, i.e., $J^\mu = 0$, $J^{\mu-1} \neq 0$. The nilpotency index μ is said to be the global index of the given DAE. If a linear DAE has global index μ, then it is index-μ-tractable . On the other hand, using transformations of lower smoothness $F \in C_N^1$, $E \in C$, each index-1-tractable DAE can be reduced to its Kronecker normal form ([15]). The corresponding assertion concerning index-2-tractable DAEs is under preparation.

Interesting particular results may be proved for DAEs with periodic coefficients (e.g. [15]).

Definition: Two DAEs with T-periodic coefficients are called (periodically) equivalent if there is a T-periodic transform $F \in C_N^1$ and a periodic scaling $E \in C$ that connect the coefficients by (3.16). As in the regular ODE case we call $X(T)$ the monodromy matrix. Indeed, it plays the expected role.

With considerable effort it is possible to generalize the well-known results of Lyapunov for DAEs having index-1 or index-2 with $Q_1 \in C_N^1$.

THEOREM 3.5 *("Lyapunov for DAEs"):*

(i) If linear T-periodic DAEs are equivalent, then their monodromy matrices are similar.

(ii) If the monodromy matrices of linear T-periodic DAEs are similar, then these DAEs are equivalent.

(iii) A linear T-periodic DAE is equivalent to a $2T$-periodic real (T-periodic complex) DAE in constant coefficient Kronecker normal form.

Surprisingly, apart from the technical amount of the proof, this assertion sounds as simple and transparent as the original reduction theorem of Lyapunov, proved more than 100 years ago. Note that the representation theorem of Floquet also holds true for DAEs.

4. Nonlinear DAEs

Considering nonlinear equations

$$f(x'(t), x(t), t) = 0 \tag{4.1}$$

we may try to form an analogous chain of subspaces, projectors and matrices by using the partial Jacobians $f_y'(y, x, t)$, $f_x'(y, x, t)$ pointwise instead of $A(t)$, $B(t)$ in

Section 3 (cf. e.g. [14], [16]). Roughly speaking, we are aiming at the following situation:

"*The DAE* (4.1) *has index* μ *if the linearized DAE has, and vice versa.*"

However, except for the index-1 case, which is well-understood , there remain a number of open questions on how to take into account the different kinds of rotating subspaces.

From the analytical point of view, the behaviour of the leading nullspace of (4.1), that is, $\ker f'_y(y,x,t)$, is of great importance - even in the index-1 case. In [17], Ch. Lubich pointed out that, if the leading nullspace varies with (y,x), the solutions involve certain derivatives, similarly to the solutions of the linear index-2 DAEs (cf. Section 3). On the other hand, an index-1 DAE (4.1) whose leading nullspace is invariant of y and x does not show this unpleasant feature.

To measure the sensitivity of solutions with respect to inherent differentiations the so-called perturbation index (e.g. [18]) has been introduced. It provides essential information on the difficulty of the problem from the numerical point of view. One might conjecture that just rotating subspaces causes the perturbation index to be higher than the (formal) differentiation index.

For simplicity, only quasilinear DAEs of the form

$$A(t)x'(t) + g(x(t),t) = 0 \tag{4.2}$$

are considered in the following. $A(t)$, $g(x,t)$, $g'_x(x,t)$ are supposed to be continuous on $J_0 \subset \mathbb{R}$ and $\mathcal{D}_0 \times J_0 \subseteq \mathbb{R}^m \times \mathbb{R}$, respectively. The nullspace $N(t) := \ker A(t)$ is assumed to be smooth (cf. Section 3). $Q(t)$ again denotes a C^1 projector function onto $N(t)$, $P(t) := I - Q(t)$.

In Appendix B it is shown how the results obtained for (4.2) immediately apply to the general DAE (4.1) provided that the leading nullspace is invariant of y and x. In order to treat the problems (4.1), whose leading nullspace $\ker f'_y(y,x,t)$ varies with (y,x), we propose to change to the trivially enlarged system having a constant leading nullspace (Appendix C), but emphasize once more that the differentiation index increases by that transformation.

Clearly, all solutions of (4.2) should belong to the class C^1_N introduced in Section 3.

Definition: The DAE (4.2) is index-1 tractable on $J \times \mathcal{D} \subseteq J_0 \times \mathcal{D}_0$ if, and only if,

$$N(t) \cap S(x,t) = \{0\}, \quad t \in J,\, x \in \mathcal{D} \tag{4.3}$$

holds, where

$$S(x,t) := \{z \in \mathbb{R}^m : g'_x(x,t)z \in imA(t)\}\ .$$

Further, introduce the set

$$\mathcal{M}_{(1)}(t) := \{w \in \mathcal{D}_0 : g(w,t) \in imA(t)\},\ t \in J_0\ ,$$

containing all solutions of (4.1). Now, the subspace $S(x,t)$ manifests its geometrical meaning

$$S(x,t) = T_x\mathcal{M}_{(1)}(t) \quad \text{for} \quad x \in \mathcal{M}_{(1)}(t) \, .$$

Restricted to $t \in J, x \in \mathcal{M}_{(1)}(t)$, the index-1 condition (4.3) says that the leading nullspace $N(t)$ and the tangent space $T_x\mathcal{M}_{(1)}(t)$ have to intersect transversally. However, condition (4.3) applies also to elements outside of $\mathcal{M}_{(1)}(t)$.

By the basic linear algebra (cf. Appendix A), the index-1 condition is satisfied if and only if the matrix function

$$G_1(x,t) := A(t) + g'_x(x,t)Q(t) \tag{4.4}$$

remains nonsingular. Hence, we have got a nice criterion for checking index-1 and for detecting singularities, respectively. Using the decoupling technique described above and applying the implicit function theorem we are able to prove the next local solvability assertion, which is a straightforward generalization of the classical regular ODE case.

THEOREM 4.1 *Given an index-1 DAE* (4.1), $t_0 \in J$, $x_0 \in \mathcal{M}_{(1)}(t_0)$, $(x_0,t_0) \in \mathcal{D} \times J$.

(i) Exactly one solution passes through (x_0, t_0).

(ii) For sufficiently small $\tau > 0$ *the IVPs*

$$\begin{aligned} &A(t)x'(t) + g(x(t),t),t) = 0, \quad P(t_0)(x(t_0) - x^0) = 0 \, , \\ &|P(t_0)(x_0 - x^0)| \le \tau, \;\; x^0 \in \mathbb{R}^m \, , \end{aligned} \tag{4.5}$$

are locally uniquely solvable.

(iii) The solution $x(\cdot\,, t_0, x^0)$ *of* (4.5) *depends in a continuously differentiable way on* x^0.

It seems that the IVP solutions provided by Theorem 4.1 can be continued as long as they do not leave the index-1 domain. Recall that impasse points and bifurcations may occur at points where the index-1 condition is lost.

If the index-1 condition (4.3) fails uniformly, then we may expect a higher index problem. This happens in example (1.2) above, which can be characterized to have index 2. Note that for (1.2) it becomes characteristic that $S_1(x) \cap N_1(x) = \{0\}$ holds, where $S_1(x) := \{z \in \mathbb{R}^3 : x_1 z_1 + x_2 z_2 = 0\} = S(x)$, $N_1(x) := \{z \in \mathbb{R}^3 : z_1 = 0, z_2 + z_3/x_2 = 0\}$, but $N := \{z \in \mathbb{R}^3 : z_1 = 0, z_2 = 0\} \subset S(x)$, $\dim(N \cap S(x)) = 1$.

Next we turn to the linearizations along a given solution, say $x_*(\cdot) : J_* \to \mathbb{R}^m$ of (4.2). Denote $B(t) := g'_x(x_*(t),t)$ such that $A(t)$, $B(t)$ are the coefficients of the linearized along x_* DAE. Further, we will now use the projectors, subspaces etc. related to the linear DAE with the coefficients $A(t)$, $B(t)$, which we introduced in Section 3. In particular, let

$$\Pi_{(\mu)}(t) := \begin{cases} P(t) & \text{in case of } \mu = 1 \\ P(t)P_1(t) & \text{in case of } \mu = 2 \end{cases}$$

Moreover, we use the canonical projector function

$$\Pi_{\mathrm{can}(\mu)}(t) := \begin{cases} P_{\mathrm{can}} & \text{for } \mu = 1 \\ \Pi_{\mathrm{can}} & \text{for } \mu = 2 \,. \end{cases}$$

As we pointed out above, the relation

$$im\Pi_{\mathrm{can}(1)}(t) = T_{x_*(t)}\mathcal{M}_{(1)}(t)$$

becomes true in the index-1 case. Such a relation is expected to be true also for $\mu = 2$, but then we would have to describe the corresponding complicated subset $\mathcal{M}_{(2)}(t) \subset \mathcal{M}_{(1)}(t)$ in some detail.

As we saw for linear DAEs, the canonical projector functions $\Pi_{\mathrm{can}(\mu)}$ are rather complex, whereas the projector functions $\Pi_{(\mu)}$ are fairly simple. Furthermore, the $\Pi_{(\mu)}$ belong to the class C^1, whereas the $\Pi_{\mathrm{can}(\mu)}$ belong to C^1_N exclusively.

In this sense, $\Pi_{(\mu)}$ can be regarded as a practicable substitute of $\Pi_{\mathrm{can}(\mu)}$. Essential preconditions for this are the following inherent properties

$$\begin{aligned} ker\Pi_{\mathrm{can}(\mu)}(t) &= ker\Pi_{(\mu)}(t) \,, \\ P(t)\Pi_{\mathrm{can}(\mu)}(t) &= P(t)\Pi_{(\mu)}(t) \,. \end{aligned}$$

Now, $im\Pi_{(\mu)}(t)$ can also be considered as a substitute for the tangent space $T_{x_*(t)}\mathcal{M}_{(\mu)}(t)$.

Taking into consideration the solvability statements for linear DAEs and the description of admissible righthand-sides by Theorems 3.1 – 3.3, we now introduce the function spaces $C_{(\mu)}$, $\mu = 1, 2$, equipped with the norms $\|\cdot\|_{(\mu)}$ by

$$\begin{aligned} C_{(1)} &:= C, \quad \|\cdot\|_{(1)} := \|\cdot\|_\infty \\ C_{(2)} &:= \{q \in C : PQ_1G_2^{-1}q \in C^1\}, \|q\|_{(2)} := \|q\|_\infty + \|(PQ_1G_2^{-1}q)'\|_\infty \end{aligned}$$

THEOREM 4.2 *Let $x_* \in C^1_N(J_*, \mathbb{R}^m)$ solve the DAE* (4.2) *on the compact interval $J_* \ni t_0$. Let the DAE linearized at x_* have index $\mu \in \{1, 2\}$.*

If $\mu = 2$, certain additional structural and smoothness conditions are supposed.

(i) Then, for sufficiently small $\tau > 0$, $\sigma > 0$ the IVPs

$$A(t)x'(t) + g(x(t), t) = q(t) \,, \tag{4.6}$$

$$\Pi_{(\mu)}(t_0)(x(t_0) - x^0) = 0 \,, \tag{4.7}$$

$$|\Pi_{(\mu)}(t_0)(x_*(t_0) - x^0)| \le \tau \,, \quad \|q\|_{(\mu)} \le \sigma$$

are locally uniquely solvable on J_*.

(ii) The solution $x(\cdot, t_0, x^0)$ of the IVP (4.2), (4.7) *depends in a continuously differentiable way on x^0, and $\frac{\partial x}{\partial x^0}$ satisfies the variational DAE.*

(iii) The solution of (4.6), (4.7) depends continuously on $(x^0, q) \in \mathbb{R}^m \times C_{(\mu)}$.

This assertion is proved by applying the implicit function theorem after carefully preparing the properties of the maps involved (cf. [4],[19]).

The additional smoothness and structural conditions mentioned in Theorem 4.2 allow, on the one hand, the necessary inherent differentiation in the index-2 case. On the other hand, they guarantee that the index-2 property of the linearization can be extended to a neighbourhood by certain restrictions of the structure. For linear DAEs and DAEs in Hessenberg form of size 2 (e.g. [6]) these structural conditions are always fulfilled. For more general classes of DAEs, however, their description is technically rather expensive.

It should be mentioned that, up to now, the *maximal index-2 structure* has not been clarified. C. Tischendorf ([20]) has obtained a structural criterion that is very useful in applications and generalizes the results in [19].

Note that Theorem 4.2 proves its value with respect to the numerical treatment of DAEs. There are analogous "discretized" versions (e.g. [20]) which show how to manage the computations well.

It should be mentioned once more that Theorem 4.2 (iii) means in fact that the DAE (4.2) has perturbation index μ if it is index-μ tractable.

On the other hand, we are now motivated to consider also solutions on infinite intervals, say $J_* = [t_0, \infty)$.

Definition: Given a solution $x_* \in C_N^1([t_0,\infty), \mathbb{R}^m)$ of (4.2). Let (4.2) have index $\mu \in \{1,2\}$. x_* is said to be stable in Lyapunov's sense if there is a $\tau > 0$, and, moreover, there exists a $\delta(\varepsilon) > 0$ to each $\varepsilon > 0$ such that

(i) the IVPs

$$A(t)x'(t)+g(x(t),t) = 0, \Pi_{(\mu)}(t_0)(x(t_0)-x^0) = 0, |\Pi_{(\mu)}(t_0)(x_*(t_0)-x^0)| \leq \tau$$

are solvable on $[t_0, \infty)$.

(ii) $|\Pi_{(\mu)}(t_0)(x_*(t_0) - x^0)| \leq \delta(\varepsilon)$ implies $|x(t) - x_*(t)| \leq \varepsilon \quad$ for $t \geq t_0$.

x_* is called asymptotically stable if, additionally, $|\Pi_{(\mu)}(t_0)(x_*(t_0) - x^0)| \leq \tau_0$ implies $x(t) - x_*(t) \to 0 (t \to \infty)$ for sufficiently small $\tau_0 \leq \tau$.

Our definition reflects the geometrical meaning of Lyapunov stability properly. By means of that special statement of initial conditions, without evaluating the implicit state manifold, the neighbouring solutions on that manifold are caught properly to be compared with x_*.

Autonomous DAEs

$$Ax' + g(x) = 0 \tag{4.8}$$

seem to be essentially simpler than nonautonomous ones. In particular, in linear homogeneous constant coefficient DAEs all critical parts disappear, and the solution is smooth.

Let $x_* \in \mathcal{D}_0 \subseteq \mathbb{R}^m$ represent a stationary solution of (4.8), i.e., $g(x_*) = 0$. In this case, the equation linearized at x_* has constant coefficients A and $B := g'(x_*)$. The well-known Lyapunov-Theorem of asymptotic stability of a stationary solution for DAEs sounds as simple as it does for regualr ODEs (cf. [21]).

Theorem 4.3 *Let* $x_* \in \mathcal{D}_0$, $g(x_*) = 0$, $g \in C^2(\mathcal{D}_0, \mathbb{R}^m)$. *Let the matrix pencil* $\{A, B\}$ *be regular with index* $\mu \in \{1,2,3\}$, $\sigma\{A, B\} \subset \mathbb{C}^-$. *If* $\mu > 1$, *let some additional structural restrictions be satisfied. Then,* x_* *is asymptotically stable.*

Note that now all solutions have additional regularity, namely, they are C^1.

Next, let $x_* \in C^1_N$ be a T-periodic solution of (4.2) whose stability properties are to be considered. Then, supposed (4.2) is T-periodic, the coefficients $A(t), B(t)$ of the linearization are also T-periodic, and in the consequence all our subspaces, projectors etc. are so.

Now we can follow Lyapunov's way of showing stability via linearization, transformation of the linear part into constant coefficient form, and using a Lyapunov function for that simple case. As an auxiliary problem we have to consider the stability of stationary solutions of DAEs

$$Ax'(t) + Bx(t) + h(x'(t), x(t), t) = 0 \tag{4.9}$$

with small nonlinearities. More precisely, assume $h(0,0,t) \equiv 0$ and consider the trivial solution $x_* = 0$. Next, collect some assumptions to be satisfied for (4.9):

Let $\{A, B\}$ be regular with index $\mu \in \{1,2\}$ and let $\ker A \subseteq \ker h'_y(y,x,t)$. To each $\varepsilon > 0$, there is a $\delta(\varepsilon) > 0$ such that

$$\left.\begin{array}{l} |h'_x(y,x,t)| \le \varepsilon \\ |h'_y(y,x,t)| \le \varepsilon \end{array}\right\} \quad \text{for} \quad |x| \le \delta(\varepsilon),\ |y| \le \delta(\varepsilon),\ t \in [t_0, \infty)\,.$$

In case of $\mu = 2$, certain additional assumptions have to be satisfied. Then we obtain the result we are searching for; it sounds as nice as the original by Lyapunov for regular ODEs.

Lemma 4.4 *If* $\sigma\{A, B\} \subset \mathbb{C}^-$, *then the origin is an asymptotically stable point of* (4.9).

Finally, return to the T-periodic case of (4.2) with a T-periodic solution $x_*(\cdot)$.

Theorem 4.5 *Let* g *have an additional continuous derivative* g''_{xx}, *and let the DAE linearized at* x_* *have index 1.*

If the monodromy matrix $X(T)$ *has all its eigenvalues in* $\{z \in \mathbb{C} : |z| < 1\}$, *then the periodic solution* $x_*(\cdot)$ *is asymptotically stable in the sense of Lyapunov.*

The proof is given in [15]. It is hoped that a version for the index-2 case will be completed soon.

5. Conclusion

Modelling with DAEs is growing in importance, and consequently so too is their numerical treatment. But the direct numerical treatment of lower index DAEs for practical applications needs further analysis (asymptotic behaviour, practicable index criteria, role of rotating subspaces, index changes, small parameters, singularities...). From this point of view, ODEs should be revisited and generalized to include DAEs. There are grounds to be optimistic in this respect.

Appendix A: Basic linear algebra lemma

A basic connection between the spaces appearing at the tractability index and the choice of the corresponding projectors is given by the following lemma, which may be directly obtained from Theorem A.13 and Lemma A.14 of [11].

LEMMA A.1 *Let* $\bar{A}, \bar{B}, \bar{Q} \in L(\mathbb{R}^m)$ *be given,* $\bar{Q}^2 = \bar{Q}, im(\bar{Q}) = \ker(\bar{A})$*, i.e., let* $\bar{Q}$ *be a projector onto* $\ker(\bar{A})$*. Denote* $\bar{S} := \{z \in \mathbb{R}^m : \bar{B}z \in im(\bar{A})\}$*. Then the following conditions are equivalent:*

(i) The matrix $\bar{G} := \bar{A} + \bar{B}\bar{Q}$ *is nonsingular.*

(ii) $\mathbb{R}^m = \bar{S} \oplus \ker(\bar{A})$.

(iii) $\bar{S} \cap \ker(\bar{A}) = \{0\}$.

If $\bar{G}$ *is nonsingular, then the relation*

$$\bar{Q}_s = \bar{Q}\bar{G}^{-1}\bar{B}$$

holds for the canonical projector $\bar{Q}_s$ *(canonical means:* $\bar{Q}_s$ *projects* $\mathbb{R}^m$ *onto* $\ker(\bar{A})$ *along* $\bar{S}$*).*

Proof:

(i) $\to$ (ii) First, the space $\mathbb{R}^m$ can be described as $\bar{S} + \ker(\bar{A})$ because

$$z = (I - \bar{Q}\bar{G}^{-1}\bar{B})z + \bar{Q}\bar{G}^{-1}\bar{B}z =: z_1 + z_2 \qquad (*)$$

holds for any $z \in \mathbb{R}^m$.
Now z_2 obviously lies in $\ker(\bar{A})$ because $\bar{Q}$ is a projector onto $\ker(\bar{A})$.
For z_1 we obtain

$$\bar{B}z_1 = (I - \bar{B}\bar{Q}\bar{G}^{-1})\bar{B}z = \bar{A}\bar{G}^{-1}\bar{B}z \in im(\bar{A}),$$

i.e., $z_1 \in \bar{S}$.

It remains to show that $\bar{S} \cap \ker(\bar{A}) = \{0\}$. For that, let $x \in \bar{S} \cap \ker(\bar{A})$. Then $x = \bar{Q}x$ holds and there exists a $z \in \mathbb{R}^m$ such that

$$\bar{A}z = \bar{B}x \quad \text{and so} \quad \bar{G}^{-1}\bar{A}z = \bar{G}^{-1}\bar{B}\bar{Q}x,$$

i.e., $(I - \bar{Q})z = \bar{Q}x$, thus $0 = \bar{Q}x = x$.

(ii) $\to$ (iii) This holds trivially because of the definition.

(iii) $\to$ (i) Let $x \in \mathbb{R}^m$ be chosen such that $\bar{G}x = 0$, i.e., $\bar{B}\bar{Q}x = -\bar{A}x$ and so $\bar{Q}x \in \bar{S}$. On the other hand, $\bar{Q}x$ lies in $\ker(\bar{A})$. Thus, $x \in \ker(\bar{Q})$ holds because of the assumption. That means $\bar{A}x = 0$, hence $x \in im(\bar{Q})$. Therefore, $x = 0$ must hold and $\bar{G}$ is nonsingular.

Due to the uniqueness of the partition $(*)$, the latter assertion follows immediately.

Appendix B: Enlarged system/solution invariant leading nullspace

Consider the DAE system with m equations

$$f(x'(t), x(t), t) = 0, \tag{B.1}$$

where $f(y, x, t)$ and the partial Jacobians $f'_y(y, x, t), f'_x(y, x, t)$ are continuous in (y, x, t).

Let the leading nullspace of (B.1), that is, the nullspace of the leading Jacobian $f'_y(y, x, t)$, be invariant of y and x. Denote

$$N(t) := \ker f'_y(y, x, t).$$

Additionally, let $N(t)$ be smooth (cf. Section 2), and let $Q \in C^1$ denote a projector function onto that nullspace, $P := I - Q$.

For characterizing the tractability index of (B.1) we introduce the following subspaces and matrices

$$\begin{array}{lcl} S(y,x,t) & := & \{z : f'_x(y,x,t)z \in im f'_y(y,x,t)\}, \\ G_1(y,x,t) & = & f'_y(y,x,t) + f'_x(y,x,t)Q(t), \\ A_1(y,x,t) & = & G_1(y,x,t)(I - P(t)P'(t)Q(t)), \\ S_1(y,x,t) & = & \{z : f'_x(y,x,t)P(t)z \in im A_1(y,x,t)\} \\ N_1(y,x,t) & = & \ker A_1(y,x,t). \end{array}$$

Besides (B.1), we consider the enlarged system with $2m$ equations

$$\left.\begin{array}{lcl} P(t)x'(t) - y(t) & = & 0 \\ f(y(t), x(t), t) & = & 0 \end{array}\right\} \tag{B.2}$$

with the unknown function $\binom{x(t)}{y(t)} =: \bar{x}(t)$. Obviously, (B.2) has the form of (4.2). Writing (B.2) in a compact way as

$$\bar{f}(\bar{x}'(t), \bar{x}(t), t) = 0$$

we find the respective subspaces and matrices for (B.2) to be

$$\bar{N}(t) = \left\{ \binom{z_1}{z_2} : z_1 \in N(t) \right\},$$

and then, dropping the arguments,

$$\begin{array}{lcl} \bar{f}'_{\bar{y}} & = & \begin{pmatrix} P & 0 \\ 0 & 0 \end{pmatrix}, \quad \bar{f}'_{\bar{x}} = \begin{pmatrix} 0 & -I \\ f'_x & f'_y \end{pmatrix}, \quad \bar{Q} = \begin{pmatrix} Q & 0 \\ 0 & I \end{pmatrix}, \\ \bar{G}_1 & = & \begin{pmatrix} P & -I \\ f'_x Q & f'_y \end{pmatrix}, \bar{A}_1 = \bar{G}_1 \begin{pmatrix} I - PP'Q & 0 \\ 0 & I \end{pmatrix}, \\ \bar{S} & = & \left\{ \binom{z_1}{z_2} : Qz_2 = 0, z_1 \in S, f'_x z_1 + f'_y z_2 = 0 \right\} \end{array}$$

$$\begin{aligned}
\bar{S}_1 &= \left\{ \binom{z_1}{z_2} : z_1 \in S_1 \right\}, \\
\bar{S} \cap \bar{N} &= \left\{ \binom{z_1}{z_2} : z_1 \in N \cap S, z_2 = -f_y'^{+} f_x' z_1 \right\}, \\
\bar{S}_1 \cap \bar{N}_1 &= \left\{ \binom{z_1}{z_2} : z_1 \in N_1 \cap S_1, z_2 = (P - PP'Q) z_1 \right\}.
\end{aligned}$$

Now, the following assertion becomes obvious.

LEMMA B.1: It holds that

$$\begin{aligned}
dim\bar{N} &= dimN + m, \\
dim\bar{N} \cap \bar{S} &= dimN \cap S, \\
dim\bar{N}_1 \cap \bar{S}_1 &= dimN_1 \cap S_1.
\end{aligned}$$

Trivially, Lemma B.1 makes clear that (B.2) is an index-1 DAE with $dim\bar{N} = m$ if and only if (B.1) represents a regular implicit ODE (which corresponds to $dimN = 0$).
The more interesting case is $dimN > 0$.

COROLLARY: Let the nullspace be nontrivial and $\mu \in \{1, 2\}$. Then (B.1) is index-μ tractable if and only if (B.2) is so.

Hence, for the cases we are interested in, enlarging the given DAE does not change the index. Therefore, the results obtained for (4.2) in Section 4 can be carried over to the general equation (B.1) via (B.2). For that, we also give the projectors

$$\bar{Q}_1 = \begin{pmatrix} Q_1 & 0 \\ PQ_1 - PP'QQ_1 & 0 \end{pmatrix}, \quad \bar{P}\bar{Q}_1 = \begin{pmatrix} PQ_1 & 0 \\ 0 & 0 \end{pmatrix}.$$

Appendix C: Enlarged system/solution dependent leading nullspace

As in Appendix B, we consider the DAE

$$f(x'(t), x(t), t) = 0, \tag{C.1}$$

but now we allow the leading nullspace to depend on all given variables. Denote

$$N(y, x, t) := \ker f_y'(y, x, t).$$

The leading Jacobian $f_y'(y, x, t)$ is assumed to have constant rank. Thus, the subspace $N(y, x, t)$ has constant dimension. Introduce further the subspace $S(y, x, t)$ as described in Appendix B. The index-1 condition on a region $\mathcal{G}$ for (C.1) reads now

$$N(y, x, t) \cap S(y, x, t) = \{0\}, \quad (y, x, t) \in \mathcal{G}.$$

However, the theory given in Section 4 does not apply and a respective enlarged system (B.2) makes no sense. Instead of (B.2) we consider the enlarged system

$$\left.\begin{aligned} x'(t) - y(t) &= 0 \\ f(y(t), x(t), t) &= 0 \end{aligned}\right\} \tag{C.2}$$

which contains also $2m$ equations. Reformulating (C.2) as $\bar{f}(\bar{x}'(t), \bar{x}(t), t) = 0$, $\bar{x} := \binom{x}{y}$, we have (again dropping the arguments)

$$\begin{aligned}
\bar{f}'_{\bar{y}} &= \begin{pmatrix} I & 0 \\ 0 & 0 \end{pmatrix}, \quad \bar{f}'_{\bar{x}} = \begin{pmatrix} 0 & -I \\ f'_x & f'_y \end{pmatrix}, \quad \bar{Q} = \begin{pmatrix} Q & 0 \\ 0 & I \end{pmatrix}, \\
\bar{S} &= \left\{ \binom{z_1}{z_2} : f'_x z_1 + f'_y z_2 = 0 \right\}, \\
\bar{N} &= \left\{ \binom{z_1}{z_2} : z_1 = 0 \right\}, \quad \bar{N}_1 = \left\{ \binom{z_1}{z_2} : z_1 = z_2 = 0 \right\}, \\
\bar{S}_1 &= \left\{ \binom{z_1}{z_2} : f'_x z_1 \in im f'_y \right\}, \\
\bar{S} \cap \bar{N} &= \left\{ \binom{z_1}{z_2} : z_1 = 0, z_2 \in \bar{N} \right\}, \\
\bar{S}_1 \cap \bar{N}_1 &= \left\{ \binom{z_1}{z_2} : z_1 = z_2 \in N \cap S, \right\}.
\end{aligned}$$

Hence, the following assertion is verified.

LEMMA C.1: It holds that

$$\begin{aligned}
dim \bar{N} &= m, \\
dim \bar{N} \cap \bar{S} &= dim N, \\
dim \bar{N}_1 \cap \bar{S}_1 &= dim N \cap S.
\end{aligned}$$

The advantage of the system (C.2) consist in its semi-explicit form and its constant leading nullspace. However, (C.2) has index 1 only for regular ODEs (C.1).
In general, the enlarged form (C.2) has a higher index than (C.1). More precisely, Lemma C.1 implies the following:

COROLLARY: (C.1) fulfils the index-1 condition if and only if (C.2) is index-2 tractable.

To illustrate the role of the solution dependent leading nullspace, let us quote the simple but characteristic example from [17] mentioned above.

Example: Given the DAE with $m = 3$

$$\left. \begin{aligned}
x'_1 - x_3 x'_2 + x_2 x'_3 &= 0 \\
x_2 &= q_2(t) \\
x_3 &= q_3(t)
\end{aligned} \right\} \tag{C.3}$$

The nullspace of the leading Jacobian

$$f'_y(y, x, t) = \begin{pmatrix} 1 & -x_3 & x_2 \\ 0 & 0 & 0 \\ 0 & 0 & 0 \end{pmatrix}$$

rotates if x_2, x_3 vary. The index-1 condition is satisfied, i.e.,

$$\begin{aligned} N(y,x,t) \cap S(y,x,t) &= \{z : z_1 - x_3 z_2 + x_2 z_3 = 0, z_2 = z_3 = 0\} \\ &= \{0\}. \end{aligned}$$

The IVP solution of (C.3) with $x_1(0) = 0$ has the component

$$x_1(t) = \int_0^t \{q_3(s)q_2'(s) - q_2(s)q_3'(s)\}\mathrm{d}s,$$

which depends on certain derivatives, in fact. Although the index-1 condition is fulfilled, we have a perturbation index greater than one here.

Since DAEs (C.1) with solution dependent leading nullspace that satisfy the index-1 condition behave analytically like index-2 tractable DAEs of the form (4.2), the transition from (C.1) to (C.2) is reasonable in this case. Now, the results from Section 4 can be applied to index-2 tractable DAEs (4.2) and reformulated for (C.1) via (C.2) (cf. [19]).

References

[1] C.W. Gear: The simultaneous numerical solution of differential-algebraic equations. IEEE Trans. Circuit Theory, CT-18, 89–95, 1971

[2] C.W. Gear, H.H. Hsu and L.R. Petzold: Differential-algebraic equations revisited. Proc. ODE Meeting, Oberwolfach. Inst. f. Geom. und Praktische Mathem., Techn. Hochschule Aachen, Germany, Bericht 9, 1981

[3] L.R. Petzold: Differential/algebraic equations are not ODEs. SIAM J. Sci. Stat. Comput. 3(1982), 367–384

[4] R. März: On correctness and numerical treatment of boundary value problems in differential-algebraic equations. Zhurnal vychisl. matem. i. matem. fiziki 26(1), 50–64, 1986

[5] O.-P. Piirilä and J. Tuomela: Differential-algebraic systems and formal integrability. Helsinki Univ. of Technology, Inst. of Math., Report A 326, 1993

[6] K.E. Brenan, S.L. Campbell and L.R. Petzold: Numerical solution of initial value problems in differential-algebraic equations. Elsevier Science Publ. Co. Inc., 1989

[7] E. Griepentrog: Index reduction methods for differential-algebraic equations. Humboldt-Univ. Berlin, Inst. für Mathem., Seminarbericht Nr. 92-1, 1992, 14–29

[8] P.J. Rabier and W.C. Rheinboldt: A geometric treatment of implicit differential-algebraic equations. J. Diff. Equations 109 (1994), 110–146

[9] C.W. Gear: Maintaining solution invariants in the numerical solution of ODEs. SIAM J. Sci. Stat. Comp. 7 (1986), 734–743

[10] V. Dolezal: Generalized solutions of semistate equations and stability. Circuit Systems Signal Process 5(4), 391–403, 1986

[11] E. Griepentrog and R. März: Differential-algebraic equations and their numerical treatment. Teubner-Texte zur Mathematik 88, 1986

[12] F. Takens: Constrained equations: A study of implicit differential equations and their discontinuous solutions. Lecture Notes in Math. 525, 143–234, 1976

[13] M. Lentini and R. März: Conditioning and dichotomy in differential-algebraic equations. SIAM J. Num. Anal. 27(6), 15 1519–1526, 1990

[14] R. März: Numerical methods for differential-algebraic equations. Acta Numerica 1992, 141–198

[15] R. Lamour, R. März and R. Winkler: How Floquet-theory applies to differential-algebraic equations. Humboldt-Univ. Berlin, Inst. für Mathem. Preprint 96–15, 1996, to appear in J. Mathem. Analysis and Applications

[16] R. März: Some new results concerning index-2 differential-algebraic equations. J. Mathem. Analysis and Applications 140 (1), 177–199, 1989

[17] Ch. Lubich: Linearly implicit extrapolation methods for differential-algebraic systems. Numer. Math. 55, 197–211, 1989

[18] E. Hairer and G. Wanner: Solving ordinary differential equations II. Springer-Verlag, Berlin, 1991

[19] R. März: On linear differential-algebraic equations and linearizations. Applied Numerical Mathematics 18, 267–292, 1995

[20] C. Tischendorf: Solution of index-2 differential-algebraic equations and its application in circuit simulation. Dissertation, Humboldt-Univ. Berlin, Inst. für Mathem., April 1996

[21] R. März: Practical Lyapunov stability criteria for differential-algebraic equations. Banach Center Publications 29, 245–266, 1994

[22] C. W. Gear and L.R. Petzold: ODE methods for the solution of differential/algebraic systems. SIAM J. Num. Anal. 21(4), 716–728, 1984

Progress in Mathematics

Edited by:

Progress in Mathematics *is a series of books intended for professional mathematicians and scientists, encompassing all areas of pure mathematics. This distinguished series, which began in 1979, includes authored monographs, and edited collections of papers on important research developments as well as expositions of particular subject areas.*

We encourage preparation of manuscripts in such form of TeX for delivery in camera-ready copy which leads to rapid publication, or in electronic form for interfacing with laser printers or typesetters.

Proposals should be sent directly to the editors or to:
Birkhäuser Boston, 675 Massachusetts Avenue, Cambridge, MA 02139, U.S.A.

92 Connes/Duflo/Joseph/Rentschler. Operator Algebras, Unitary Representations, Enveloping Algebras, and Invariant Theory. A Collection of Articles in Honor of the 65th Birthday of Jacques Dixmier
93 Audin. The Topology of Torus Actions on Symplectic Manifolds
94 Mora/Traverso (Eds). Effective Methods in Algebraic Geometry
95 Michler/Ringel (Eds). Representation Theory of Finite Groups and Finite–Dimensional Algebras
96 Malgrange. Equations Différentielles à Coefficients Polynomiaux
97 Mumford/Norman/Nori. Tata Lectures on Theta III
98 Godbillon. Feuilletages, Etudes géométriques
99 Donato/Duval/Elhadad/ Tuynman. Symplectic Geometry and Mathematical Physics. A Collection of Articles in Honor of J.-M. Souriau
100 Taylor. Pseudodifferential Operators and Nonlinear PDE
101 Barker/Sally. Harmonic Analysis on Reductive Groups
102 David. Séminaire de Théorie des Nombres, Paris 1989-90
103 Anger/Portenier. Radon Integrals
104 Adams/Barbasch/Vogan. The Langlands Classification and Irreducible Characters for Real Reductive Groups
105 Tirao/Wallach. New Developments in Lie Theory and Their Applications
106 Buser. Geometry and Spectra of Compact Riemann Surfaces
107 Brylinski. Loop Spaces,Characteristic Classes and Geometric Quantization
108 David. Séminaire de Théorie des Nombres, Paris 1990-91
109 Eyssette/Galligo. Computational Algebraic Geometry
110 Lusztig. Introduction to Quantum Groups
111 Schwarz. Morse Homology
112 Dong/Lepowsky. Generalized Vertex-Algebras and Relative Vertex Operators
113 Moeglin/Waldspurger. Décomposition Spectrale et Series d'Eisenstein
114 Berenstein/Gay/Vidras/Yger. Residue Currents and Bezout Identities
115 Babelon/Cartier/Kosmann-Schwarzbach (Eds). Integrable Systems. The Verdier Memorial Conference
116 David (Ed.). Séminaire de Théorie des Nombres, Paris, 1991-1992
117 Audin/Lafontaine (Eds). Holomorphic Curves in Symplectic Geometry
118 Vaisman. Lectures on the Geometry of Poisson Manifolds
119 Joseph/Mignot/Murat/Prum/Rentschler (Eds) First European Congress of Mathematics (Paris, July 6-10, 1992). Volume I: Invited Lectures (Part 1)
120 Joseph/Mignot/Murat/Prum/Rentschler (Eds). First European Congress of Mathematics (Paris, July 6-10, 1992). Volume II: Invited Lectures (Part 2)

121 JOSEPH/MIGNOT/MURAT/PRUM/RENTSCHLER (Eds). First European Congress of Mathematics (Paris, July 6-10, 1992). Volume III: Round Tables
122 GUILLEMIN. Moment Maps and Combinatorial Invariants of Hamiltonian T^n-spaces
123 BRYLINSKI/BRYLINSKI/GUILLEMIN/KAC (Eds). Lie Theory and Geometry: In Honor of Bertram Kostant
124 AEBISCHER/BORER/KÄLIN/LEUENBERGER/REIMANN. Symplectic Geometry. An Introduction based on the Seminar in Bern, 1992
125 LUBOTZKY. Discrete Groups, Expanding Graphs and Invariant Measures
126 RIESEL. Prime Numbers and Computer Methods for Factorization
127 HÖRMANDER. Notions of Convexity
128 SCHMIDT. Dynamical Systems of Algebraic Origin
129 DIJKGRAAF/FABER/VAN DER GEER (Eds). The Moduli Space of Curves
130 DUISTERMAAT. Fourier Integral Operators
131 GINDIKIN/LEPOWSKY/WILSON. Functional Analysis on the Eve of the 21st Century in Honor of the 80th Birthday of I.M. Gelfand. Vol. I
132 GINDIKIN/LEPOWSKY/WILSON. Functional Analysis on the Eve of the 21st Century in Honor of the 80th Birthday of I.M. Gelfand. Vol. II
131/132 Set Vols I+II
133 HOFER/TAUBES/WEINSTEIN/ZEHNDER (Eds). The Floer Memorial Volume
134 CAMPILLO LÓPEZ/ NARVÁEZ MACARRO (Eds). Algebraic Geometry and Singularities
135 Amrein/Boutet de Monvel/Georgescu. C_0-Groups, Commutator Methods and Spectral Theory of *N*-Body Hamiltonians
136 BROTO/CASACUBERTA/MISLIN (Eds). Algebraic Topology. New Trends in Localization and Periodicity
137 VIGNÉRAS. Représentations *l*-modulaires d'un groupe réductif *p*-adique avec $l \neq p$
140 KNAPP. Lie Groups Beyond an Introduction
141 CABANES (Ed.). Finite Reductive Groups: Related Structures and Representations
142 MONK. Cardinal Invariants on Boolean Algebras
143 GONZÁLEZ-VEGA/RECIO (Eds). Algorithms in Algebraic Geometry and Applications
144 BELLAÏCHE / RISLER (Eds). Sub-Riemannian Geometry
145 ALBERT / BROUZET / DUFOUR (Eds). Integrable Systems and Foliations Feuilletages et Systemes Integrables
146 JARDINE. Generalized Etale Cohomology Theories
147 DI BIASE. Fatou Type Theorems Maximal Functions and Approach Regions
148 HUANG. Two-Dimensional Conformal Geometry and Vertex Operator Algebras
149 SOURIAU. Structure of Dynamical Systems. A Symplectic View of Physics
150 SHIOTA. Geometry of Subanalytic and Semialgebraic Sets
151 HUMMEL. Gromov's Compactness Theorem for Pseudo-holomorphic Curves
153 BUESCU. Exotic Attractors. From Liapunov Stability to Riddled Basins
154 BÖTTCHER/KARLOVICH. Carleson Curves, Muckenhoupt Weights, and Toeplitz Operators
155 DRAGOMIR/ORNEA. Locally Conformal Kähler Geometry
156 GUIVARC'H/JI/TAYLOR. Compactifications of Symmetric Spaces
157 MURTY/MURTY. Non-vanishing of *L*-functions and Applications
158 TIRAO/VOGAN/WOLF (Eds). Geometry and Representation Theory of Real and *p*-Adic Groups
159 THANGAVELU. Harmonic Analysis on the Heisenberg Group
161 SAGAN. Mathematical Essays in Honor of Gian-Carlo Rota
162 ARNOLD/GREUEL/STEENBRINK. Singularities. The Brieskorn Anniversary Volume
163 BERNDT/SCHMIDT. Elements of the Representation Theory of the Jacobi Group
164 ROUSSARIE.. Bifurcations of Planar Vector Fields and Hilbert's Sixteenth Problem
165 MIGLIORE. Introduction to Liaison Theory and Deficiency Modules
166 ELIAS/GIRAL/MIRÓ-ROIG/ZARZUELA (Eds). Six Lectures on Commutative Algebra
167 FACCHINI.Module Theory
168 BALOG / KATONA / SZA'SZ / RECSKI (Eds). European Congress of Mathematics, Budapest, July 22-26, 1996. Vol. I
169 BALOG / KATONA / SZA'SZ / RECSKI (Eds). European Congress of Mathematics, Budapest, July 22-26, 1996. Vol. II
168/169 Set Vols I+II